CONFOCAL LASER MICROSCOPY - PRINCIPLES AND APPLICATIONS IN MEDICINE, BIOLOGY, AND THE FOOD SCIENCES

Edited by **Neil Lagali**

Confocal Laser Microscopy - Principles and Applications in Medicine, Biology, and the Food Sciences
http://dx.doi.org/10.5772/50821
Edited by Neil Lagali

Contributors

Neil Lagali, Beatrice Bourghardt Peebo, Johan Germundsson, Ulla Eden, Reza Danyali, Per Fagerholm, Marcus Rinaldo, Rita Marchi, Emi Nishijima Sakanashi, Herrera, Marc Navarro, Jun Fujita, Natsuko Hemmi, Shugo Tohyama, Tomohisa Seki, Yuuichi Tamura, Keiichi Fukuda, Akira Kobayashi, Enzo Di Iorio, Gary Chinga-Carrasco, Magnus B. Lilledahl, Catharina Davies, M. Vitoria Bentley, Anjali Basil, Wahid Wassef

Published by InTech

Janeza Trdine 9, 51000 Rijeka, Croatia

Copyright © 2013 InTech

Notice

Publishing Process Manager Oliver Kurelic
Technical Editor InTech DTP team
Cover InTech Design team

Additional hard copies can be obtained from orders@intechopen.com

Confocal Laser Microscopy - Principles and Applications in Medicine, Biology, and the Food Sciences,
Edited by Neil Lagali
p. cm.
ISBN 978-953-51-1056-9

Meet the editor

Neil S. Lagali, PhD, obtained undergraduate and graduate degrees in Canada. He has had several years of industry experience and held postdoctoral fellowships in Canada, The Netherlands, and Sweden. He has published more than 35 peer-reviewed articles in international journals in the fields of engineering, biomedical optics, biomaterials, ophthalmology, and translational medicine. He has co-authored several book chapters, and holds patents in the fields of optical devices, biosensors, and non-invasive imaging methods. He gives numerous lectures and courses on microscopy and clinical imaging, is a regular invited speaker and session organizer at international conferences, and is associate editor for the journal BMC Ophthalmology. Dr. Lagali is with the Department of Clinical and Experimental Medicine, Linköping University, Sweden.

Contents

Preface

Science often grows exponentially from certain key insights, and the insight Marvin Minsky had in 1955 is no exception. While working as a postdoctoral fellow at HarvardUniversity, Minsky sought a method to reduce the high level of light scatter that made high contrast imaging of thick specimens impossible by standard light microscopy. He realized that an arrangement of lenses having conjugate focal planes (hence the term 'confocal'), with a specimen and a pinhole placed at these opposite planes, could serve to reject the light that was scattered out-of-plane at the specimen. The result was a conceptually simple yet powerful method that not only improved the contrast of an image, but that also enabled adjacent serial optical sections of tissue to be analysed. Although these benefits were readily apparent in early confocal microscope systems employing white light sources, it was the advent of inexpensive diode lasers emitting at various wavelengths that dramatically improved the image contrast, fluorophore targeting, and the axial resolution of the confocal system. Today, laser-scanning confocal microscopy is ubiquitous in biomedical, biological, and non-biological research. Its versatility has ensured its widespread popularity, and the confocal technique has spawned many variants, aiming to improve detection, resolution, and contrast.

Laser-based confocal microscopy has provided the basic platform upon which more exotic laser microscopy techniques have been developed. These techniques improve on some of the limitations of confocal microscopy, by exploiting non-linear effects of laser light to provide advantages such as thinner optical sectioning, non-damaging low-energy excitation, and imaging with intrinsic molecular contrast. While non-linear techniques undoubtedly improve our ability to extract information from specimens and can provide unmatched image quality and information in certain situations, laser confocal microscopy is still the most widely applicable laser microscopy technique in use today. Indeed, this volume is a testament to this broad applicability, describing the use of laser confocal techniques to address diverse questions in medicine, biology, and the non-biological sciences.

In the biomedical field, confocal microscopy is crucial for examining single cells labelled with multiple fluorescent probes, by means of serial confocal examination with multiple excitation wavelengths. This technique is eloquently described in Chapter 1, where the authors detail insights gained into stem cell-based cardiac regenerative therapy. Laser confocal techniques such as thin sectioning, 3D image reconstruction, and spectrally-resolved fluorescent detection are discussed. Chapter 2 describes the application of high resolution sectioning and multi-channel fluorescent detection to elucidate the detailed morphology and structural composition of retinal blood vessels in the mouse – a model commonly used to study eye disease and angiogenesis.

Chapters 3, 4, and 5 concern the use of laser confocal microscopy in the cornea. Because it is a thin, transparent tissue that is externally accessible, the cornea has become a model organ for examination by laser confocal microscopy. These chapters describe the use of an in vivo variant of laser confocal microscopy to study the live cornea in patients, as well as in preclinical animal models. Additionally, corneal cells and tissues can be studied in vitro and ex vivo by similar techniques, for example, to gain insights into stem cell cultivation for corneal regenerative medicine. The widespread use of laser confocal microscopy in ophthalmology has led to the development of sophisticated image acquisition and analysis techniques, which can be readily exported to other fields.

Another external tissue amenable to examination is the body's largest organ, the skin. Chapter 6 comprehensively describes the use of laser confocal microscopy to analyze skin biopsy specimens and the skin of live patients in vivo. Diagnosis of melanomas, carcinomas, dermatitis, and keratosis can be aided by the confocal microscope. Additionally, the dynamics and efficacy of drug delivery through the skin can be investigated through laser confocal microscopic detection of fluorescently-labelled drugs or tagged nanoparticles. The condition of dermatitis, however, can also occur in the oral cavity, and the problem of allergy to dental alloys can cause a particularly severe form of dermatitis. Chapter 7 addresses this topic, describing the nature and extent of the condition. Laser confocal microscopy is presented as a new tool to improve the objectivity of grading the severity of allergic skin reactions to various dental metals.

For internal organs of the body not externally accessible to in vivo confocal microscopic observation, confocal laser endomicroscopy is an emerging technique that combines traditional endoscopic imaging with the high magnification, cellular level resolution and targeted fluorescence excitation provided by confocal microscopy. In Chapter 8, in vivo monitoring and diagnosis of conditions of the gastrointestinal tract by confocal laser endomicroscopy is described. An excellent correlation between confocal images and histopathology is possible in conditions such as Barrett's esophagus.

Within the biological sciences, structural fibers such as collagen, elastin, and cellulose form the scaffold of various organs and structures in humans, animals, or plants. These fibers form a matrix that interacts with cells and provides a medium for molecular signaling. In Chapter 9, 3D visualization and quantification of structural fibers is described, along with emerging non-linear laser microscopic imaging techniques. For these sophisticated imaging techniques, powerful image analysis techniques for data extraction are also discussed, including the use of various transforms and gradient methods. In Chapter 10, another type of biological matrix, that produced by fibrin, is discussed. The fibrin matrix and fibrin-cell interactions are probed in three dimensions by the use of laser confocal microscopy, to investigate pheonmena that can have wide ranging consequences for studying the processes of clotting, inflammation, wound healing, and angiogenesis.

Finally in Chapter 11, a thorough introduction and review is given of the application of laser confocal microscopy in the food sciences. Microscopic-level information from foods and their ingredients, such as the distribution of crystal size, homogeneity of emulsions, texture, processing characteristics, dynamic changes with time and temperature, etc., are invaluable for food production, processing, and delivery. As detailed in the chapter, laser confocal microscopy is proving to be a valuable tool for monitoring and visualizing foods in their natural and processed states, and evaluating the effects of food additives at the microscopic level.

I wish to sincerely thank all the chapter authors for their valuable contributions to this book. Their knowledge and deep understanding of the most interesting and relevant developmentswithin their respective fields, combined with comprehensive reference lists, provides an ideal starting point for the reader to delve more deeply into specific areas of interest. I also wish to thank InTech Publishing, who provided me the opportunity to serve as editor, and additionally provided administrative, technical, and publishing support to ensure this volume was compiled in a timely and professional manner.

Reading carefully through all the chapters in this book, which I have been privileged to be able to do, gives one a sense that laser-based confocal microscopy is a powerful technique we are only just beginning to apply. Laser confocal microscopy furthermore defies broad generalizations; while it may be a maturing method for multi-channel fluorescent detection of laboratory-labelled specimens, it is a rapidly growing method in ophthalmology for in vivo examination of the eye, and an emerging modality in the areas of dentistry, endoscopy, and in the food sciences. Also, it is certain that the technology will expand into new fields in time. Current trends that become apparent upon reading this volume are the extension of laser confocal techniques into the non-linear optical domain, development and applications of confocal systems for in vivo clinical use, automation of image acquisition and image analysis, and 3D visualization and data extraction.

The range of applications described in this book within seemingly disparate fields provides us with the impetus to venture beyond our specific domain of interest to learn the tools and techniques used in other domains. It is hoped that this volume will inspire a cross-fertilization of ideas within the community that utilizes laser confocal microscopy. It is these ideas that have the potential to push forward the boundaries of what is possible with laser confocal microscopy – and may one day lead to key insights such as the one Minsky had more than half a century ago.

Neil Lagali
Department of Clinical and Experimental Medicine,
Faculty of Health Sciences, Linköping University,
Linköping, Sweden

Section 1

Applications in Medicine

Chapter 1

Practical Application of Confocal Laser Scanning Microscopy for Cardiac Regenerative Medicine

Jun Fujita, Natsuko Hemmi, Shugo Tohyama,
Tomohisa Seki, Yuuichi Tamura and Keiichi Fukuda

Additional information is available at the end of the chapter

http://dx.doi.org/10.5772/55864

1. Introduction

Heart failure (HF) is an insidious disease in developed countries. Despite recent medical progress, the number of patients with HF continues to increase, with the mortality of HF as high as that of cancer. The only radical treatment for HF is cardiac transplantation, although the shortage of donor hearts poses a serious problem [1]. To overcome this unmet medical need, innovative technology is required. Specifically, cell transplantation therapy with regenerative cardiomyocytes is expected to eventually replace cardiac transplantation as the treatment for severe HF.

It was believed that, after the neonatal stage, heart cells could no longer proliferate and regenerate. However, recent evidence demonstrates the regenerative capacity of cardiomyocytes obtained from several different cell sources, such as mesenchymal stem cells (MSCs), cardiac progenitor cells (CPCs), and neural crest derived stem cells (NCSCs) [2-5]. In addition, pluripotent stem cells (PSCs), such as human embryonic stem cells (hESCs) and human induced pluripotent stem cells (hiPSCs), seem to be potential cell sources of regenerative cardiomyocytes. Thus, basic in vivo and in vitro studies have evolved into translational research focused on stem cell therapy for severe HF.

Without the development of innovative scientific technology enabling the precise observation and analysis of individual cells, these recent advances in regenerative medicine would not have been possible. In such basic studies, the cells are often marked with green fluorescent protein (GFP) or red fluorescent protein, and co-stained with various cell-specific markers, such as α-actinin, MF20, and cardiac troponin in the case of cardiomyocytes. Fluorescence-activated cell sorting (FACS) enables population analysis of both stem cells and differentiated cells, and fluorescent microscopy visualizes regenerative cardiomyocytes in culture, as well as in tissues. However, for either technique to be useful, it is necessary to collect detailed information at the level of individual cells.

Confocal laser scanning microscopy (CLSM) has emerged as a new high-tech method for exploring the stem cell field in cardiovascular medicine. CLSM enables the kinetics of single stem cells and differentiated cells to be studied both in vivo and in vitro, and so has opened up a new world, revealing the regenerative potential of stem cells. In this chapter we explain how CLSM has contributed to new scientific findings in cardiac regenerative medicine.

2. Technical advantages of CLSM for the investigation of stem cells in cardiovascular medicine

2.1. Serial optical thin sections

Fluorescent immunohistochemistry is important for studies into the topography of stem cells. Samples can be double stained with different markers, with the resulting fluorescent images enabling visualization of the co-localization of the different signals. It is impossible to distinguish overlapping signals using conventional fluorescent microscopy because this technique detects signals in both the field of focus and all the unfocused signals. The distinctive feature of CLSM is a pinhole that permits focusing on a small focal point compared with conventional microscopy. This technology underpins one of the advantages of CLSM, namely spatial resolution via the acquisition of a series of images called the Z-stack (Figure 1). CLSM will detect signals only at the focal point in thick samples, and can thus distinguish overlapping signals that cannot be differentiated using conventional microscopy. Another advantage of CLSM is multiple track detection, which contributes to the exclusion of signal crosstalk. In single track detection, multiple lasers excite multiple fluorescent probes simultaneously and all the fluorescent signals are emitted at the same time. In this case, each signal from each of the fluorescent probes cannot be completely delineated because the spectral wavelengths of the probes overlap. In contrast, in the case of multiple track detection, the excitation lasers stimulate the sample sequentially, eliminating signal crosstalk among fluorescent signals. This technology has made a considerable contribution to stem cell research.

2.2. Three dimensional imaging and multidimensional views

The acquisition of Z-stack images using the CLSM enables reconstruction of three-dimensional (3D) images, which make it easier to sterically analyze an object [6]. Using CLSM, signals from multiple cells can be distinguished from overlapping signals within single cells. The multidimensional view afforded by CLSM also helps researchers to understand tissue organization, particularly in vivo.

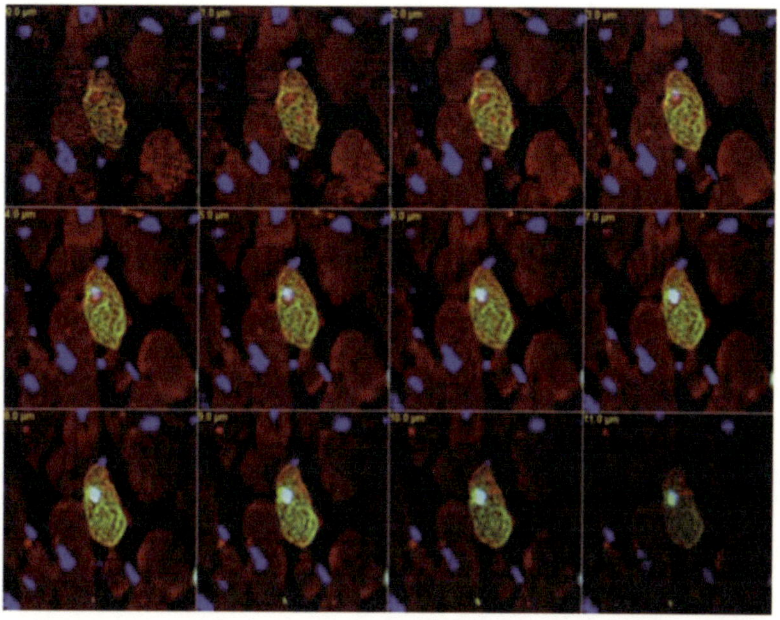

Figure 1. Sequential images of a green fluorescent protein (GFP)-positive bone marrow stem cells (BMSC)-derived cardiomyocyte were acquired with confocal laser scanning microscopy (CLSM). Red, α-actinin; blue, nuclei; green, GFP.

2.3. Region of interest scanning

By using CLSM, it is easy to focus on a region of interest (ROI). If the ROI cannot be scanned specifically, as may be the case using conventional microscopy, an all-field picture must be taken and the ROI analyzed later using imaging software. In this case, the image acquired contains too much extra information and the file that needs to be saved is very big. If the field of focus is a single cell or a part of a cell, such as the nuclear membrane or organelles, it is often difficult to obtain clear images using conventional microscopy. Focusing on the ROI also helps prevent the loss of signals in other parts of the field.

2.4. Emission fingerprinting and multifluorescence imaging

The spectral imaging (SI-) CLSM system developed by Carl Zeiss is the most innovative technology in this field. The SI-CLSM system simultaneously detects spectral curves on the fluorescence wavelength (λ). Conventional filter systems cannot distinguish closely adjacent signals, such as those of GFP (peak emission wavelength; 509 nm) and fluorescein isothiocyanate (FITC) (peak emission wavelength; 525 nm), but the SI-CLSM system uses a grating mirror and a 32-channel array detector to separate close emission spectra. In addition, the maximum number of available signals using conventional filter systems is usually four; in contrast, more than four colors are available in the SI-CLSM system. Another advantage of the SI-CLSM system is its ability to distinguish specific wavelengths of an object against non-specific background signals. The SI-CLSM system has significantly increased the reliability of data obtained in regenerative medicine [7-10].

2.5. Two-photon laser scanning microscopy and time-lapse imaging

One of the major disadvantages of CLSM is sample damage caused by the laser. Furthermore, if the signal is very weak, the laser power must be high. This can result in photobleaching, and the subsequent disappearance of signal. It is also difficult to take pictures of living cells using CLSM. These problems have been overcome by two-photon laser scanning microscopy. The infrared laser used in two-photon laser scanning microscopy causes less damage to living cells than visible light lasers, allowing time-lapse images of living cells to be acquired using LSM [11]. Time-lapse images are important for analyzing stem cell behavior in vitro and the differentiation process of pluripotent stem cells [12]. Furthermore, two-photon laser scanning microscopy enables the detection of signals from deeper within tissues because the infrared laser tends to reach greater depths within specimens compared with ultraviolet and visible light [13].

Overall, the development of CLSM and two-photon laser scanning microscopy has been essential for advances in cardiovascular regenerative medicine.

3. Bone marrow stem cells

3.1. Bone marrow stem cell-derived cardiomyocytes

Bone marrow stem cells (BMSCs) consist of hematopoietic stem cells (HSCs) and mesenchymal stem cells (MSCs), and MSCs have been shown to have the potential to develop into cardiomyocytes both in vitro and in vivo [2, 8, 14].

In an early study, Makino et al. established an MSC cell line (cardiomyogenic [CMG] cells) that stably developed into cardiomyocytes [2]. The cardiomyocytes derived from this cell line exhibited the same functional properties as native cardiomyocytes [15]. In a later study, CMG cells with cardiac-specific promoter (myosin light chain-2v [MLC-2v])-derived GFP were generated. Transplantation of MLC-2v-GFP CMG cells in vivo demonstrated the successful delivery of MSC-derived cardiomyocytes into the murine heart [7], with CLSM clearly showing the GFP signals of the donor cardiomyocytes.

However, the origin of the bone marrow (BM)-derived cardiomyocytes (i.e., HSC or MSC) remained contentious. To investigate this issue, we generated BM transplantation models with HSCs, whole BM, and MSCs [14]. Myocardial infarction (MI) was induced in these BM-transplanted mice, and the BM cells were mobilized with granulocyte colony-stimulating factor (G-CSF). In contrast with results obtained following transplantation of whole BM, HSC-derived cardiomyocytes were very rare and, on the basis of these observations, it was concluded that BMSC-derived cardiomyocytes were of MSC origin [14].

In pressure overloaded HF models (i.e. hypoxia-induced pulmonary hypertension-induced right ventricular hypertrophy and transverse aortic constriction-induced left ventricular hypertrophy), many BMSC-derived cardiomyocytes were mobilized with ventricular pressure by both cell fusion and transdifferentiation [8]. These GFP-labeled BMSC-derived cardiomyocytes are clearly visible using CLSM (Figure 2), with their signals clearly distinguished from non-specific background signals (Figure 3).

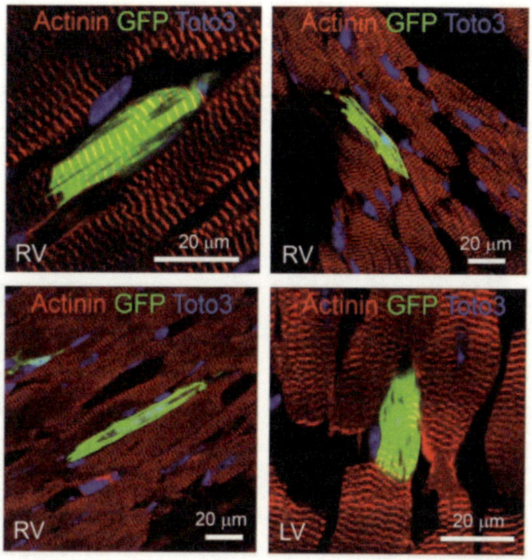

Figure 2. Mobilization of GFP-positive BMSC-derived cardiomyocytes after transplantation in the host heart. The red periodic striations represent expression of the myocyte marker, α-actinin, in cardiac muscle. RV, right ventricle; LV, left ventricle; Toto3; nuclear marker. (Reproduced with permission from Endo et al. [8].)

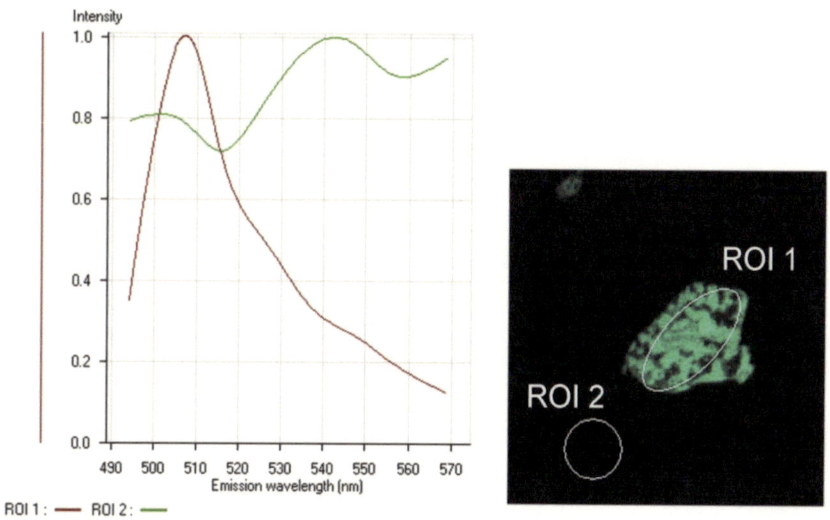

Figure 3. Emission profile of the GFP signal in BMSC-derived cardiomyocytes. The image was acquired with the Zeiss spectral imaging (SI)-CLSM system. Note that the cardiomyocyte on the right clearly shows the emission wavelength of GFP (left panel). ROI, region of interest. (Figures are modified from Endo et al. [8].)

3.2. BMSC-derived vascular cells

It has been reported that vascular progenitor cells (VPCs) can also be derived from BMSCs [16, 17]. Cytokine therapy is a useful method of mobilizing BMSC-derived VPCs to ischemic areas and inducing angiogenesis in ischemic limbs. The combination of G-CSF and hepatocyte growth factor (HGF) was shown to increase the number of BMSC-derived endothelial and smooth muscle cells, and to promote angiogenesis [18]. The induction of angiogenesis was greater following G-CSF than HGF treatment. CLSM clearly showed the colocalization of endothelial markers, such as von Willebrand Factor, CD31, and α-smooth muscle actin (Figure 4).

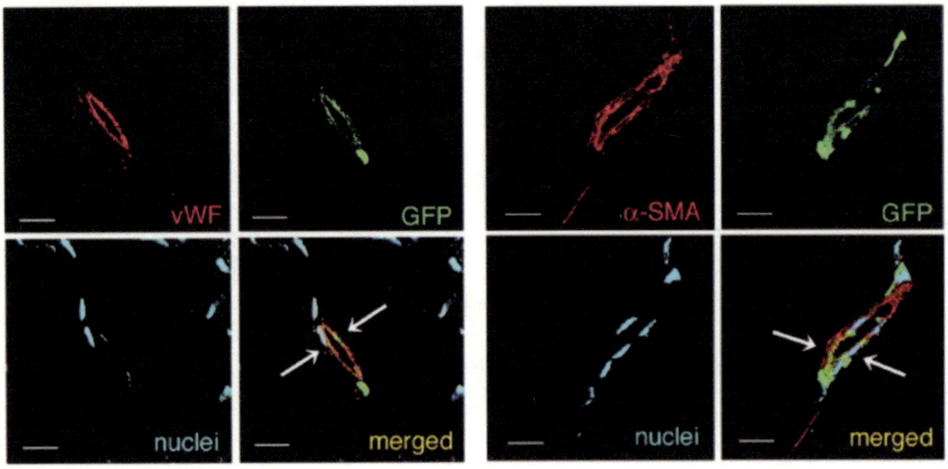

Figure 4. BMSC-derived endothelial and smooth muscle cells were recruited to the ischemic limb with cytokine therapy. vWF, von Willebrand factor. Bars, 10 μm. (Reproduced with permission from Ieda et al. [18].)

3.3. BMSC-derived cells in pulmonary hypertension

BMSC-derived cells are involved in the vascular remodeling of pulmonary arteries and the progression of pulmonary hypertension. In one study, whole BM cells from GFP transgenic mice were transplanted into wild-type mice [9] and pulmonary hypertension was induced in the BM-transplanted mice by placing them in a hypoxic chamber. A considerable number of GFP-positive BMSC-derived cells were found to be involved in pulmonary artery remodeling [9], which was confirmed by CLSM using a grating mirror and a 32-channel array detector (Figure 5).

3.4. BMSC-derived myofibroblasts in MI

In wild-type mice transplanted with GFP-positive BM, the administration of G-CSF improved cardiac function, prevented cardiac remodeling, and improved survival after MI [14], although the presence of BMSC-derived cardiomyocytes was not enough to explain the beneficial effects of G-CSF therapy after MI. In a subsequent study, we found that G-CSF mobilized a considerable number of BMSC-derived myofibroblasts in the MI scar [10]. The BM-derived GFP-

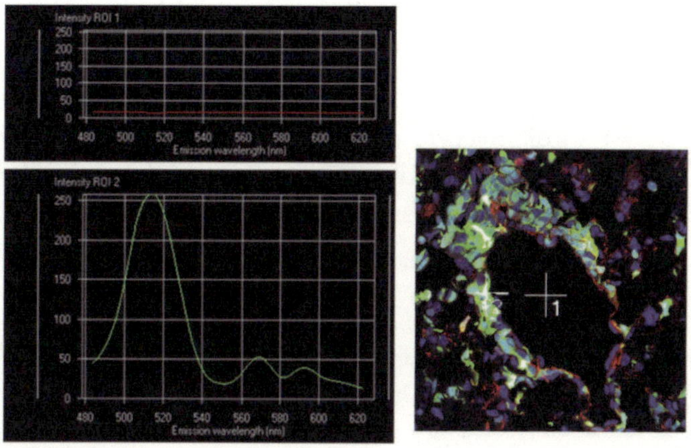

Figure 5. CLSM confirmation of the contribution of BMSC-derived cells to pulmonary artery remodeling in pulmonary hypertension. The image was acquired with the Zeiss SI-CLSM system. The GFP signal from the BMSC presented in the right micrograph (bottom left graph) was clearly distinguished with non-specific background (top left graph). (Reproduced with permission from Hayashida et al. [9].)

positive cells were co-stained with vimentin, α-smooth muscle actin, and a nuclear marker. Fluorescent signals were detected using a 32-channel array detector and, although the emission signals for Toto3 and Alexa 660 were highly overlapping, they were clearly separated using a grating mirror and a 32-channel array detector (Figure 6).

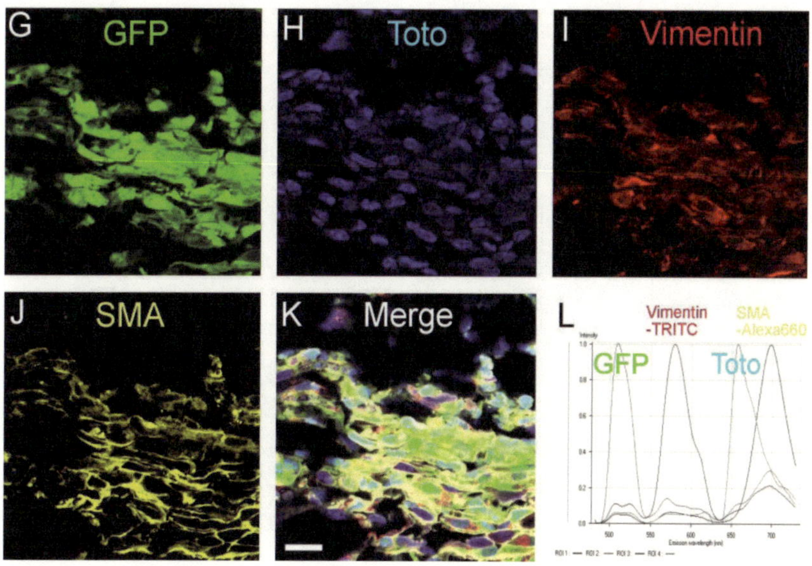

Figure 6. Migration of BMSC-derived cells into the infarcted area and their differentiation into myofibroblasts after myocardial infarction (MI). SMA, α-smooth muscle actin; GFP, green fluorescent protein; TRITC, tetramethylrhodamine-5-(and 6)-isothiocyanate. Bar, 20 μm. (Reproduced with permission from Fujita et al. [10].)

4. Cardiac development and cardiac neural crest stem cells

Embryonic development is instructive for regenerative medicine. Specifically, heart development is a textbook for cardiomyocyte differentiation from stem cells, because effective differentiation depends on a precise process. CLSM is a powerful tool with which the localization of small groups of cells in small tissues (e.g. murine embryonic hearts) can be observed.

4.1. Cardiac neural crest-derived cardiomyocytes

During development of the mammalian heart, neural crest-derived stem cells (NCSCs) migrate to the developing heart and differentiate into several types of cells, including cardiomyocytes [4]. The number of NCSC-derived cardiomyocytes increases during postnatal growth. The NCSCs in a heart can be cultured as a cardiosphere and will develop into neurons, smooth muscle cells, and cardiomyocytes in vitro. They can also migrate into a heart after the induction of MI and develop into cardiomyocytes [19]. CLSM has contributed to observations of NCSC-derived cells in the heart (Figure 7).

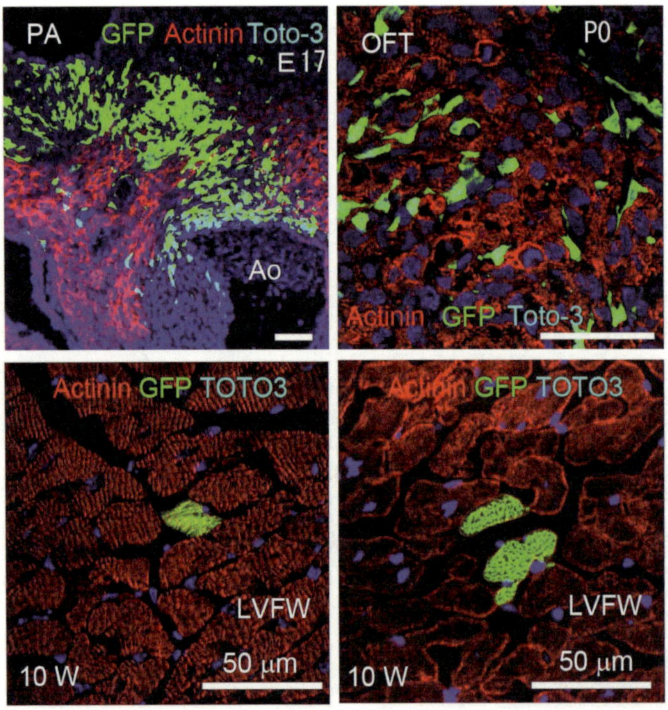

Figure 7. Neural crest stem cells migrate into the developing heart and differentiate into cardiomyocytes after MI in adult mice. PA, pulmonary artery; Ao, aorta; E17, embryonic day 17; OFT, outflow tract; P0, postnatal day 0; LVFW, left ventricular free wall; 10, 10 weeks postnatally; GFP, green fluorescent protein. Bars, 50 μm on top panels. (Reproduced with permission from Tamura et al. [19])

http://dx.doi.org/10.5772/55864

4.2. Heart valve development and chondromodulin-I

Heart valve formation is controlled by anti-angiogenic activity, such as that of chondromodulin-I. Downregulation of chondromodulin-I leads to neovascularization of the cardiac valves, resulting in valvular heart diseases [20]. CLSM has been used to clarify the localization of chondromodulin-expressing cells in developing embryonic hearts (Figure 8).

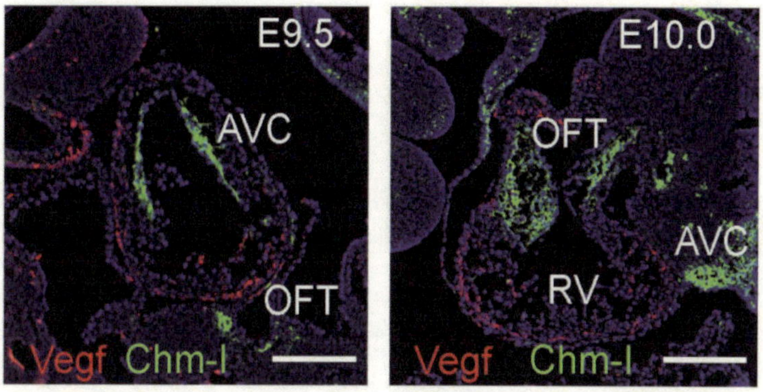

Figure 8. Expression of chondromodulin-I at the atrioventricular canal and outflow tract in the developing murine heart. AVC, atrioventricular canal; Vegf, vascular endothelial growth factor; Chm-1, chondromudulin-I; E, embryonic day; RV, right ventricle; OFT, outflow tract. Bars, 200 μm. (Reproduced with permission from Yoshioka et al. [20].)

5. Pluripotent stem cells

In 1998, hESCs were reported as true PSCs [21]. Although the clinical application of hESCs has been hindered by ethical considerations, tumor formation, and immunological rejection, the clinical potential of hESCs as a cell source for regenerative medicine is undeniable [22]. In addition, hiPSCs were developed in 2007 following the transfection of four pluripotent factors into fibroblasts to yield PSCs with the same differentiation capacity as hESCs [23, 24]. The advantage of hiPSCs is that immunosuppressive therapy and ethical issues are not limiting factors in their clinical application (as opposed to hESCs) because hiPSCs are generated from individual patients. Both hESCs and hiPSCs have good potential to differentiate into all components of the heart, including endothelial cells, smooth muscle cells, and cardiomyocytes. Two-photon laser scanning microscopy has proved useful in observing PSCs-derived cells (Figure 9), which are expected to become a future cell source for human regenerative cardiomyocytes. Nevertheless, there are still some issues that need to be resolved before the application of cell therapy using PSC-derived cardiomyocytes. For example, teratoma formation as a result of contamination by residual PSCs is the most pressing issue, highlighting the need to purify the differentiated cardiomyocytes. To this end, GFP-labeled hESCs or iPSCs are extremely valuable in studies investigating the differentiation of undifferentiated PSCs to yield pure cardiomyocytes (Figure 10).

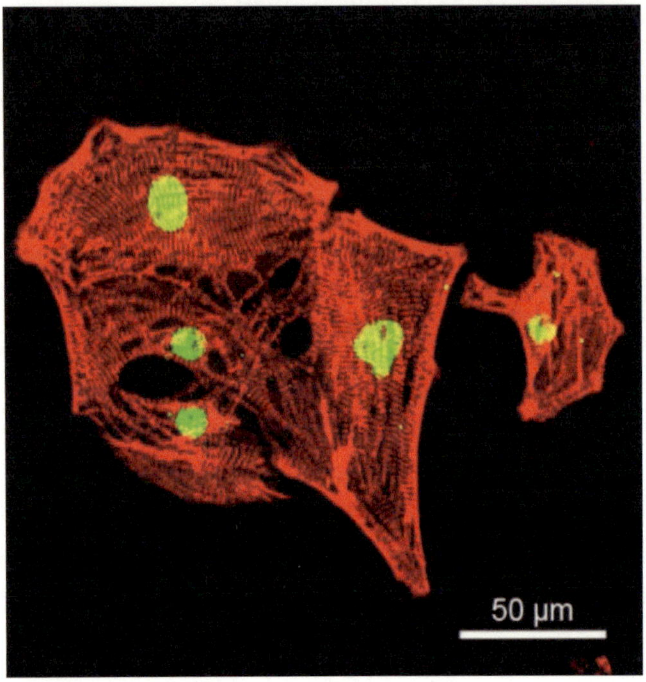

Figure 9. Human induced pluripotent stem cell (253 G4, [25])-derived cardiomyocytes photographed using two-photon CLSM. Green, Nkx2-5 (cardiac-specific transcription factor); red, α-actinin.

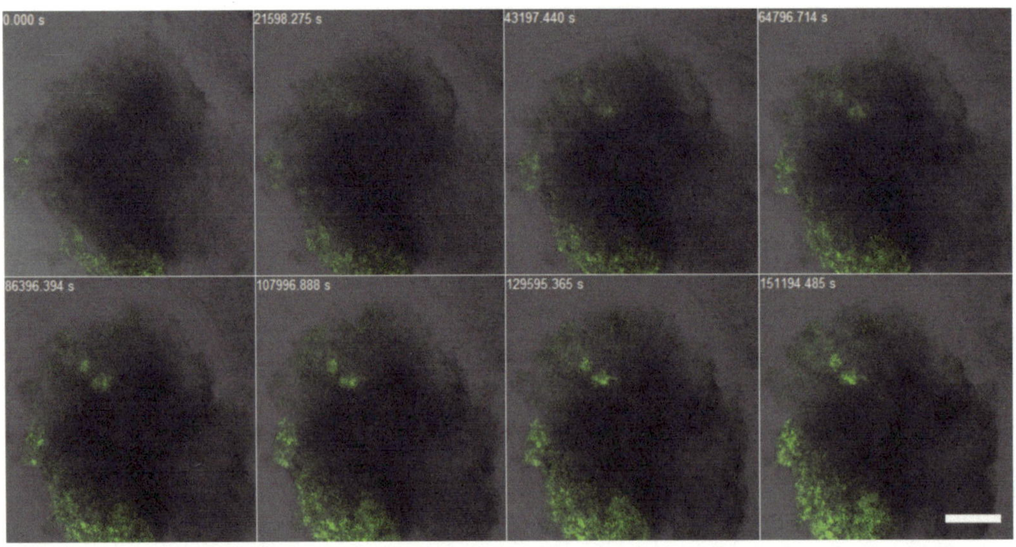

Figure 10. Time-lapse imaging of human embryonic stem cell-derived cardiomyocytes showing the formation of embryoid bodies by the H9-hTnnTZ pGZ-D2 embryonic stem cell line and increased expression of the cardiac marker troponin T-GFP in sequential time frame. Bar, 200 μm.

6. In vivo assessment of tissue engineering of heart diseases

There are three major methods for the transplantation of regenerative cardiomyocytes: direct cell injection, the use of a biological scaffold, and the use of a cell sheet [26]. The status of transplanted cells is the most critical issue before cardiomyocyte cell therapy can be realized. For example, although direct cell transplantation has traditionally been the most common way to transplant cells, < 15% of transplanted cardiomyocytes survive due to aggregation and necrosis of the grafted cells. [7, 27]. CLSM has advanced analyses of the status of transplanted cells in vivo. Following direct cell injection, CLSM could readily distinguish transplanted cells from host tissue, and the GFP signal was confirmed using a 32-channel array detector [7].

The 3D reconstruction of the myocardium is a challenge for tissue engineering applications in the field of cardiovascular therapy. Many biomaterials are available for the construction of 3D scaffolds for regenerative therapy [28]. Although creating both aligned donor cardiomyocytes and dense myocardial tissues is difficult, cell sheet technology enables the construction of myocardial tissue with aligned cardiomyocytes. Numerous basic studies have shown that Myocardial cell sheets (MCSs) effectively restore cardiac function [26]. In MCSs, the densely aligned myocardium was clearly shown by CLSM (Figure 11). The advantages associated with high-resolution CLSM aid in the analysis of dense myocardial tissue.

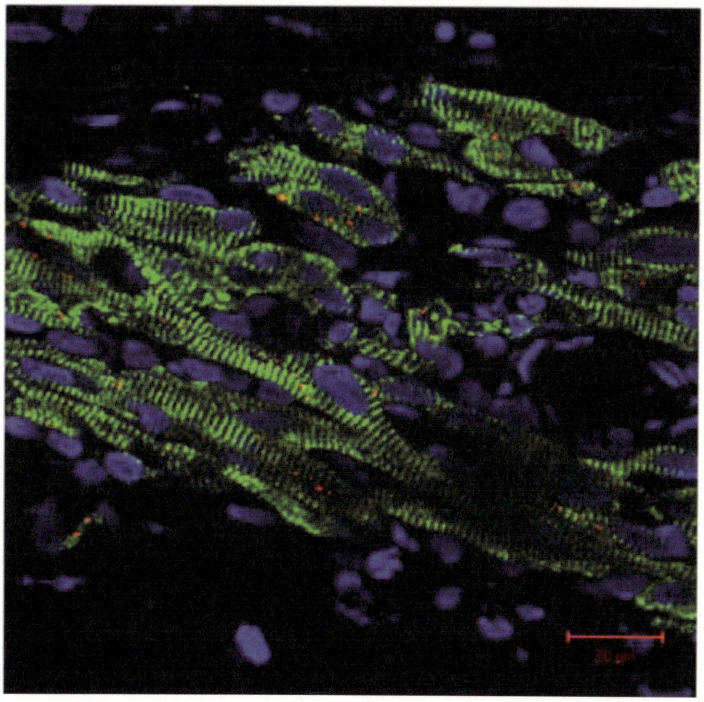

Figure 11. Graft cardiomyocytes constitute the functional myocardial cell sheets. Green, GFP; red, connexin 43; blue, nuclei. Bar, 20 μm. (Figures modified from Itabashi et al. [29, 30].)

7. Conclusion

CLSM has contributed to the advances in cardiac regenerative medicine. The precise data obtained using CLSM have been extremely valuable in confirming the potential of stem cells to differentiate into cardiomyocytes. Based on the investigations completed thus far, translational research in large, animal models and clinical studies have started. It is expected that the findings of basic research into cell therapies will be transferred successfully into clinical practice, with cell therapy ultimately becoming the standard treatment for severe HF. CLSM has thus opened up a new therapeutic field for HF with cardiac regenerative medicine.

Acknowledgements

The authors thank Toshiyuki Watanabe (Carl Zeiss Japan) for valuable suggestions and critical comments. The authors thank RIKEN BioResource Center for the 253 G4 iPS cells and the WiCell research institute for the H9-hTnnTZ pGZ-D2 ES cells.

Author details

Jun Fujita*, Natsuko Hemmi, Shugo Tohyama, Tomohisa Seki, Yuuichi Tamura and Keiichi Fukuda

*Address all correspondence to: jfujita@a6.keio.jp

Department of Cardiology, Keio University School of Medicine, Shinanomachi Shinjuku-ku, Tokyo, Japan

References

[1] Stehlik, J, Edwards, L. B, Kucheryavaya, A. Y, Benden, C, Christie, J. D, Dobbels, F, et al. The Registry of the International Society for Heart and Lung Transplantation: Twenty-Eighth Adult Heart Transplant Report--2011. J Heart Lung Transplant. (2011). Epub 2011/10/04., 30(10), 1078-94.

[2] Makino, S, Fukuda, K, Miyoshi, S, Konishi, F, Kodama, H, Pan, J, et al. Cardiomyocytes Can Be Generated from Marrow Stromal Cells in Vitro. J Clin Invest. (1999). Epub 1999/03/13., 103(5), 697-705.

[3] Beltrami, A. P, Barlucchi, L, Torella, D, Baker, M, Limana, F, Chimenti, S, et al. Adult Cardiac Stem Cells Are Multipotent and Support Myocardial Regeneration. Cell. (2003)., 114(6), 763-76.

[4] Tomita, Y, Matsumura, K, Wakamatsu, Y, Matsuzaki, Y, Shibuya, I, Kawaguchi, H, et al. Cardiac Neural Crest Cells Contribute to the Dormant Multipotent Stem Cell in the Mammalian Heart. J Cell Biol. (2005). Epub 2005/09/28., 170(7), 1135-46.

[5] Oh, H, Bradfute, S. B, Gallardo, T. D, Nakamura, T, Gaussin, V, Mishina, Y, et al. Cardiac Progenitor Cells from Adult Myocardium: Homing, Differentiation, and Fusion after Infarction. Proceedings of the National Academy of Sciences. (2003). , 100(21), 12313-8.

[6] Malliaras, K, Zhang, Y, Seinfeld, J, Galang, G, Tseliou, E, Cheng, K, et al. Cardiomyocyte Proliferation and Progenitor Cell Recruitment Underlie Therapeutic Regeneration after Myocardial Infarction in the Adult Mouse Heart. EMBO Mol Med. (2012). Epub 2012/12/21.

[7] Hattan, N, Kawaguchi, H, Ando, K, Kuwabara, E, Fujita, J, Murata, M, et al. Purified Cardiomyocytes from Bone Marrow Mesenchymal Stem Cells Produce Stable Intracardiac Grafts in Mice. Cardiovasc Res. (2005). Epub 2005/01/11., 65(2), 334-44.

[8] Endo, J, Sano, M, Fujita, J, Hayashida, K, Yuasa, S, Aoyama, N, et al. Bone Marrow Derived Cells Are Involved in the Pathogenesis of Cardiac Hypertrophy in Response to Pressure Overload. Circulation. (2007). Epub 2007/08/19., 116(10), 1176-84.

[9] Hayashida, K, Fujita, J, Miyake, Y, Kawada, H, Ando, K, Ogawa, S, et al. Bone Marrow-Derived Cells Contribute to Pulmonary Vascular Remodeling in Hypoxia-Induced Pulmonary Hypertension. Chest. (2005). Epub 2005/05/13., 127(5), 1793-8.

[10] Fujita, J, Mori, M, Kawada, H, Ieda, Y, Tsuma, M, Matsuzaki, Y, et al. Administration of Granulocyte Colony-Stimulating Factor after Myocardial Infarction Enhances the Recruitment of Hematopoietic Stem Cell-Derived Myofibroblasts and Contributes to Cardiac Repair. Stem Cells. (2007). Epub 2007/08/11., 25(11), 2750-9.

[11] Liu, H, Shao, Y, Qin, W, Runyan, R. B, Xu, M, Ma, Z, et al. Myosin Filament Assembly onto Myofibrils in Live Neonatal Cardiomyocytes Observed by Tpef-Shg Microscopy. Cardiovasc Res. (2012). Epub 2012/11/03.

[12] Nakano, T, Ando, S, Takata, N, Kawada, M, Muguruma, K, Sekiguchi, K, et al. Self-Formation of Optic Cups and Storable Stratified Neural Retina from Human Escs. Cell Stem Cell. (2012). Epub 2012/06/19., 10(6), 771-85.

[13] Rubart, M, Pasumarthi, K. B, Nakajima, H, Soonpaa, M. H, Nakajima, H. O, & Field, L. J. Physiological Coupling of Donor and Host Cardiomyocytes after Cellular Transplantation. Circ Res. (2003). Epub 2003/05/06., 92(11), 1217-24.

[14] Kawada, H, Fujita, J, Kinjo, K, Matsuzaki, Y, Tsuma, M, Miyatake, H, et al. Nonhematopoietic Mesenchymal Stem Cells Can Be Mobilized and Differentiate into Cardiomyocytes after Myocardial Infarction. Blood. (2004). Epub 2004/08/07., 104(12), 3581-7.

[15] Hakuno, D, Fukuda, K, Makino, S, Konishi, F, Tomita, Y, Manabe, T, et al. Bone Marrow-Derived Regenerated Cardiomyocytes (Cmg Cells) Express Functional Adrenergic and Muscarinic Receptors. Circulation. (2002). Epub 2002/01/24., 105(3), 380-6.

[16] Asahara, T, Murohara, T, Sullivan, A, Silver, M, Van Der Zee, R, Li, T, et al. Isolation of Putative Progenitor Endothelial Cells for Angiogenesis. Science. (1997). Epub 1997/02/14., 275(5302), 964-7.

[17] Saiura, A, Sata, M, Hirata, Y, Nagai, R, & Makuuchi, M. Circulating Smooth Muscle Progenitor Cells Contribute to Atherosclerosis. Nat Med. (2001). Epub 2001/04/03., 7(4), 382-3.

[18] Ieda, Y, Fujita, J, Ieda, M, Yagi, T, Kawada, H, Ando, K, et al. G-Csf and Hgf: Combination of Vasculogenesis and Angiogenesis Synergistically Improves Recovery in Murine Hind Limb Ischemia. J Mol Cell Cardiol. (2007). Epub 2007/01/16., 42(3), 540-8.

[19] Tamura, Y, Matsumura, K, Sano, M, Tabata, H, Kimura, K, Ieda, M, et al. Neural Crest-Derived Stem Cells Migrate and Differentiate into Cardiomyocytes after Myocardial Infarction. Arterioscler Thromb Vasc Biol. (2011). Epub 2011/01/08., 31(3), 582-9.

[20] Yoshioka, M, Yuasa, S, Matsumura, K, Kimura, K, Shiomi, T, Kimura, N, et al. Chondromodulin-I Maintains Cardiac Valvular Function by Preventing Angiogenesis. Nat Med. (2006). Epub 2006/09/19., 12(10), 1151-9.

[21] Thomson, J. A, Itskovitz-eldor, J, Shapiro, S. S, Waknitz, M. A, Swiergiel, J. J, Marshall, V. S, et al. Embryonic Stem Cell Lines Derived from Human Blastocysts. Science. (1998). , 282(5391), 1145-7.

[22] Schwartz, S. D, Hubschman, J. P, Heilwell, G, Franco-cardenas, V, Pan, C. K, Ostrick, R. M, et al. Embryonic Stem Cell Trials for Macular Degeneration: A Preliminary Report. Lancet. (2012). Epub 2012/01/28., 379(9817), 713-20.

[23] Takahashi, K, Tanabe, K, Ohnuki, M, Narita, M, Ichisaka, T, Tomoda, K, et al. Induction of Pluripotent Stem Cells from Adult Human Fibroblasts by Defined Factors. Cell. (2007). , 131(5), 861-72.

[24] Yu, J, Vodyanik, M. A, Smuga-otto, K, Antosiewicz-bourget, J, Frane, J. L, Tian, S, et al. Induced Pluripotent Stem Cell Lines Derived from Human Somatic Cells. Science. (2007). , 318(5858), 1917-20.

[25] Nakagawa, M, Koyanagi, M, Tanabe, K, Takahashi, K, Ichisaka, T, Aoi, T, et al. Generation of Induced Pluripotent Stem Cells without Myc from Mouse and Human Fibroblasts. Nat Biotechnol. (2008). Epub 2007/12/07., 26(1), 101-6.

[26] Fujita, J, Itabashi, Y, Seki, T, Tohyama, S, Tamura, Y, Sano, M, et al. Myocardial Cell Sheet Therapy and Cardiac Function. Am J Physiol Heart Circ Physiol. (2012). Epub 2012/09/25.

[27] Van Laake, L. W, Passier, R, Monshouwer-kloots, J, Verkleij, A. J, Lips, D. J, Freund, C, et al. Human Embryonic Stem Cell-Derived Cardiomyocytes Survive and Mature in the Mouse Heart and Transiently Improve Function after Myocardial Infarction. Stem Cell Res. (2007). Epub 2007/10/01., 1(1), 9-24.

[28] Rane, A. A, & Christman, K. L. Biomaterials for the Treatment of Myocardial Infarction a 5-Year Update. J Am Coll Cardiol. (2011). Epub 2011/12/14., 58(25), 2615-29.

[29] Itabashi, Y, Miyoshi, S, Kawaguchi, H, Yuasa, S, Tanimoto, K, Furuta, A, et al. A New Method for Manufacturing Cardiac Cell Sheets Using Fibrin-Coated Dishes and Its Electrophysiological Studies by Optical Mapping. Artif Organs. (2005). Epub 2005/01/27., 29(2), 95-103.

[30] Itabashi, Y, Miyoshi, S, Yuasa, S, Fujita, J, Shimizu, T, Okano, T, et al. Analysis of the Electrophysiological Properties and Arrhythmias in Directly Contacted Skeletal and Cardiac Muscle Cell Sheets. Cardiovasc Res. (2005). Epub 2005/05/21., 67(3), 561-70.

Chapter 2

The Use of Confocal Laser Microscopy to Analyze Mouse Retinal Blood Vessels

David Ramos, Marc Navarro, Luísa Mendes-Jorge,
Ana Carretero, Mariana López-Luppo,
Víctor Nacher, Alfonso Rodríguez-Baeza and
Jesús Ruberte

Additional information is available at the end of the chapter

http://dx.doi.org/10.5772/56131

1. Introduction

Until recently, the house mouse (*Mus musculus*) was not a prefered model to study the mammalian visual system [1]. However, the power of transgenic and knockout mice as tools to analyze the genetic basis and the pathophysiology of human eye diseases, have become the mouse one of the most used animals for the study of retinopathy [2].

In the retina there is a compromise between transparency and optimal oxygenation [3]. Thus, retinal vasculature must show special characteristics in order to minimize their interference with the light path. Retinal capillaries are sparse and small [4], representing only 5% of the total retinal mass [5]. Hence, retinal blood volume is relatively low [6]. This feature, together with an extremely active cellular metabolism, 10% of resting body energy expenditure is consumed by retinal tissue [7], makes retina very susceptible to hypoxia.

The study of retinal vasculature has an increasing relevance, since vascular alterations are one of the earliest events observed during retinopathy [8]. Vascular alterations compromise blood flow, diminish oxygen supply, and neovascularization develops in response to hypoxia. This neovascularization is the most common cause of blindness, with a growing social impact in the world [9].

The structure of the mouse retina has been extensively studied anatomically using silver impregnations [10], Nissl staining [11], electron microscopy [12, 13], differential interference contrast microscopy [14] and confocal laser microscopy [15]. More specifically, mouse retinal blood vessels have been analyzed by angiography using fluorescent dyes [16], vascular corrosion cast [17], trypsin/pepsin digestion [18] and confocal laser microscopy (CLM) [19-23]. However, retinal whole-mount observation by confocal laser technology is the only method that allows a three-dimensional microscopical analysis of the entire retina, combining the use of fluorescent markers for proteins, signaling molecules, etc.

The visual organ, the eye, is a structure that transforms light into electrical impulses, which are sent to the brain. The visual organ is formed by the eyeball and the accessory ocular organs. Lids, lacrimal glands and extraocular muscles provide protection and help to the visual function (Figure 1A).

The adult mouse is a very small nocturnal mammal with a relatively small eyeball having an axial length from anterior cornea to choroid of about 3.4 mm [24]. As is typical for nocturnal mammals, the mouse eyeball resembles a hollow sphere with a relatively large cornea. The eyeball is formed by three layers or tunicae, which contains the eye chambers and a very large lens that represents approximately 65% of the axial length (Figure 1B). The anterior chamber is placed between the cornea and the iris. The posterior chamber is the space situated between the iris and the lens. The vitreous chamber of the eyeball is placed behind the lens, surrounded by the retina (Figure 1B).The three tunicae of the eyeball are concentrically placed and, from most internal to most external, are: the nervous layer, formed by the retina; the vascular layer, where can be found choroid, ciliary body and iris; and the external layer, which is formed by cornea and sclera [25] (Figure 1B).

The retina is the most complex part of the eye. Its structure and function is similar to those of the cerebral cortex. In fact, retina can be considered as an outpouching of central nervous system during embryonic development. The retina comprises a blind part, insensitive to light, associated with the ciliary body and the iris; and an optical part, containing photoreceptors. In turn, optical part is formed by two sheets: the neuroepithelial stratum, composed by neurons, and the retinal pigmentary epithelium. Mouse neuroretina is composed by eight layers (Figure 1C): the nerve fiber layer, hardly distinguishable in equatorial retina; the ganglion cell layer; the inner plexiform layer; the inner nuclear layer, containing bipolar, amacrine, horizontal and the nuclei of Müller cells; the outer plexiform layer; the outer nuclear layer, formed by photoreceptors nuclei; and the layers of internal and external segments of photoreceptors. Two limiting membranes can also be distinguished: the internal limiting membrane, placed between the vitreous and the retina; and external limiting membrane, found between the outer nuclear layer and the external segment of photoreceptors [25].

Mice and humans have holoangiotic retinas [26]. In these species the entire retina is vascularized, in contrast with anangiotic retinas, such as the avian retinas, where there are not blood vessels inside the retina. In holoangiotic retinas blood flow is directed from the optic disc radially to the periphery of the retina, and vasculature consists of arteries, veins and a wide network of capillaries (Figures 2A and 2B). The retinal circulation develops from the hyaloid artery that regresses after birth. Hyaloid blood vessels following a template of astrocytes growth superficially and deeply forming the retinal capillary plexi [27]. In mouse retina, as happens in most of the mammals including man, blood supply is carried out by two different

http://dx.doi.org/10.5772/56131

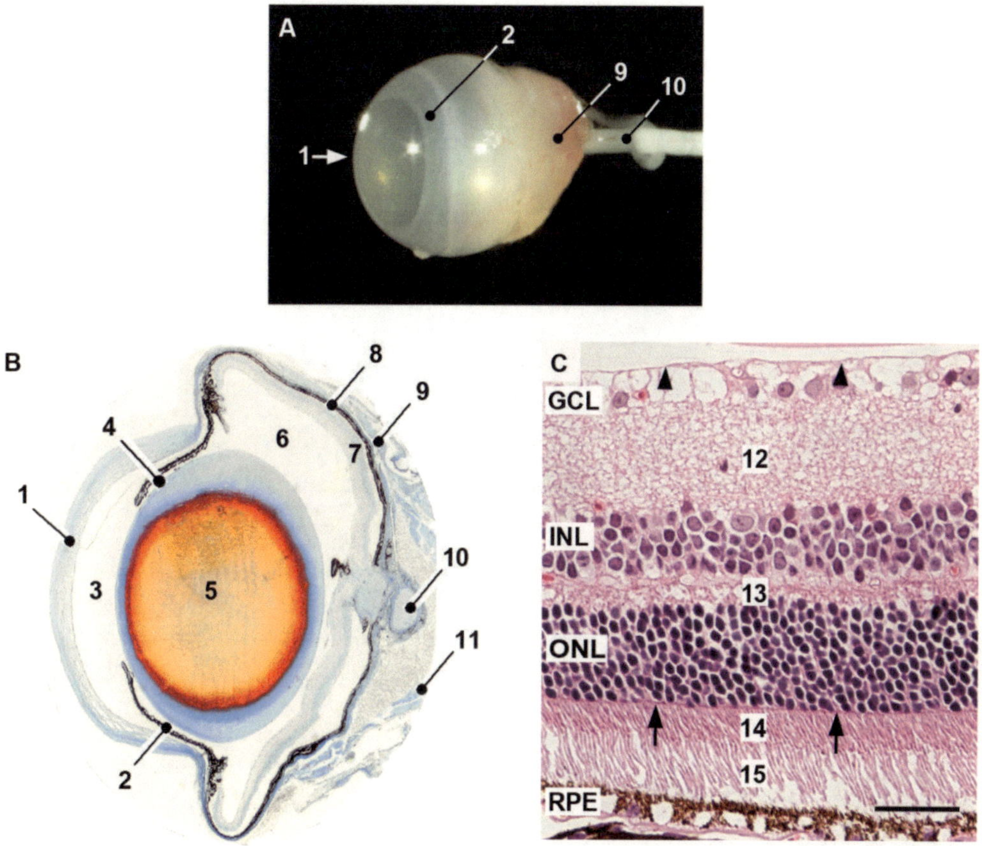

Figure 1. Topography and structure of the retina in the mouse eye. Enucleated eye (A). Paraffin section of an eye stained with Azan trichrome (B). Hematoxylin/eosin stained paraffin section of retina (C). 1: cornea; 2: iris; 3: anterior chamber; 4: posterior chamber; 5: lens; 6: vitreous chamber; 7: retina; 8: choroid; 9: sclera; 10: optic nerve; 11: extraocular muscles; 12: inner plexiform layer; 13: outer plexiform layer; 14: photoreceptor inner segment; 15: photoreceptor outer segment; GCL: ganglion cell layer; INL: inner nuclear layer; ONL: outer nuclear layer; RPE: retinal pigmentary epithelium; arrow head: internal limiting membrane; arrow: external limiting membrane. Scale bar: 34 μm.

vascular systems: retinal vessels, which irrigate from the internal limiting membrane to the inner nuclear layer; and choroidal vessels that supply the rest of the retina [25] (Figure 2F).

The main source for retinal blood supply is the internal carotid artery that gives rise to the ophthalmic artery. This artery goes along with the optic nerve and internally is the origin of the central retinal artery [25, 28, 29]. At the level of the optic disc, the central retinal artery branch in four to eight retinal arterioles, depending on mouse strain. Arterioles run towards retinal periphery, where retinal venules are originated. (Figures 2C and 2D). Retinal arterioles are the origin of precapillary arterioles, which give rise to a capillary network settled between retinal arterioles and venules (Figure 2E). Capillaries are placed in the retina forming two plexi: internal vascular plexus, at the level of ganglion cells and inner plexiform layers; and external vascular plexus, between inner and outer nuclear layers (Figure 2F). The figure 3 shows a

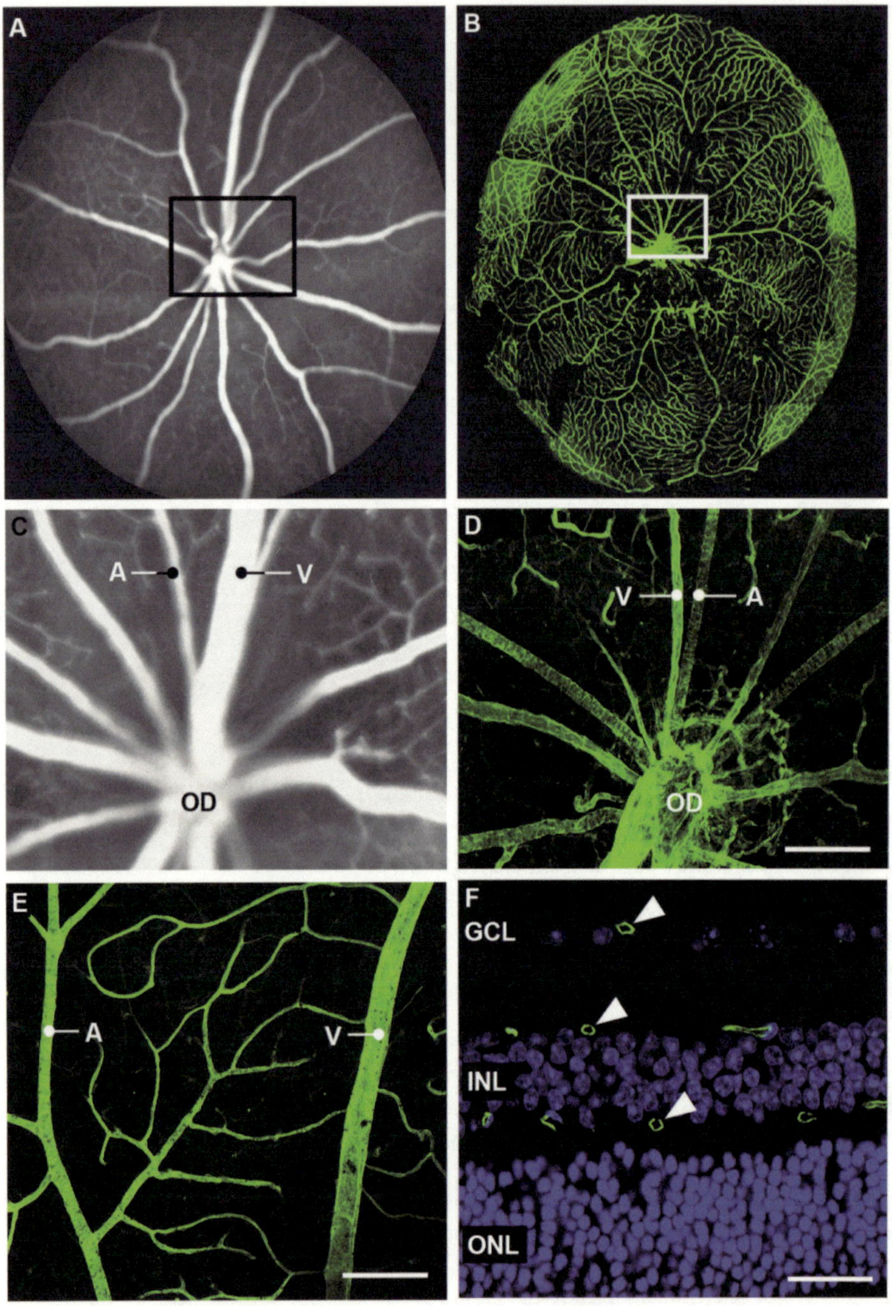

Figure 2. Blood vessel distribution in mouse retina. To show the topography of blood vessels in the mouse retina, scan laser ophtalmoscope images (A and C), collagen IV antibody immunohistochemistry (green) of whole-mount (B,D and E) and paraffin embedded (F) mouse retinas are presented. Nuclei counterstained with ToPro-3 (blue). A: arteriole; V: venule; OD: optic disc; GCL: ganglion cell layer; INL: inner nuclear layer; ONL: outer nuclear layer; arrowhead: blood vessels. Scale bars: 108 μm (D), 122 μm (E) and 34 μm (F).

http://dx.doi.org/10.5772/56131

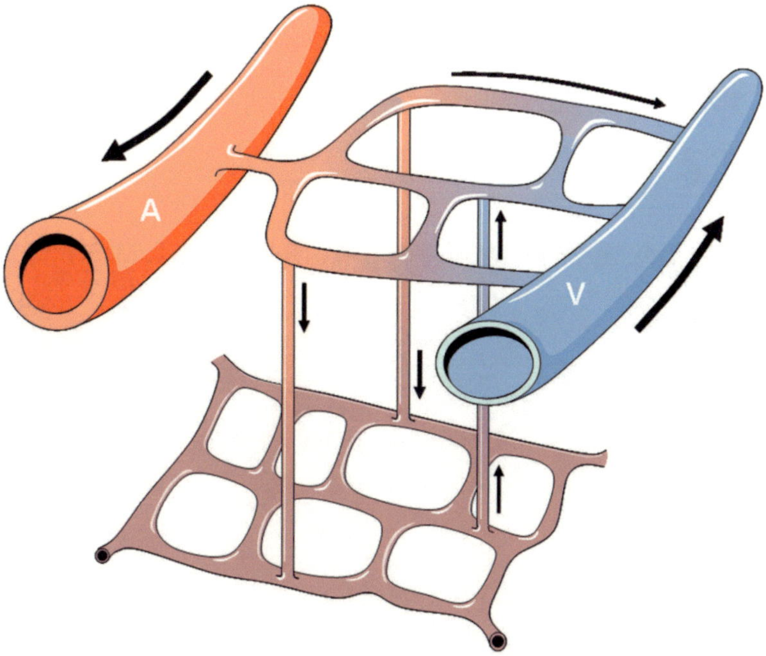

Figure 3. Representative schema of vascular architecture and blood flow in mouse retina. Blood flows from arterioles to venules through a capillary network. Capillaries are placed forming two plexi, the most superficial situated in the outer plexiform layer and the deepest localized in the ganglion cell and inner plexiform layers. Hypoxic blood is collected from capillaries by retinal venules. A: arteriole; V: venule; arrows: blood flow direction.

schematic representation of capillary retinal plexi (adapted from [19]). Retinal capillaries converge into retinal venules, which course parallel to arterioles and drive hypoxic blood to the central retinal vein (Figures 2C and 2D).

The structure of retinal blood vessels is similar to other localizations of the body. The blood vessel wall can be divided in three layers or tunicae: the adventitia layer, the most external, is formed by connective tissue; the media layer, where can be found smooth muscle cells; and the intima layer, consisting in a monolayer of endothelial cells [30]. Retinal arterioles show a tunica adventitia, mainly formed by collagen IV, surrounding all cellular components of blood vessel wall (Figure 4A). Retinal arterioles have a well-developed tunica media, formed by one layer of smooth muscle cells placed perpendicularly to vascular axis (Figures 4B and 4C). Smooth muscle cell number diminishes when arterioles branch in precapillary arterioles, forming a non-continuous layer of sparse smooth muscle cells. Finally, the tunica intima is made of endothelial cells placed parallel to the vessel axis (Figure 4D).

Retinal capillaries are formed by pericytes and endothelial cells surrounded by basement membrane (Figure 5A). Pericytes are a contractile cell population positive in retina for β-actin (Figure 5B), nestin (Figure 5C), NG2 (Figure 5D) and PDGF-Rβ (Figure 5E), among others [31-34]. Endothelial cells are placed in the most internal part of capillaries, in direct contact with blood stream. These cells show an elongated morphology with a big nucleus that protrudes to the

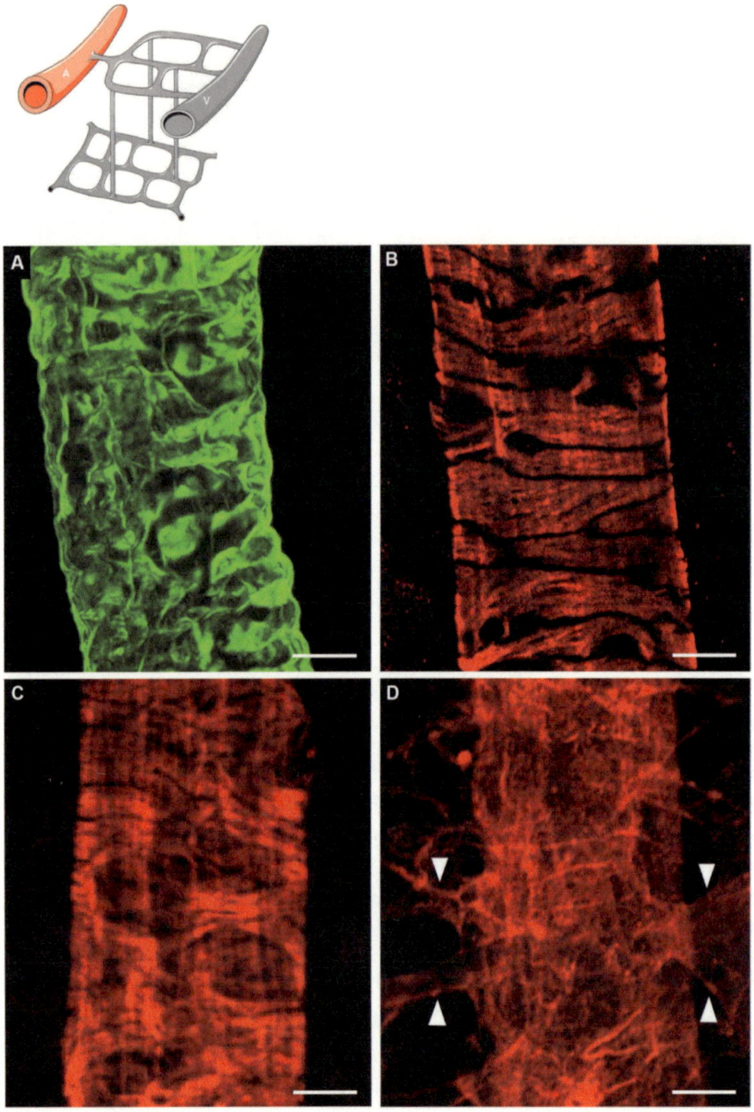

Figure 4. Morphology and composition of retinal arterioles. Different markers were used in order to show the components of the arteriole wall. (A) collagen IV antibody (green) was used to specifically stain basement membrane. Smooth muscle cells (red) were evidenced by means of α-smooth muscle actin antibody (B) and phalloidin (C). (D) Lectin from *Lycopersicon sculentum* allowed the analysis of both endothelial glycocalyx and microglial cells (arrowhead). Scale bars: 7 µm (A,B and C) and 8 µm (D).

vascular lumen (Figure 5). Different markers stain specifically endothelial cells, among others: Von Willebrand factor (Figure 6A), PECAM-1 (Figure 6B) and CD34 (Figure 6C). As happens in the brain, endothelial cells are connected by tight junctions (*zonula ocludens*) (Figure 6D). These tight junctions are an important component of blood-retinal barrier, which prevents the free pass of blood borne molecules to the retinal parenchyma [8].

http://dx.doi.org/10.5772/56131

Retinal venules show, as arterioles do, three concentrically placed layers: a tunica adventitia mainly formed of collagen IV (Figure 7A); a tunica media, consisting of a non-continuous layer of sparse smooth muscle cells (Figure 7B); and a monolayer of endothelial cells, the tunica intima (Figure 7C).

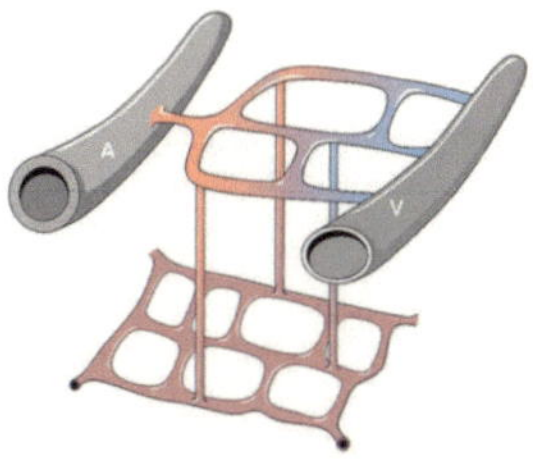

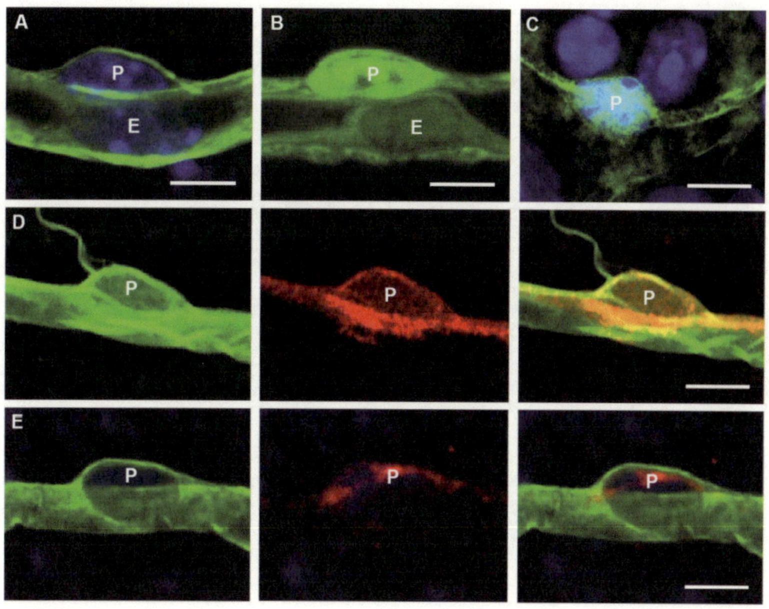

Figure 5. Morphology and composition of retinal capillaries. (A, D and E) Blood basement membrane was marked with anti-collagen IV antibody (green). (B) Capillary morphology in a β-actin/EGFP (green) transgenic mouse. Note that pericytes expressed more β-actin than endothelial cells. (C) Pericytes expressed nestin in the retinal capillaries of a nestin/EGFP (green) transgenic mouse. Specific pericyte markers, such as NG2 (D) and PDGF-Rβ (E), has also been employed to show the distribution and morphology of pericytes. Nuclei counterstained with ToPro-3 (blue). E: endothelial cell; P: pericyte. Scale bars: 9.5 µm (A and B) and 7.3 µm (C,D and E).

In addition to neurons, retinal blood vessels are surrounded by glia that seems to play a role in the formation of blood-retinal barrier [5, 35-38] and the control of retinal blood flow [38]. The term glia encloses two components: neuroglia and microglia. Retinal neuroglia is formed by astrocytes and Müller cells (Figure 8). Astrocytes are only placed in the internal part of the retina, nerve fiber and ganglion cell layers, in close relation with arterioles and venules [39] (Figures 8A and 8B). The principal markers for astrocytes are glial fibrillary acidic protein

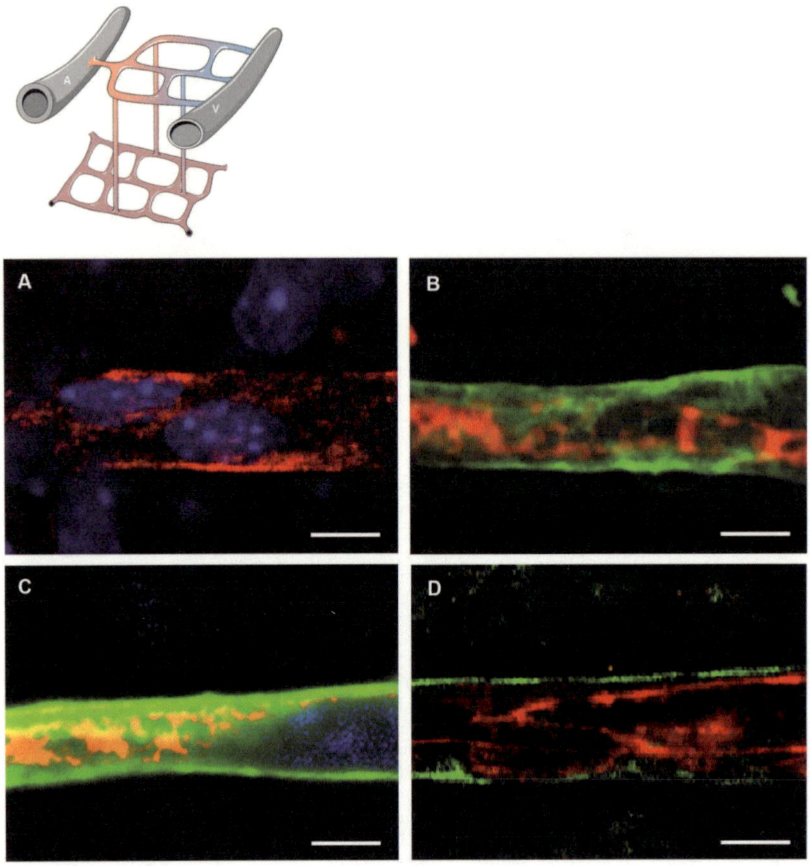

Figure 6. Morphology and composition of retinal capillaries. (B,C and D) Blood basement membrane was marked with anti-collagen IV antibody (green). Endothelial cells were specifically marked with anti-Von Willebrand factor (red) (A), anti-PECAM-1 (red) (B) and anti-CD34 (red) (C) antibodies. (D) Endothelial cell contribution to blood retinal barrier was evidenced using anti-ocludin antibody, a specific marker for endothelial tight junctions. Nuclei counterstained with ToPro-3 (blue). Scale bars: 5.5 μm (A,B and C) and 6.8 μm (D).

(GFAP) (Figure 8A) and desmin (Figure 8B). The nuclei of Müller cells are localized in the inner nuclear layer and their cytoplasmic prolongations extend practically to the entire retina forming the inner and outer limiting membranes (Figure 1C). Müller cells are very easily distinguished using the PDGF-Rα (Figure 8C). Cytoplasmic prolongations of neuroglia, called vascular end-feet, contact with retinal blood vessels (Figures 8A, 8B and 8C).

Retinal microglia originates from hemopoietic cells and invade the retina from the blood vessels of the ciliary body, iris and retinal vasculature [40]. Resting microglial cells are scattered troughout the retina forming a network of potential immunoephector cells, easily marked with Iba1 (Figure 8D). Several studies show that microglial cells have characteristics of dendritic antigen-presenting cells, while others resemble macrophages [41]. During retinopathy, activated microglial cells participate in phagocytosis of debris and facilitate the regenerative processes. Microglial cells are also in contact with blood vessels, forming a special subtype of

http://dx.doi.org/10.5772/56131

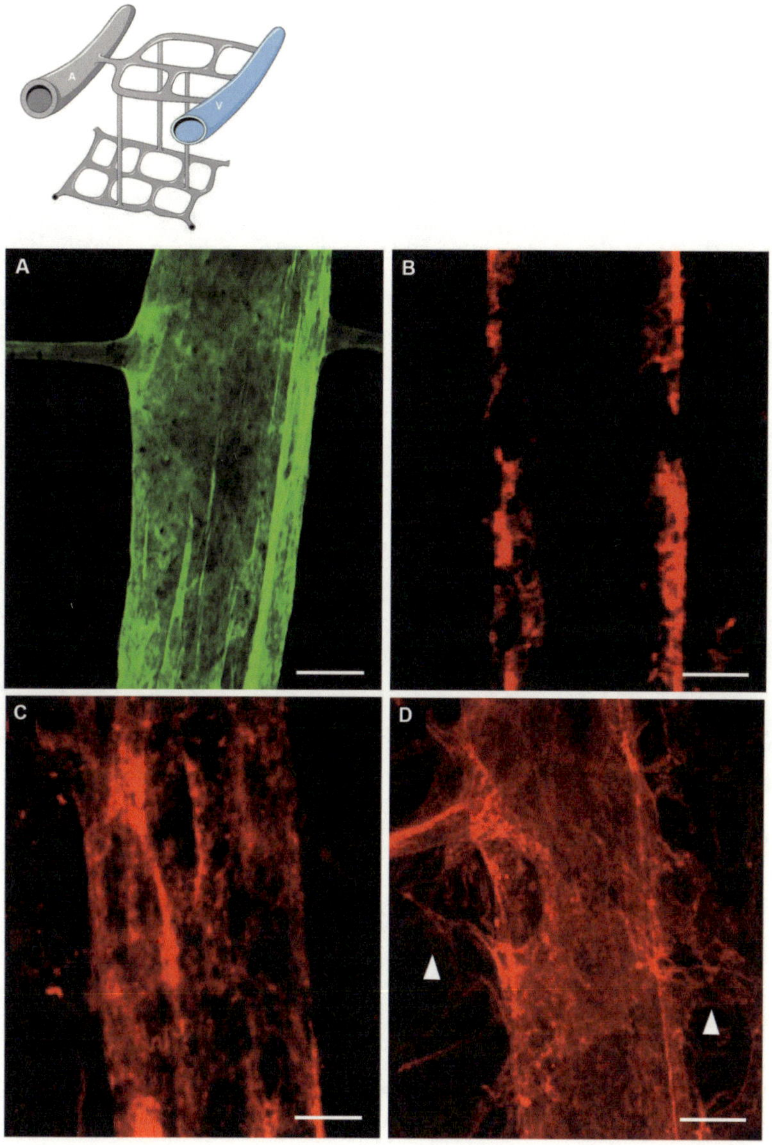

Figure 7. Morphology and composition of retinal venules. Different markers were used in order to show the components of the venule wall. (A) Collagen IV antibody (green) was used to specifically stain basement membrane. (B) Smooth muscle cells were evidenced by phalloidin binding (red). (C) Immunohistochemistry against Von Willebrand factor (red) showed endothelial cell morphology. (D) Lectin from *Lycopersicon sculentum* allowed the analysis of both endothelial glycocalyx and microglial cells (arrowhead). Scale bars: 14.5 μm (A) and 11.5 μm (B,C and D).

perivascular microglial cells, localized in the perivascular space of Virchow-Robin (Figures 4D and 7D).

During the examination of retinal vasculature labeled with two different fluorescent markers, emission signals can often overlap in the final image. This effect, known as colocalization,

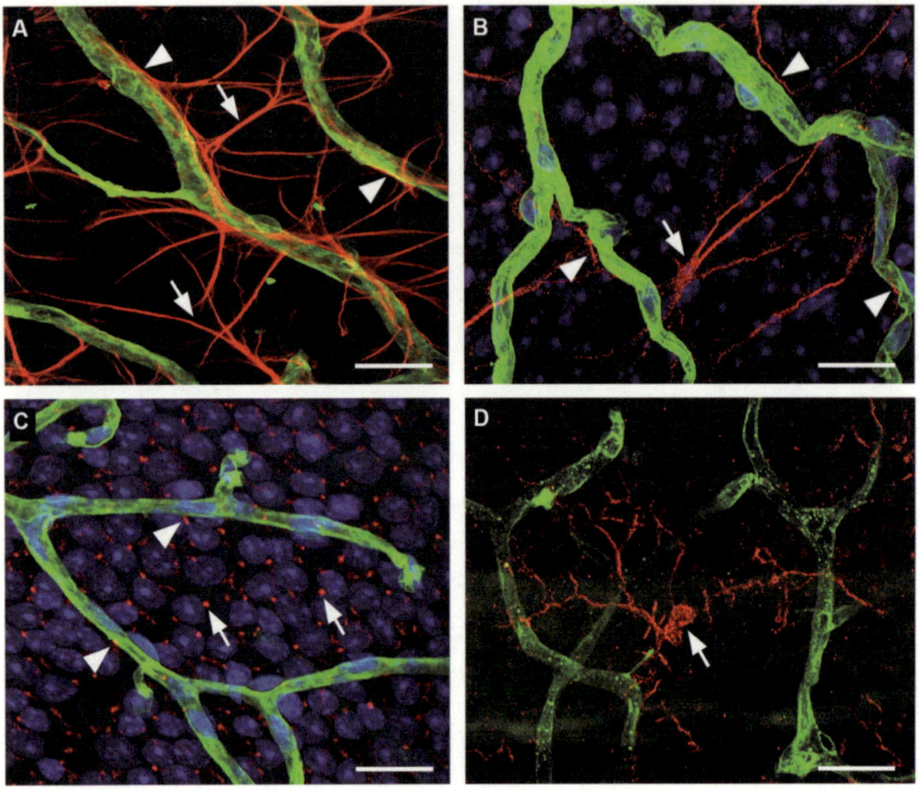

Figure 8. Perivascular glia in mouse retina. Different cell markers were used in order to show perivascular neuroglia and microglia. GFAP (A) and desmin (B) mark astrocytes (arrow), PDGF-Rα (C) stain Müller cells (arrow) and Iba1 (D) is expressed by microglial cells (arrow). (A,B,C and D) Blood basement membrane was marked with anti-collagen IV antibody (green). Nuclei counterstained with ToPro-3 (blue). Arrowhead: vascular end-foot. Scale bars: 23 μm (A,B and C) and 20 μm (D).

occurs when fluorescent dyes bind to molecules residing in a very close spatial position in the tissue [42]. Although colocalization is getting more relevance in modern cell and molecular biological studies, it is probably one of the most misrepresented and misunderstood phenomena. In this way, proteins continue to be described as more or less colocalized with no quantitative justification. This lack of information prevents researchers to analyze protein dynamics or protein-protein interactions [43].

In Figure 9 we can observe a retinal arteriole with the blood vessel basement membrane stained with anti-collagen IV (green) and anti-matrix metalloproteinase 2 (MMP2) (red). MMP2 is a constitutive gelatinase protein that can be observed in a wide variety of healthy mice tissues [44]. One of the main substrates of MMP2 is collagen IV, so MMP2 colocalize with collagen IV in retinal arterioles (yellow). An accurate colocalization analysis is only possible if fluorescent emission spectra are well separated between fluorophores and a correct filter setting is used (Figure 9). When a high degree of emission spectra overlap and/or filter combinations are not well defined the resulting colocalization will be meaningless [42].

http://dx.doi.org/10.5772/56131

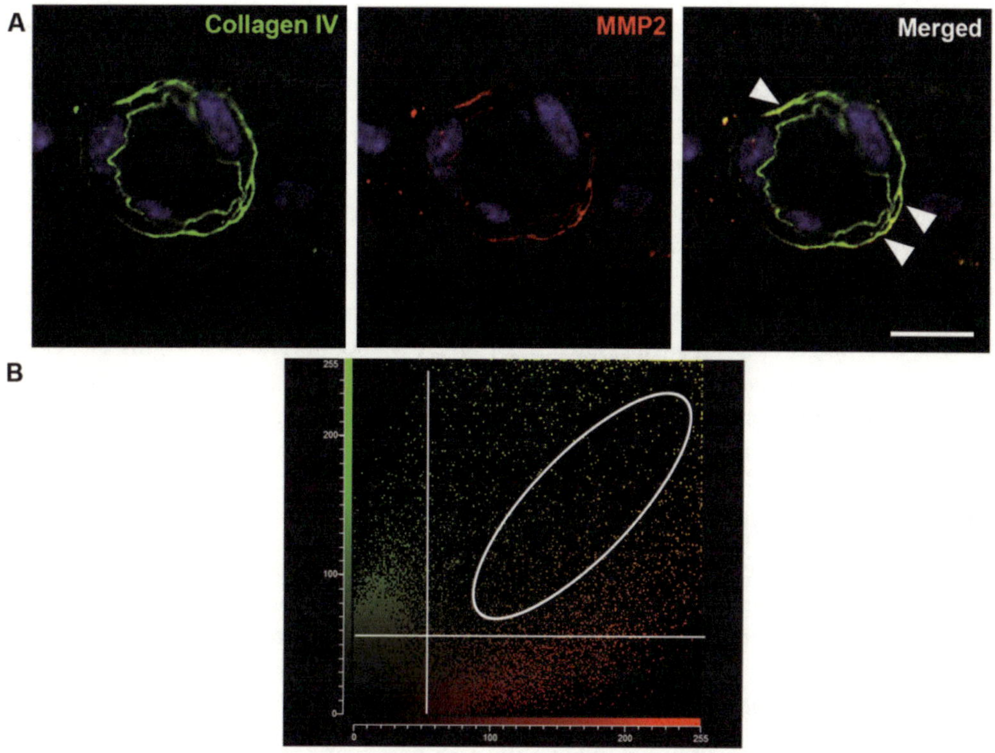

C

$$R_r = \frac{\sum_i (S1_i - S1_{aver}) \cdot (S2_i - S2_{aver})}{\sqrt{\sum_i (S1_i - S1_{aver})^2 \cdot \sum_i (S2_i - S2_{aver})^2}}$$

D

$$R^2 = k_1 \cdot k_2 \qquad k_1 = \frac{\sum_i S1_i \cdot S2_i}{\sum_i (S1_i)^2} \qquad k_2 = \frac{\sum_i S1_i \cdot S2_i}{\sum_i (S2_i)^2}$$

Figure 9. Colocalization of collagen IV with MMP2 in the arteriolar basement membrane. Double inmunohistochemistry was performed in retinal paraffin sections using antibodies against collagen IV (green) and matrix MMP2 (red). (A) Digital images of green (left image) and red (central image) channels showing colocalization (arrowheads) in the right image. Nuclei counterstained with ToPro-3 (blue). (B) Graphic representation in a scatterplot, where pure red and green pixels are between abscissa/ordinate and white lines. Colocalizating pixels are found inside white elliptic region. (C) Pearson's correlation coefficient, where S1 and S2 are pixel intensities in channels red and green respectively; and $S1_{aver}$ ($S2_{aver}$) is the average value of pixels in the first (second) channel. (D) Overlap coefficient, with k_1 being sensitive to the differences in the intensities of channel 2 and k_2 depending on the intensity of channel 1 pixels. Scale bar: 10.2 μm.

Graphical display for colocalization analysis is well represented by a fluorogram: a scatterplot which graphs the intensity of one color versus another on a two-dimensional histogram (Figure

9B). Along the y-axis is plotted green channel, while red channel is graphed on the x-axis. Thus, having each pixel a pair of fluorescent intensities in a Cartesian system. In the scatterplot, pixels having lower fluorescent intensities are close to the origin of abscissa and ordinate, while brighter pixels are dispersed along the graph (Figure 9B). Pure green and red pixels cluster close to the axes of the graph, while colocalized pixels are localized in the center and in the upper right hand of scatterplot (Figure 9B).

As discussed above, a quantitative assessment of colocalization is important in order to analyze protein dynamics and association. Using the information given by the scatterplot several values can be generated. Pearson's correlation coefficient (R_r) is used as standard technique for image pattern recognition. This coefficient is employed to describe the degree of overlap between two images and can be calculated according to the equation seen in Figure 9 C. Values of this coefficient ranges from -1 to 1. The value -1 correspond to a complete lack of overlap between images and 1 a total match of pixels in the two images. Pearson's coefficient takes into account only similarity among pixels in the two images, and does not consider information of pixel intensities. Thus, Pearson's correlation coefficient can overestimate colocalization when the degree of colocalization is low [45].

Another standard value used to quantify colocalization is the overlap coefficient (R^2) (Figure 9D). This coefficient uses two values (k_1 and k_2) in order to characterize colocalization in both channels. This coefficient avoids negative values, which have a harder interpretation. Some authors find this coefficient less reliable than Pearson's correlation coefficient, since overlap coefficient is only applicable in images with similar intensities in the two channels [45].

Diabetic retinopathy is a common and specific microvascular complication of diabetes, and remains the leading cause of blindness in working-aged people [46]. Recent metadata studies established that in the world there are 93 million people with diabetic retinopathy [47]. Nearly all individuals with type 1 diabetes and more than 60% of individuals with type 2 diabetes have some degree of retinopathy after 20 years of disease. There are two phases in diabetic retinopathy. Early phase is known as non-proliferative diabetic retinopathy, and is characterized by thickening of capillary basement membrane, pericyte and vascular smooth muscle cell loss, capillary occlusion and formation of microaneurysms [48]. Proliferative retinopathy, the second phase of the disease, is characterized by the formation of new vessels that pas through the inner limiting retinal membrane and penetrate in the vitreous chamber. New vessels are surrounded by fibrous tissue that may contract, leading to retinal detachment and sudden visual loss. Neovascularization is a consequence of retinal increase of cytokines and growth factors produced in ischemic conditions. Proliferative retinopathy appears in approximately 50% of patients with type 1 diabetes and in about 15% of patients with type 2 diabetes [49].

Confocal laser microscopy allows the study of retinal blood vessels in diabetic mouse models (Fig. 10). Non-obese diabetic (NOD) mice develop type 1 diabetes by autoimmune destruction of pancreatic β cells [50].

The analysis of 8 months-old NOD mice whole-mount flat retinas marked with anti-collagen IV antibody showed basement membrane alterations in venules (Fig. 10). Similarly, db/db mice also showed alterations in basement membrane of retinal venules (Fig. 10). Db/db mice are homozygous for a mutation in the leptin receptor, and spontaneously develop type 2 diabetes

http://dx.doi.org/10.5772/56131

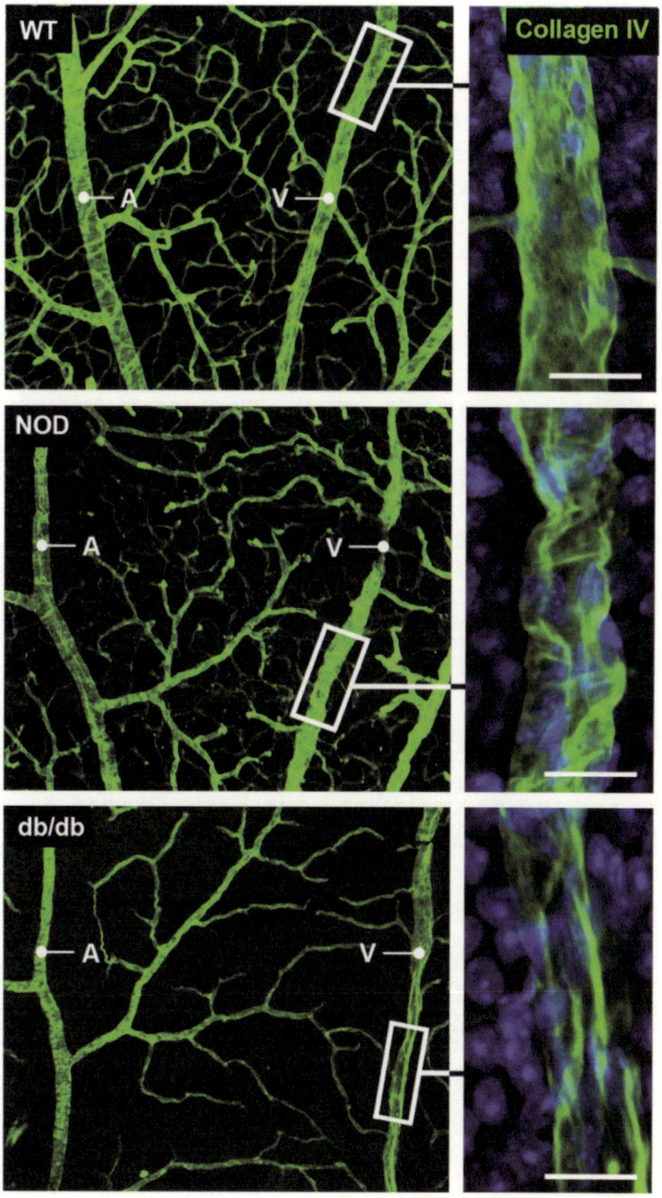

Figure 10. Venule basement membrane alterations in diabetic retinopathy mouse models. Compare the morphology of basement membrane (green) between wild type (WT) mice and Non-obese diabetic (NOD) mice, and db/db mice. Nuclei counterstained with ToPro3 (blue). Blood basement membrane was marked with anti-collagen IV antibody. A: arteriole, V: venule. Scale bars: 21,2 μm (WT), 15 μm (NOD) and 16,2 μm (db/db).

[51]. These result suggested that, both in diabetes type 1 and 2, normal functions of blood basement membrane are altered in venules during diabetes. Venule basement membrane separate endothelial cells and smooth muscle cells from underlying connective tissue provid-

ing structural support, a selective barrier of filtration, and a substrate for molecular adhesion that modulates cells of the vascular wall.

By the above, we can conclude that the analysis of whole-mount retinas by laser confocal technology is a reliable method that allows a complete three-dimensional microscopical analysis of retinal blood vessels in health and disease. Combining the use of fluorescent markers for proteins, signaling molecules, etc, it will be possible study the topography and the structure of blood vessels. Thus, the use of confocal laser microscopy together with new mouse eye disease models may provide basis to fully understand the alterations of retinal vasculature during retinopathies.

Acknowledgements

We thank V. Melgarejo, L. Noya and A. Vazquez for technical assistance.

This work was supported by grants PS09/01152 from the Instituto de Salud Carlos III, Ministerio de Ciencia e Innovación, Spain and PTDC/SAU-ORG/110856/2009 from the Fundação para a Ciência e a Tecnologia do Ministério da Ciência, Tecnologia e Ensino Superior, Portugal.

Author details

David Ramos[1,2], Marc Navarro[1,2], Luísa Mendes-Jorge[1,3], Ana Carretero[1,2], Mariana López-Luppo[1,2], Víctor Nacher[1,2], Alfonso Rodríguez-Baeza[4] and Jesús Ruberte[1,2]

*Address all correspondence to: jesus.ruberte@uab.cat

1 CBATEG, Autonomous University of Barcelona, E-08193-Bellaterra, Spain

2 Department of Animal Health and Anatomy, School of Veterinary Medicine, Autonomous University of Barcelona, Spain

3 CIISA, Faculty of Veterinary Medicine, Technical University of Lisbon, Av. Da Universidade Técnica, Lisbon, Portugal

4 Department of Morphological Sciences, Faculty of Medicine, Autonomous University of Barcelona, Spain

References

[1] Leamey CA, Protti DA, Dreher B. Comparative survey of the mammalian visual system with reference to the mouse. In: Chalupa LM and Williams RW (Eds.) Eye,

Retina, and Visual System of the Mouse. The Mit Press. Cambridge, Massachusetts; 2008.

[2] Stahl A, Connor KM, Sapieha P, Chen J, Dennison RJ, Krah NM, Seaward MR, Willett KL, Aderman CM, Guerin KL, Hua J, Löfqvist C, Hellström A, Smith A. The mouse retina as an angiogenesis model. Investigative Ophthalmology and Visual Science 2010;51(6): 2813-26. http://www.iovs.org/content/51/6/2813.full.pdf+html

[3] Funk RH. Blood supply of the retina. Ophtalmic Research 1997;29(5): 320-5.

[4] Wangsa-Wirawan ND, Linsenmeier RA. Retinal oxygen: Fundamental and Clinical Aspects. Archives of ophthalmology 2003;121(4): 547-57. http://archopht.jamanetwork.com/article.aspx?articleid=415259

[5] Gartner LP, Hiatt JL. Texto Atlas de Histología. McGraw-Hill (ed.) Interamericana; 2002.

[6] Alm A, Bill A. Ocular and normal blood flow at normal and increased intraocular pressure in monkeys (Macaca irus): a study with radioactively labelled microspheres including flow determinations in brain and some other tissues. Experimental Eye Research 1973;15: 15-29. http://www.sciencedirect.com/science/article/pii/0014483573901851

[7] Bristow EA, Griffiths PG, Andrews RM, Johnson MA and Turnbull DM. The distribution of mitochondrial activity in relation to optic nerve structure. Archives of Ophtalmology 2002;120(6): 791-6. http://archopht.jamanetwork.com/article.aspx?articleid=271059

[8] Knott RM, Forrester JV. Pathogenesis of diabetic eye disease. In: Pickup J and Williams G (eds.) Textbook of Diabetes. Blackwell; 2003. p. 48.1-48.17.

[9] Cheung N, Mitchell P, Wong TY. Diabetic retinopathy. Lancet 2010;376(9735):124-36.

[10] Ramon y Cajal S. Histologie du Système Nerveux de l'Homme et des Vertébrés. Madrid: Consejo Superior de Investigaciones Científicas 1952.

[11] Dräger UC, Olsen JF. Ganglion cell distribution in the retina of the mouse. Investigative Ophtalmology and Visual Science 1981;20(3):285-93. http://www.iovs.org/content/20/3/285.full.pdf+html

[12] Carter-Dawson LD, LaVAil MM. Rods and cones in the mouse retina. I. Structural analysis using light and electron microscopy. The Journal of Comparative Neurology 1979;188(2): 245-62. http://onlinelibrary.wiley.com/doi/10.1002/cne.901880204/abstract

[13] Tsukamoto Y, Morigiwa K, Ueda M, Sterling P. Microcircuits for night vision in mouse retina. The Journal of Neuroscience 2001;21(21): 8616-23. http://www.jneurosci.org/content/21/21/8616.full.pdf+html

[14] Jeon CJ, Strettoi E, Masland RH. The major cell populations of the mouse retina. The Journal of Neuroscience 1998;18(21): 8936-46. http://www.jneurosci.org/content/18/21/8936.full.pdf+html

[15] Haverkamp S, Wässle H. Immunocytochemical analysis of the mouse retina. The Journal of Comparative Neurology 2000;424(1): 1-23. http://onlinelibrary.wiley.com/doi/10.1002/1096-9861%2820000814%29424:1%3C1::AID-CNE1%3E3.0.CO;2-V/pdf

[16] Paques SM, Simonutti M, Roux MJ, Picaud S, Levavasseur E, Bellman C, Sahel JA. High resolution fundus imaging by confocal scanning laser ophthalmoscopy in the mouse. Vision Research 2006;46(8-9): 1336-45.

[17] Richter M, Gottanka J, May CA, Welge-Lüssen U, Lütjen-Drecoll E. Retinal vasculature changes in Norrie disease mice. Investigative Ophtalmology and Visual Science 1998;39(12): 2450-7. http://www.iovs.org/content/39/12/2450.full.pdf+html

[18] Kuwabara T, Cogan DC. Studies of retinal vascular patterns. I. Normal architecture. Archives of Ophtalmology 1960;64: 904-11.

[19] Paques M, Tadayoni R, Sercombe R, Laurent P, Genevois O, Gaudric A, Vicaut E. Structural and hemodynamic analysis of the mouse retinal microcirculation. Investigative Ophtalmology and Visual Science 2003;44: 4960-67. http://www.iovs.org/content/44/11/4960.full.pdf+html

[20] Ruberte J, Ayuso E, Navarro M, Carretero A, Nacher V, Haurigot V, George M, Llombart C, Casellas A, Costa C, Bosch A, Bosch F. Increased ocular levels of IGF-1 in transgenic mice lead to diabetes-like eye disease. The Journal of Clinical Investigation 2004;113:1149–57. http://www.jci.org/articles/view/19478

[21] Van Eeden PE, Tee LBG, Lukehurst S, Lai C-M, Rakoczy EP, Beazley LD, Dunlop SA. Early Vascular and Neuronal Changes in a VEGF Transgenic Mouse Model of Retinal Neovascularization. Investigative Ophtalmology and Visual Science 2006;47(10): 4638-45. http://www.iovs.org/content/47/10/4638.full.pdf+html

[22] Mendes-Jorge L, Ramos D, Luppo M, Llombart C, Alexandre-Pires G, Nacher V, Melgarejo V, Correia M, Navarro M, Carretero A, Tafuro S, Rodríguez-Baeza A, Esperança-Pina JA, Bosch F, Ruberte J. Scavenger function of resident autofluorescent perivascular macrophages and their contribution to the maintenance of the blood-retinal barrier. Investigative Ophtalmology and Visual Science 2009;50(12): 5997-6005. http://www.iovs.org/content/50/12/5997.full.pdf+html

[23] Mendes-Jorge L, Llombart C, Ramos D, López-Luppo M, Valença A, Nacher V, Navarro M, Carretero A, Méndez-Ferrer S, Rodriguez-Baeza A, Ruberte J. Intercapillary bridging cells: immunocytochemical characteristics of cells that connect blood vessels in the retina. Experimental Eye Research 2012;98: 79-87. http://ac.els-cdn.com/S0014483512000942/1-s2.0-S0014483512000942-main.pdf?_tid=1808ed70-1de9-11e2-b382-00000aab0f27&acdnat=1351089954_3d594e6fd63d988a1b69c98e4b33fa0f

[24] Remtulla S, Hallet PE. A schematic eye of the mouse, and comparisons with the rat. Vision Research 1985;25(1): 21-31. http://www.sciencedirect.com/science/article/pii/0042698985900768

[25] Smith RS, Hawes NL, Chang B, Nishina PM. Retina. In: Smith RS (ed.) Systematic evaluation of the mouse eye. CRC Press; 2002. p 195 – 225.

[26] Germer A, Biedermann B, Wolburg H, Schuck J, Grosche J, Kuhrt H, Reichelt W, Schousboe A, Paasche G, Mack AF, Reichenbach A. Distribution of mitochondria within müller cells—I. Correlation with retinal vascularisation in different mammalian species. Journal of Neurocytology 1998;27: 329-45.

[27] Dorrell M, Aguilar E, Friedlander M. Retinal vascular development is mediated by endothelial filopodia, a pre-existing astrocytic template and specific R-cadherin adhesion. Investigative Ophthalmology and Visual Science 2002;43: 3500-10.

[28] Saint-Geniez M, D'Amore PA. Development and pathology of the hyaloid, choroidal and retinal vasculature. International Journal of Developmental Biology 2004;48: 1045-58. http://www.ijdb.ehu.es/web/paper.php?doi=10.1387/ijdb.041895ms

[29] Schaller O. Illustrated Veterinary Anatomical Nomenclature. Ferdinand Enke Verlag; 1992. http://books.google.es/books?hl=ca&lr=&id=0xIbF8NHU5kC&oi=fnd&pg=PA1&dq=Schaller+O.+Illustrated+Veterinary+Anatomical+Nomenclature&ots=hElShZ0as5&sig=G61V_ozauUAbmGH-lIdnRQQdLBgc#v=onepage&q=Schaller%20O.%20Illustrated%20Veterinary%20Anatomical%20Nomenclature&f=false

[30] Laflamme MA, Sebastian MM, Buetow BS. Cardiovascular. In: Comparative Anatomy and Histology. A Mouse and Human Atlas. Academic Press; 2012. p 135-153.

[31] Rouget CMB. Mémoire sur le développment, la structure et les propriétés physiologiques des capillaires sanguins et lymphatiques. Archives de physiologie normale et pathologique 1873;5: 603-63.

[32] Zimmermann KW. Der feinere Bau der Blutkapillaren. Z. Anat. Entwicklungsgesch 1923;68: 29-109.

[33] Hammes HP, Lin J, Renner O, Shani M, Lundqvist A, Betsholtz C, Brownlee M, Deutsch U . Pericytes and the pathogenesis of diabetic retinopathy. Diabetes 2002;51(10): 3107-12. http://diabetes.diabetesjournals.org/content/51/10/3107.full.pdf+html

[34] Armulik A, Genové G, Bestholtz C. Pericytes: Developmental, physiological and pathological perspectives, problems and promises. Developmental Cell 2011;21(2): 193-215. http://www.sciencedirect.com/science/article/pii/S1534580711002693#

[35] Newman EA. New roles for astrocytes: regulation of synaptic transmission. Trends in Neurosciences 2003;26: 536-42. http://www.sciencedirect.com/science/article/pii/S0166223603002376

[36] Antonetti DA, Barber AJ, Bronson SK, Freeman WM, Gardner TW, Jefferson LS, Kester M, Kimball SR, Krady JK, LaNoue KF, Norbury CC, Quinn PG, Sandirasegarane L, Simpson IA. Perspectives in diabetes. Diabetic retinopathy. Seeing beyond glucose-induced microvascular disease. Diabetes 2006;55: 2401-11. http://diabetes.diabetesjournals.org/content/55/9/2401.full.pdf+html

[37] Kim JH, Kim JH, Yu YS, Kim DH, Kim KW. Recruitment of pericytes and astrocytes is closely related to the formation of tight junction ion developing retinal vessels. Journal of Neuroscience Research 2009;87(3): 653-9. http://onlinelibrary.wiley.com/doi/10.1002/jnr.21884/pdf

[38] Attwell D, Buchan AM, Charpak S, Lauritzen M, Macvicar BA, Newman EA. Glial and neuronal control of brain blood flow. Nature 2010;468(7321): 232-43. http://www.nature.com/nature/journal/v468/n7321/full/nature09613.html

[39] Hollander H, Makarov F, Deher Z, van Driel D, Chang-Ling TL, Stone J. Structure of the macroglia of the retina: sharing and division of labour between astrocytes and müller cells. The Journal of Comparative Technology 1991;313: 587-603.

[40] Chen L, Yang P, Kijlstra A. Distribution, markers and function of retinal microglia. Ocular Immunology and Inflammation 2002;10(1): 27-39.

[41] Provis JM, Diaz CM, Penfold PL. Microglia in human retina: a heterogeneous population with distinct ontogenies. Perspectives on Developmental Neurobiology 1996;3(3): 213-22.

[42] Smallcombe A. Multicolor imaging: The important question of co-localization. Biotechniques 2001;30(6): 1240-6. http://www.biotechniques.com/multimedia/archive/00011/01306bt01_11243a.pdf

[43] Zinchuk V and Zinchuk O. Quantitative Colocalization Analysis of Confocal Fluorescence Microscopy Images. In: Current Protocols in cell biology. Wiley Interscience; 2008. p 4.19.1-4.19.16.

[44] Yong VW, Krekoski CA, Forsyth PA, Bell R, Edwards DR. Matrix metalloproteinases and diseases of the CNS. Trends in Neurosciences 1998;21: 75-80. http://www.sciencedirect.com/science/article/pii/S0166223697011697#

[45] Wu Y, Zinchuk V, Grossenbacher-Zinchuk O, Stefani E. Critical Evaluation of Quantitive Colocalization Analysis in Confocal Fluorescence Microscopy. Interdisciplinary sciences: Computational life sciences 2012;4: 27-37. http://www.springerlink.com/content/421887r526tp5933/fulltext.pdf

[46] Cheung N, Mitchell P, Wong TY. Diabetic Retinopathy. Lancet 2010;376(9735): 124-36. http://ac.els-cdn.com/S0140673609621243/1-s2.0-S0140673609621243-main.pdf?_tid=1665d802-752d-11e2-acc0-00000aacb362&acdnat=1360684908_0b184c5dc0521f994ef5987064d04613

[47] Yau JW, Rogers SL, Kawasaki R, Lamoureux EL, Kowalski JW, Bek T, Chen SJ, Dekker JM, Fletcher A, Grauslund J, Haffner S, Hamman RF, Ikram MK, Kayama T,

Klein BE, Klein R, Krishnaiah S, Mayurasakorn K, O'Hare JP, Orchard TJ, Porta M, Rema M, Roy MS, Sharma T, Shaw J, Taylor H, Tielsch JM, Varma R, Wang JJ, Wang N, West S, Xu L, Yasuda M, Zhang X, Mitchell P, Wong TY, Meta-Analysis for Eye Disease (META-EYE) Study Group. Global prevalence and major risk factors of diabetic retinopathy. Diabetes Care 2012;35(3): 556-64.

[48] Gardiner TA, Archer DB, Curtis TM, Stitt AW. Arteriolar involvement in the microvascular lesions of diabetic retinopathy: implications for pathogenesis. Microcirculation 2007;14(1): 25-38. http://onlinelibrary.wiley.com/doi/10.1080/10739680601072123/pdf

[49] Klein R, Klein BE, Moss SE, Davis MD, DeMets DL. The Wisconsin epidemiologic study of diabetic retinopathy. II. Prevalence and risk of diabetic retinopathy when age at diagnosis is less than 30 years. Archives of Ophtalmology 1984;102(4): 520-6. http://archopht.jamanetwork.com/article.aspx?articleid=635006

[50] Shaw SG, Boden JP, Biecker E, Reichen J, Rothen B. Endothelin Antagonism Prevents Diabetic Retinopathy in NOD Mice: a Potential Role of the Angiogenic Factor Adrenomedullin. Experimental Biology and Medicine 2006;231: 1101-5. http://ebm.rsmjournals.com/content/231/6/1101.full.pdf+html

[51] Hummel KP, Dickie MM, Coleman DL. Diabetes, a new mutation in the mouse. Science 1966;153: 1127-8. http://www.sciencemag.org/content/153/3740/1127.long

Chapter 3

In Vivo Biopsy of the Human Cornea

Akira Kobayashi, Hideaki Yokogawa and
Kazuhisa Sugiyama

Additional information is available at the end of the chapter

http://dx.doi.org/10.5772/55657

1. Introduction

The human cornea is the transparent, dome-shaped tissue that covers anterior segment of the eye. It consists of five thin layers: epithelium, Bowman's layer, stroma, Descemet's membrane and endothelium. Due to its transparency, real-time *in vivo* confocal microscopic observation of the normal and diseased cornea was developed since the early 1990s. [1,2] Since histologic-like images are obtained by the device, it is called "painless biopsy" and/or "*in vivo* biopsy". In this chapter, clinical application of *in vivo* laser confocal microscopy is demonstrated.

2. *In vivo* laser confocal microscopy

In 2005, cornea specific *in vivo* laser confocal microscopy (Heidelberg Retina Tomograph 2 Rostock Cornea Module, HRT2-RCM, Heidelberg Engineering GmbH, Dossenheim, Germany) has become available (Figure 1).[3,4] It has permitted detailed *in vivo* layer-by-layer observations of corneal microstructure with an axial resolution of nearly 1 μm. [4]

Before examination, written informed consents are obtained from all patients; this includes possible consequences of this device such as superficial punctate keratopathy. A large drop of contact gel (Comfort Gel ophthalmic ointment®, Bosch & Lomb, GmbH, Berlin, German) is applied on the front surface of the microscope lens and ensuring no air bubbles had formed, a Tomo-cap® (Heidelberg Engineering GmbH, Dossenheim, Germany) is mounted on the holder to cover the microscope lens. Then, the center and peripheral cornea were examined layer by layer by *in vivo* laser confocal microscopy.

Figure 1. Heidelberg Retina Tomograph 2 Rostock Cornea Module The device uses a 60× water-immersion objective lens (Olympus Europa GmbH, Hamburg, Germany) and utilizes a 670-nm diode laser as the light source with the area of observation 400μm square section.

3. Normal human cornea and conjunctiva

In vivo laser confocal microscopy enables to visualize normal human cornea layer by layer: superficial epithelial cells, basal epithelial cells, Bowman's layer with nerves, Bowman's layer with K-structures (Kobayashi-structures), stroma and endothelial cells (Figure 2).[5,6] K-structures are fibrous structures with a diameter of 5 to 15 μm, and are considered to be anterior collagen fiber bundles running at the posterior surface of Bowman's layer. [5] Mapping of the K-structure revealed that it showed mosaic pattern which is completely different from the whirl pattern of corneal nerves. [6]

In addition, conjuctival cells and meibomian glands are able to visualize *in vivo* (Figure 3).[7, 8]

4. Corneal infections

Acanthamoeba is a ubiquitous, free-living amoeba found in water (swimming pool, hot tubs, tap water, contact lens solutions), air and soil, but *Acanthamoeba* keratitis is a relatively newly recognized entity: the first case was reported in 1974.[9] As the use of soft contact lenses increased in the early 1980s, the incidence of reported *Acanthamoeba* keratitis increased dramatically. *Acanthamoeba* keratitis is relatively uncommon but is a potentially blinding corneal infection. Clinical diagnosis is very difficult, especially in the early phase of the disease; it is often misdiagnosed and treated as a herpes simplex infection.[10] A definite diagnosis is made by confirmation of *Acanthamoeba* in corneal lesions with direct examination, corneal

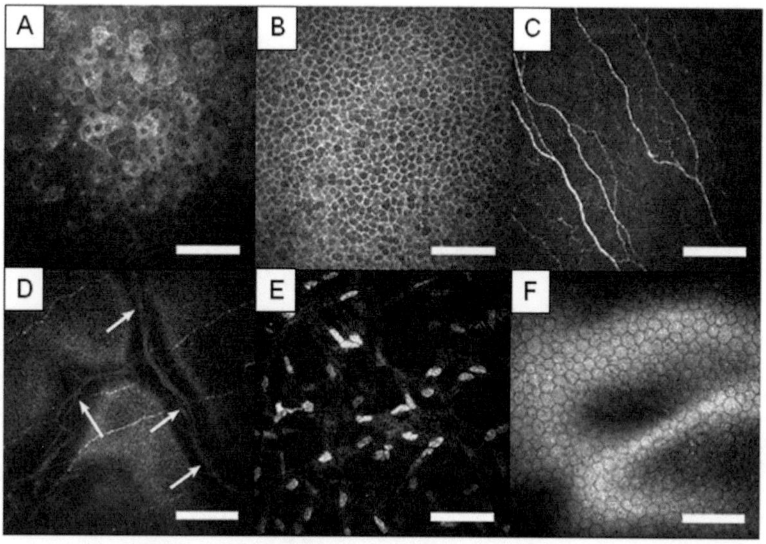

Figure 2. Normal human cornea. A. Superficial epithelial cells. B. Basal epithelial cells. C. Bowman's layer with nerves. D. Bowman's layer with K-structures (arrows). E. Stroma. F. Endothelial cells. (Bar=100μm)

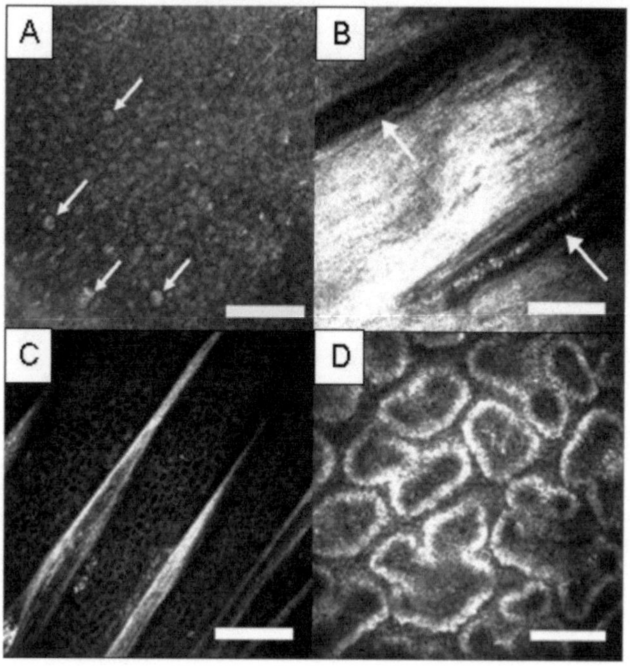

Figure 3. Normal human conjunctiva and meibomian glands. A. Conjuctival epithelium with Goblet cells (arrows). B. Sub-conjunctival fibrous tissue with conjunctival vessls (arrows). D. Palisade of Vogt of corneal limbus. D. Meibomian gland of the upper tarsus. (Bar=100μm)

biopsy or with culture; however, these methods are invasive, time-consuming, and are not always routinely available. The invasive methods are often postponed until there is a high index of suspicion for the disease and when there has been no response to treatments for bacterial, viral and/or fungal keratitis. [10] The usefulness of *in vivo* white-light confocal microscopy in diagnosis and monitoring for improvement of *Acanthamoeba* keratitis has been reported. [11] *In vivo* laser confocal microscopy has also been shown to be useful in the early diagnosis of *Acanthamoeba* keratitis (Figure 4 A-C).[12-15] It is also reported that fungal hyphae can be well visualized by *in vivo* laser confocal microscopy.[16,17]

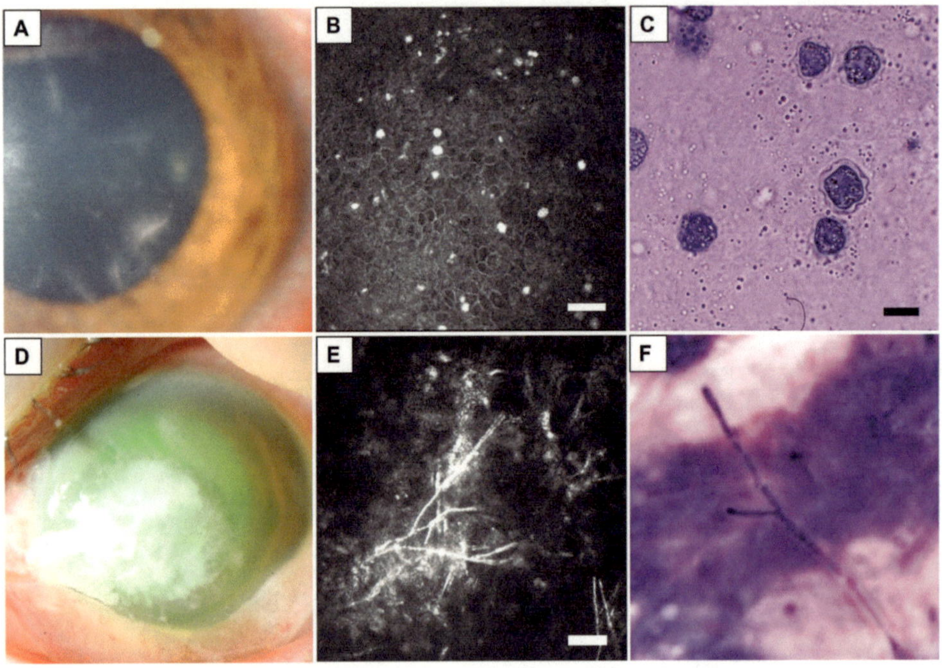

Figure 4. Corneal infections. A. Slit-lamp photograph of the cornea with *Acanthamoeba* keratitis. Subepithelial opacities and numerous radial keratoneuritis lesions were observed. B. In the epithelial basal cell layer, numerous highly reflective, high-contrast round-shaped particles 10-20μm in diameter suggestive of *Acanthamoeba* cysts were detected by *in vivo* laser confocal microscopy. (Bar=50μm) C. Direct examination of the epithelial scraping with Parker ink-potassium hydroxide revealed Acanthamoeba cysts. Note that the cysts have double walls with characteristic wrinkled outer wall. (Bar=10μm) D. Slit-lamp photograph of the cornea with *Aspergillus* keratitis. Severe corneal ulcer was observed. E. In the stormal layer, numerous highly reflective, high-contrast branching filaments suggestive of *Aspergillus* hyphae were detected by *in vivo* laser confocal microscopy. (Bar=50μm) F. Direct examination of the epithelial scraping revealed *Aspergillus* hyphae

5. Corneal dystrophies

In vivo laser confocal microscopy are proven useful to visualize pathological changes of corneal dystrophies *in vivo* (Figure 5) [18-20]. It is also useful to differentially diagnose confusing corneal dystrophies. Previously, considerable confusion exists clinically in distinguishing between Thiel-Behnke and Reis-Bücklers corneal dystrophy (Figure 5A, 5C). However, using *in vivo* laser confocal microscopy, unique and characteristic confocal images are readily obtained at the levels of Bowman's layer; relatively highly reflective deposits in Thiel-Behnke dystrophy but extremely highly reflective deposits in Reis-Bücklers corneal dystrophy. The deposits in Thiel-Behnke corneal dystrophy had round-shaped edges with dark shadows, whereas the deposits in Reis-Bücklers corneal dystrophy did not have round-shaped edges but consisted of highly reflective small granular materials without any shadows (Figure 5B, 5D)[18]. It is also possible to differentially diagnose with corneal stromal dystrophies including Avellino (Figure 5E, 5F), lattice (Figure 5G, 5H), macular (Figure 5I, 5J) and Schnyder dystrophy (Figure 5K, 5L).[19, 20]

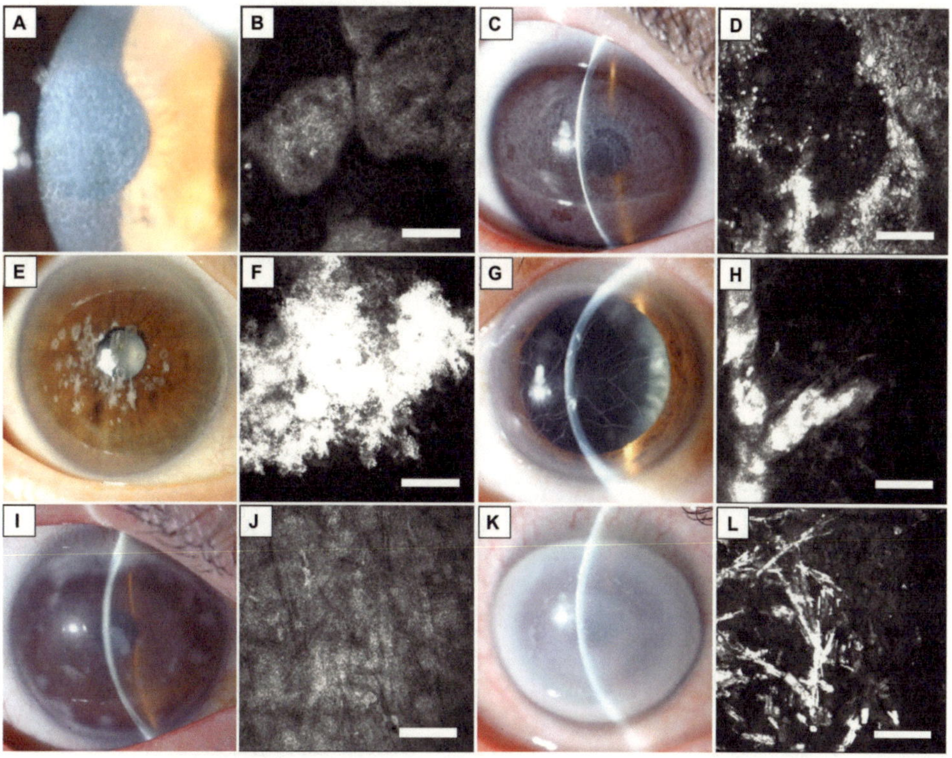

Figure 5. Corneal dystrophies. A. Slit-lamp biomicroscopic photograph of Thiel-Behnke corneal dystrophy (*TGFBI* R555Q). Honeycomb-shaped gray opacities were observed at the level of Bowman's layer. B. *In vivo* laser confocal microscopy showed focal deposition of homogeneous reflective materials with round-shaped edges in the basal epithelial layer. All deposits accompanied dark shadows. C. Slit-lamp biomicroscopic photograph of Reis-Bücklers corneal dystrophy (*TGFBI* R124L). Bilateral gray-white, amorphous opacities of various sizes at the level of Bowman's layer were observed. D. *In vivo* laser confocal microscopy showed focal deposition of highly reflective irregular and granular materials in the basal epithelial layer. No deposits accompanied dark shadows. E. Slit-lamp biomicroscopic photograph of Avellino corneal dystrophy (*TGFBI* R124H) showed round gray-white deposits and scattered stellate opacities in the superficial and mid-stroma. F. *In vivo* laser confocal microscopy showed focal deposits of extremely highly reflective material with irregular edges in the stromal layer. G. Slit-lamp biomicroscopic photograph of lattice corneal dystrophy

(*TGFBI* R124C) showed radially oriented thick lattice lines in the stroma. H. *In vivo* laser confocal microscopy showed highly reflective lattice-shaped materials in the stromal layer. I. Slit-lamp biomicroscopic photograph of macular corneal dystrophy (*CHST6* A217T) showed anterior and deep stromal opacities with indistinct borders that extend out to the corneal periphery. Some gray-white discrete deposits could be seen in the stroma. J. *In vivo* laser confocal microscopy showed homogeneous reflective materials with dark striae-like images. Normal keratocytes were not seen. K. Slit-lamp biomicroscopic photograph of Schnyder corenal dystrophy (*UBIAD1* N233H) showed dense anterior stromal disciform opacity with lipoid arcus. L. *In vivo* laser confocal microscopy showed numerous crystals with varying sizes were observed in the sub-epithelial stromal layer.

6. Cytomegalovirus corneal endotheliits/uveitis

Corneal endotheliitis, characterized by corneal edema associated with linear keratic precipitates and endothelial dysfunction, may be caused by herpes simplex virus, varicella zoster virus, or other viruses such as mumps. It often leads to irreversible corneal endothelial cell damage and severe visual disturbance. Most recently, cytomegalovirus (CMV) was recognized as a new etiologic factor for corneal endotheliitis [21-24]. Clinical manifestations of CMV endotheliitis are characterized by linear keratic precipitates associated with multiple coin-shaped lesions and local corneal stromal edema with minimal anterior chamber reactions. *In vivo* laser confocal microscopy is able to demonstrate the characteristic owl's eye cells in the corneal endothelial cell layer (Figure 6). Owl's eye cells are typically seen at autopsy or in biopsy specimens from the kidneys, lungs and other organs in cases of congenital or acquired CMV infection. By *in vivo* laser confocal microscopy, owl's eye cells are readily seen *in vivo* as large corneal endothelial cells with an area of high reflection in the nucleus surrounded by a halo of low reflection. Therefore, owl's eye cells may be a useful adjunct for the non-invasive diagnosis of CMV corneal endotheliitis [25].

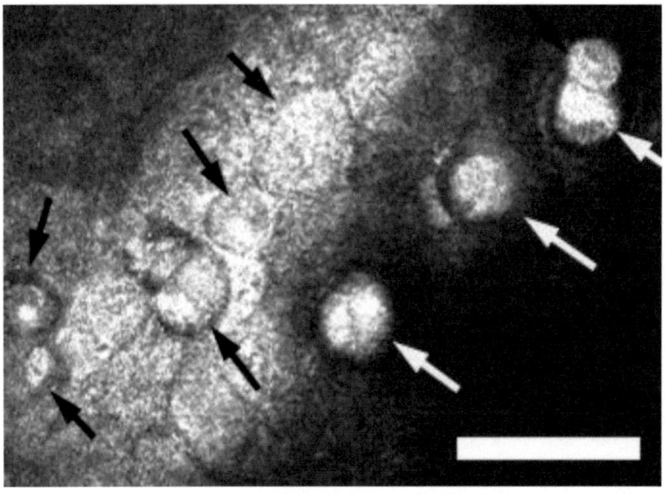

Figure 6. Owl's eye cells in cytomegalovirus corneal endotheliitis. *In vivo* laser confocal microscopic photo showed numerous owl's eye cells (arrows) in the corneal endothelial cell layer. (bar=100μm)

http://dx.doi.org/10.5772/55657

7. Post surgical anatomies

In vivo laser confocal microscopy is useful for visualization of post keratoplasty anatomies (Figure 7). Descemet's stripping automated endothelial keratoplasty (DSAEK) is a new type of keratoplasty technique in which only posterior portion of the donor cornea is transplanted inside the eye. Since the donor is attached with air, no stitches are required; this enables maintaining much of the structural integrity of the cornea and induces minimal refractive change, suggesting distinct advantages over standard penetrating keratoplasty.[26] *In vivo* laser confocal microscopy analysis identified subclinical corneal abnormality after DSAEK with high resolution; this includes subepithelial haze, donor-recipient interface haze (Figure 8), and interface particles(Figure 8).[27, 28] Quantitative analysis showed that these postoperative hazes and particles decreased significantly over follow-up. [27, 28] The influence of these hazed on vision are still under investigation.

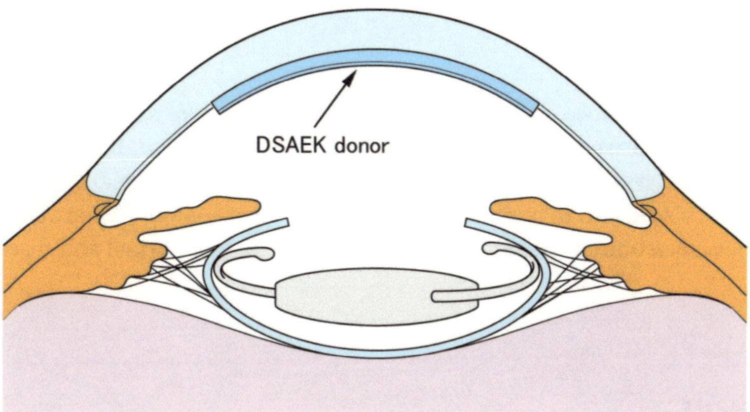

Figure 7. Schema of DSAEK. Posterior portion of the donor cornea is transplanted inside the eye.

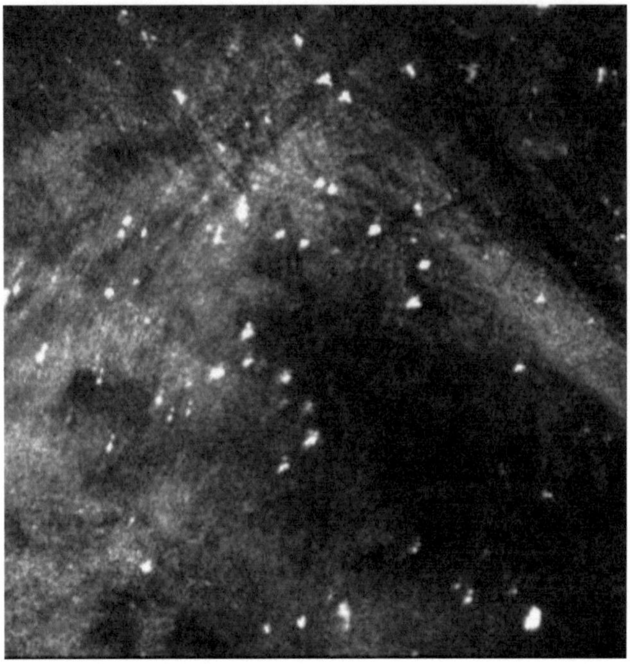

Figure 8. *In vivo* laser confocal microscopy of donor-recipient interface after DSAEK. At the level of the donor-recipient interface, highly reflective particles and interface haze were observed(400×400μm).

8. Conclusion

In vivo laser confocal microscopy is able to identify histologic-like real time optical sectioning of the normal and diseased cornea/conjuctiva with high resolution. It is clinically useful in diagnosing *Acanthamoeba* keratitis, fungal keratitis, cytomegalovirus endotheliitis and corneal dystrophies. It is also useful to observe post surgical corneal anatomies. Further investigation using this device is required to understand in vivo histology of normal and diseased cornea/ conjunctiva.

Author details

Akira Kobayashi*, Hideaki Yokogawa and Kazuhisa Sugiyama

*Address all correspondence to: kobaya@kenroku.kanazawa-u.ac.jp

Department of Ophthalmology, Kanazawa University Graduate School of Medical Science, Kanazawa, Japan

References

[1] Cavanagh, H. D, Petroll, W. M, Alizadeh, H, He, Y. G, Mcculley, J. P, & Jester, J. V. Clinical and diagnostic use of in vivo confocal microscopy in patients with corneal disease. Ophthalmology (1993). , 100(10), 1444-54.

[2] Kaufman, S. C, Musch, D. C, Belin, M. W, Cohen, E. J, Meisler, D. M, Reinhart, W. J, Udell, I. J, & Van Meter, W. S. Confocal microscopy: a report by the American Academy of Ophthalmology. Ophthalmology (2004). , 111(7), 396-406.

[3] Stave, J, Zinser, G, Grummer, G, & Guthoff, R. Der modifizierte Heidelberg-Retina-Tomograph HRT. Erste ergebnisse einer in-vivo-darstellung von kornealen strukturen. [Modified Heidelberg Retinal Tomograph HRT. Initial results of in vivo presentation of corneal structures.] Ophthalmologe (2002). In German., 99(4), 276-280.

[4] Eckard, A, Stave, J, & Guthoff, R. F. In vivo investigations of the corneal epithelium with the confocal Rostock Laser Scanning Microscope (RLSM). Cornea (2006). , 25(2), 127-131.

[5] Kobayashi, A, Yokogawa, H, & Sugiyama, K. In vivo laser confocal microscopy of Bowman's layer of the cornea. Ophthalmology. (2006). , 113(12), 2203-8.

[6] Yokogawa, H, Kobayashi, A, & Sugiyama, K. Mapping of normal corneal K-structures by in vivo laser confocal microscopy. Cornea. (2008). , 27(8), 879-883.

[7] Kobayashi, A, Yoshita, T, & Sugiyama, K. In vivo findings of the bulbar/palpebral conjunctiva and presumed meibomian glands by laser scanning confocal microscopy. Cornea. (2005). , 24(8), 985-988.

[8] Kobayashi, A, & Sugiyama, K. In vivo corneal confocal microscopic findings of palisades of Vogt and its underlying limbal stroma. Cornea. (2005). , 24(4), 435-437.

[9] Nagington, J, Watson, P. G, Playfair, T. J, Playfair, T. J, Mcgill, J, Jones, B. R, & Steele, A. D. Amoebic infection of the eye. Lancet (1974). , 2(7896), 1537-1540.

[10] Hammersmith, K. M. Diagnosis and management of Acanthamoeba keratitis. Curr Opin Ophthalmol. (2006). Review., 17(4), 327-331.

[11] Winchester, K, Mathers, W. D, Sutphin, J. E, & Daley, T. E. Diagnosis of Acanthamoeba keratitis in vivo with confocal microscopy. Cornea (1995). , 14(1), 10-17.

[12] Matsumoto, Y, Dogru, M, Sato, E. A, Katono, Y, Uchino, Y, Shimmura, S, & Tsubota, K. The application of in vivo confocal scanning laser microscopy in the management of Acanthamoeba keratitis. Mol Vis. (2007). , 13, 1319-26.

[13] Kobayashi, A, Ishibashi, Y, Oikawa, Y, Yokogawa, H, & Sugiyama, K. In vivo and ex vivo laser confocal microscopy findings in patients with early-stage acanthamoeba keratitis. Cornea. (2008). , 27(4), 439-445.

[14] Yokogawa, H, Kobayashi, A, Yamazaki, N, Ishibashi, Y, Oikawa, Y, Tokoro, M, & Sugiyama, K. Bowman's layer encystment in cases of persistent Acanthamoeba keratitis. Clin Ophthalmol. (2012). , 6, 1245-1251.

[15] Yamazaki, N, Kobayashi, A, Yokogawa, H, Ishibashi, Y, Oikawa, Y, Tokoro, M, & Sugiyama, K. Ex vivo laser confocal microscopy findings of cultured Acanthamoeba trophozoites. Clinical Ophthalmol (2012). , 6, 1365-1368.

[16] Das, S, Samant, M, Garg, P, Vaddavalli, P. K, & Vemuganti, G. K. Role of confocal microscopy in deep fungal keratitis. Cornea. (2009). , 28(1), 11-13.

[17] Takezawa, Y, Shiraishi, A, Noda, E, Hara, Y, Yamaguchi, M, Uno, T, & Ohashi, Y. Effectiveness of in vivo confocal microscopy in detecting filamentous fungi during clinical course of fungal keratitis. Cornea. (2010). , 29(12), 1346-1352.

[18] Kobayashi, A, & Sugiyama, K. In vivo laser confocal microscopy findings for Bowman's layer dystrophies (Thiel-Behnke and Reis-Bücklers corneal dystrophies). Ophthalmology. (2007). , 114(1), 69-75.

[19] Kobayashi, A, Fujiki, K, Fujimaki, T, Murakami, A, & Sugiyama, K. In vivo laser confocal microscopic findings of corneal stromal dystrophies. Arch Ophthalmol. (2007). , 125(9), 1168-1173.

[20] Kobayashi, A, Fujiki, K, Murakami, A, & Sugiyama, K. In vivo laser confocal microscopy findings and mutational analysis for Schnyder's crystalline corneal dystrophy. Ophthalmology. (2009). , 116(6), 1029-1037.

[21] Jap, A, & Chee, S. P. Viral anterior uveitis. Curr Opin Ophthalmol. (2011). Review., 22(6), 483-488.

[22] Koizumi, N, Yamasaki, K, Kawasaki, S, Sotozono, C, Inatomi, T, Mochida, C, & Kinoshita, S. Cytomegalovirus in aqueous humor from an eye with corneal endotheliitis. Am J Ophthalmol (2006). , 141(3), 564-565.

[23] Koizumi, N, Suzuki, T, Uno, T, Chihara, H, Shiraishi, A, Hara, Y, Inatomi, T, Sotozono, C, Kawasaki, S, Yamasaki, K, Mochida, C, Ohashi, Y, & Kinoshita, S. Cytomegalovirus as an etiologic factor in corneal endotheliitis. Ophthalmology (2008). , 115(2), 292-297.

[24] Chee, S. P, Bacsal, K, Jap, A, Se-thoe, S. Y, Cheng, C. L, & Tan, B. H. Clinical features of cytomegalovirus anterior uveitis in immunocompetent patients. Am J Ophthalmol. (2008). , 145(5), 834-840.

[25] Kobayashi, A, Yokogawa, H, Higashide, T, Nitta, K, & Sugiyama, K. Clinical significance of owl eye morphologic features by in vivo laser confocal microscopy in patients with cytomegalovirus corneal endotheliitis. Am J Ophthalmol. (2012). , 153(3), 445-453.

[26] Gorovoy, M. S. Descemet-Stripping Automated Endothelial Keratoplasty. Cornea (2006). , 25(8), 886-889.

[27] Kobayashi, A, Mawatari, Y, Yokogawa, H, & Sugiyama, K. In vivo laser confocal microscopy after descemet stripping with automated endothelial keratoplasty. Am J Ophthalmol (2008). , 145(6), 977-985.

[28] Kobayashi, A, Yokogawa, H, & Sugiyama, K. In vivo laser confocal microscopy after non-Descemet stripping automated endothelial keratoplasty. Ophthalmology (2009). , 116(7), 1306-1313.

Chapter 4

Laser-Scanning *in vivo* Confocal Microscopy of the Cornea: Imaging and Analysis Methods for Preclinical and Clinical Applications

Neil Lagali, Beatrice Bourghardt Peebo,
Johan Germundsson, Ulla Edén, Reza Danyali,
Marcus Rinaldo and Per Fagerholm

Additional information is available at the end of the chapter

http://dx.doi.org/10.5772/55216

1. Introduction

1.1. The cornea is a model tissue for in vivo imaging

The cornea is the most anteriorly located ocular tissue, serving principally as a transparent window for light to enter the human eye (Fig 1A). Besides the function of transparency, the curved cornea serves to refract light to deeper ocular structures, and presents a biological barrier to the outer environment. The cornea, often depicted histologically in cross-section, consists of five distinct layers: epithelium, Bowman's layer, stroma, Descemet's membrane, and endothelium (Fig 1B). For the interested reader, a detailed description of corneal anatomy and physiology can be found in Reference 1.

The cornea has long been considered to be a model tissue for microscopy, owing to the unique combination of various tissue elements (cells, extracellular matrix, nerves, vasculature, stem cells, etc.) present within a virtually transparent and relatively thin structure. The cornea transmits visible light with very little absorption, while the cellular features of interest within the cornea often strongly scatter light, resulting in the possibility of obtaining high signal-to-noise images by optical techniques. With the added advantages of confocal microscopy and the exterior anatomical location of the cornea, it has become a model tissue for high-contrast microscopic imaging in vivo.

1.2. Endogenous-contrast darkfield confocal imaging of the living cornea

Early confocal microscopy systems employed white light and tandem spinning discs to illuminate and collect light from a sample, while modern confocal systems use slit-scanning

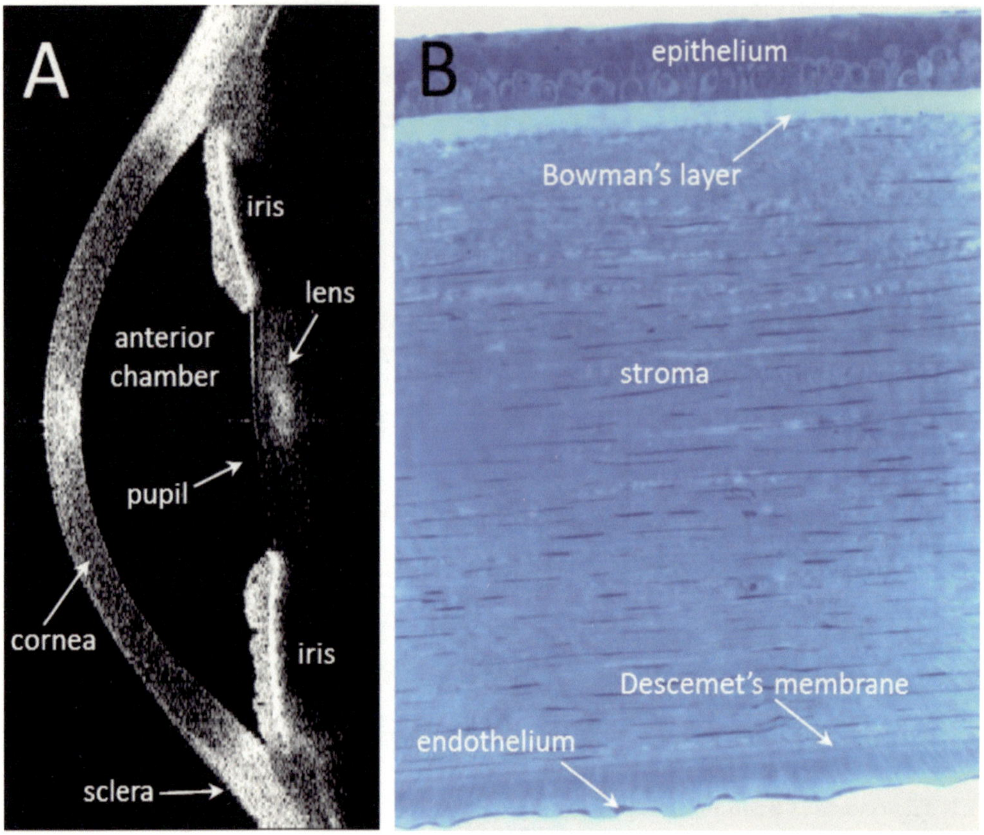

Figure 1. The human corneal anatomy. (A) In vivo cross-section of the cornea and anterior eye by anterior segment optical coherence tomography. The bright areas indicate light scatter while the dark areas are transparent. The bright area in the central cornea is an artifact. (B) Histologic view of the human cornea with toluidine blue staining, indicating the five principal layers of the normal cornea.

(white light) or point-scanning (laser) configurations. Regardless of the system, however, all confocal microscopes used to examine the cornea in vivo utilize the darkfield microscopy principle. In darkfield microscopy, only light scattered by the sample reaches the detector, while transmitted light is blocked, rendering transparent tissue or structures as 'dark'. Darkfield imaging has the advantage of achieving a high contrast of scattering structures against a dark background, and the potential for producing high-resolution images that approach or even exceed [2] the diffraction limit.

Another benefit is the endogenous contrast provided by structures in the tissue, which scatter visible light generally independent of wavelength and without the need for additional contrast-enhancing means. Practically, this endogenous contrast enables in vivo imaging of the inner structures of the cornea without the use of dyes, in a non-selective manner, since all scattering elements in the probed volume will be imaged, regardless of other biological properties. The clear advantage is that the tissue can be examined in vivo in its native state, with all cell types present. The challenge that then emerges, however, is to interpret in vivo images solely by morphology, without further cell-specific information. Some strategies for overcoming this limitation will be discussed later in this chapter.

1.3. Equipment and procedures

At present, there are two commercially-available systems for performing in vivo confocal imaging of the cornea. Both systems are manufactured for clinical use as medical devices. The Nidek Confoscan 4 from Nidek Technologies is a white-light, slit-scanning confocal microscope, while the Heidelberg Retinal Tomograph 3 with Rostock Cornea Module (HRT3-RCM) from Heidelberg Engineering is a red laser point-scanning confocal microscope (Fig 2A).

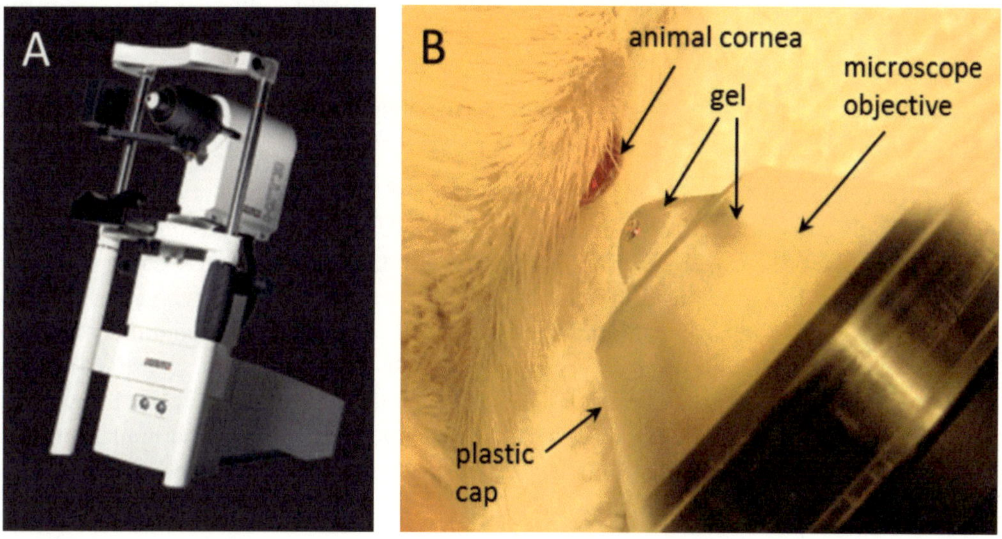

Figure 2. The HRT3-RCM system for laser-scanning in vivo confocal microscopy of the cornea. (A) The laser-scanning interface. (B) Magnified view of the scanning head with plastic cap and transparent tear gel applied. The gel and cap make contact with the ocular surface of the locally anesthetized animal (or human) cornea during imaging.

Consistent with the theme of this book, the remainder of this chapter will describe data obtained with the HRT3-RCM laser-scanning system. The system consists of a point-scanning 670nm diode laser and horizontally-mounted optics to enable access to the cornea of a seated subject. The system and its operation are described in detail elsewhere. [3] The objective lens most commonly used is a water immersion lens with high numerical aperture (63x/0.95 NA, Zeiss), upon which a disposable sterile plastic cap is placed. The cap comes into physical contact with the corneal surface through a refractive index-matching ophthalmic gel (eg., ViscoTears, Novartis). An in vivo confocal microscopy (IVCM) examination consists of one or several scans, where each scan consists of 100 consecutive images acquired at a selectable rate from 1 to 30 frames/sec. During the scan (which lasts from 3.3 to 100 sec, depending on the acquisition rate), the operator manually adjusts the position of the objective relative to the cornea in a plane parallel to the corneal surface. Additionally, the operator adjusts the depth of focus in the axial direction, by manual or motor-driven [4] means. A depth range of approximately 1200 μm is available, which is ample for examination of human corneas typically 550μm thick. In this manner, the entire corneal volume can be accessed for in vivo imaging. Practically, access to corneal regions further from the apex is achieved by manually placing a movable fixation target into which the patient is instructed to view.

The images acquired represent a field of view of either 300 × 300μm or 400 × 400μm, depending on the choice of internal field lens (which is user-interchangeable). The lateral resolution is about 1-2 μm, but can be submicron depending on contrast [2], while the axial resolution is about 4 μm. [5] Comparable figures for the slit-scanning microscope are 1-2 μm lateral and 25-27 μm axial resolution. [6,7]

Patient examinations in the clinic are conducted upon a topically anesthetized cornea, and depending on the clinical and/or scientific questions, a number of scans are completed during a given session. Upwards of 20 scans per eye (2000 single images per eye) is not uncommon. Typical examinations may last from 5 to 15 minutes, with patients rarely experiencing discomfort. For patients, an image acquisition rate of 8 frames/sec is typical, however in certain cases (eg., a subject with nystagmus) the frame rate is increased.

For preclinical animal imaging, the animal is examined under general anesthesia and with topically applied anesthetic drops. Animals are examined in the prone position, while resting on a movable stand (Figure 2B). During examination, care is taken to keep the cornea hydrated with saline drops. Image acquisition is similar as for patients, but typically with a slower frame rate (5 frames/sec) due to the minimal motion of the animal during examination. Often anatomic considerations limit the accessible region of the animal cornea. In order to obtain good in vivo images, the animal must be oriented such that the corneal surface is presented tangentially to the microscope objective lens. This often requires manual manipulation of the head and eyelids, which in practice means at least two persons are needed to conduct the examination.

1.4. Application of IVCM to clinical and animal research

The use of IVCM in scientific research in ophthalmology has rapidly expanded over the past decade. The interested reader is referred to several informative review articles describing the use of IVCM in humans, [5], [8]- [10] however, reviews are barely keeping pace with the latest developments in the field. The use of IVCM in preclinical animal research is more recent, and is slowly gaining momentum. Animal studies have detailed the animal corneal anatomy [11], the effect of various surgical techniques, [12]- [14] and phenomena such as inflammation [15] and angiogenesis. [16], [17]

Besides research, IVCM has been used for clinical diagnosis of pathology of the cornea, or as a screening/monitoring tool for patients undergoing treatment. [18]- [22] The number and scope of clinical applications of IVCM will undoubtedly increase in the coming years.

It is pertinent to note that ethical considerations in human and animal research can be influenced by the use of IVCM. In human studies, the principle of informed consent is relevant, and the non-invasive nature and short-duration of IVCM examination can facilitate patient and ethical review board acceptance. It must also be noted that the laser is non-damaging to

http://dx.doi.org/10.5772/55216

the eye, and the instrumentation is approved for clinical use in humans. In animal research, IVCM facilitates repeated, longitudinal examination of the same animal corneas over periods of hours, days, months, or years. [12], [13], [16] While the technique can therefore reduce the number of animals required for a study (since repeated non-invasive cellular-level examination is possible in the same animal), care must be taken to administer anesthesia in a humane manner. Typically, for small animals such as mice or rats, short-duration or time-lapse studies can be performed by repeated general anesthesia over a period of several hours. It is not advisable for animals to remain under general anesthesia for longer time periods, and it is advised to administer an antidote (such as atipamezole) where feasible, to rapidly reverse the effects of the anesthetic.

2. Feature recognition in IVCM

2.1. IVCM appearance of the normal human cornea

A summary of the different layers and features of the normal human cornea is presented in Figure 3. The different layers are imaged by adjusting the axial alignment to select a given focal depth in the cornea.

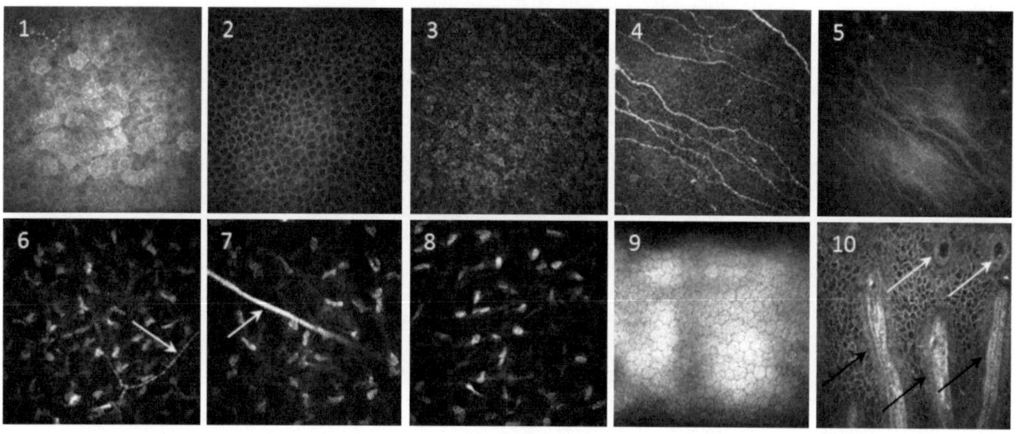

Figure 3. IVCM appearance of the corneal layers in a normal human subject. Depths from the corneal surface are given. All images are 400 × 400µm. 1. Depth 0 µm, superficial epithelium; 2. Depth 20 µm, epithelial wing cell layer; 3. Depth 30 µm, basal epithelium; 4. Depth 40 µm, subbasal nerve plexus; 5. Depth 45 µm, Bowman's layer; 6. Depth 150 µm, anterior stroma with nerve (arrow); 7. Depth 200 µm, mid stroma with nerve trunk (arrow); 8. Depth 400 µm, posterior stroma; 9. Depth 530 µm, endothelium; 10. Inferior limbal palisade ridges (black arrows) and focal stromal projections (round protrusions, white arrows) indicating the presumed limbal epithelial stem cell niche.

The *en face* images parallel to the corneal surface shown in Figure 3 are the most widely used; however, the instrument offers the possibility of oblique imaging when the microscope objective contacts the cornea off-axis. Although the scale of the image is difficult to ascertain in oblique images, the information yielded by such images is useful for localizing structures or pathology within the cornrea in a picture similar to a histological cross section (Figure 4).

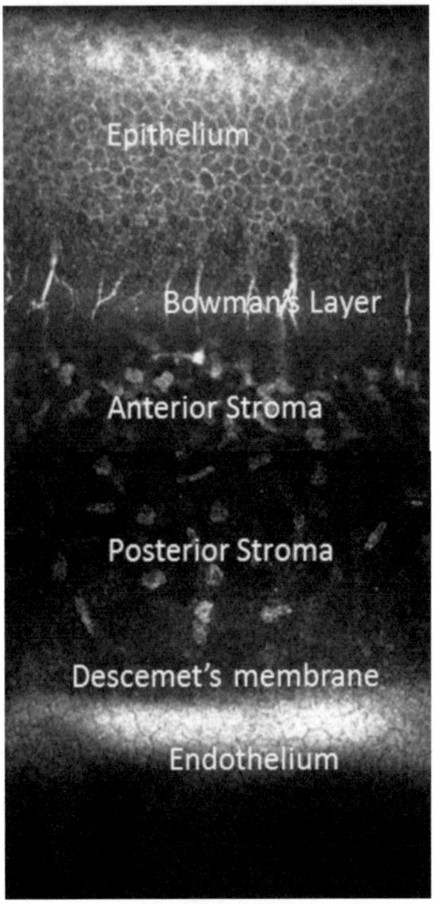

Figure 4. Live histological cross-section representation of a normal cornea by IVCM. Compare this in vivo view with that of the cross-section in Fig 1B.

2.2. Application: Distinguishing different inflammatory cell subtypes in corneal inflammation

IVCM can be used to identify the presence of inflammatory cells in the cornea, despite their invisibility by traditional methods of clinical observation such as slit lamp biomicroscopy. Moreover, the morphological appearance of the various inflammatory cell types in vivo enables the distinction between certain cell subtypes including neutrophil-granulocyte, immature/mature dendritic, and macrophage (Figure 5). Although these cells of leukocytic origin can invade the cornea from the peripheral limbus and conjunctiva in cases of overt, active inflammation, it is also important to note that IVCM can detect early inflammatory cell influx prior to the appearance of external signs of inflammation. Additionally, inflammatory cell presence can be detected in vivo in cases of chronic, sub-clinical inflammation as observed in some corneal transplant patients (Figure 5). Dendritic cells in the cornea have been the most widely studied inflammatory cell subtype by IVCM. While populations of dendritic cells are resident in the peripheral human cornea under normal conditions, [23] in an inflammatory setting these cells migrate to the central cornea and mature into MHC class II$^+$ cells capable of antigen presentation. [24]

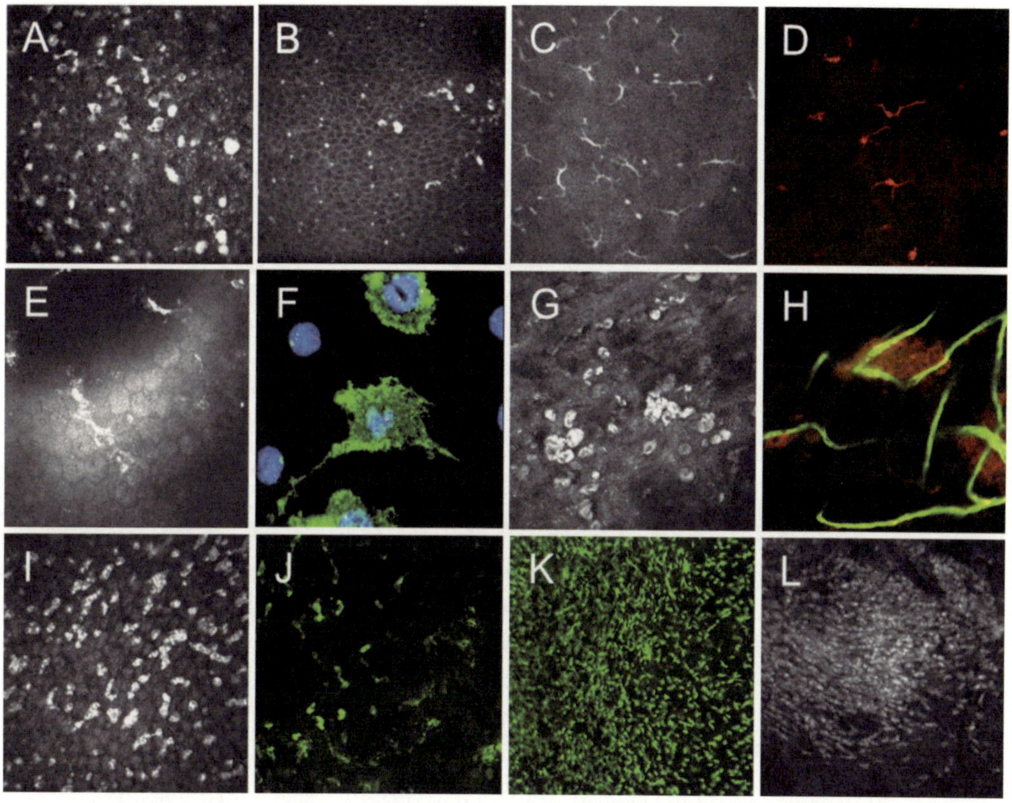

Figure 5. Inflammatory cell imaging by IVCM. (A-B) Leukocyte invasion of the epithelium one year after corneal transplantation in two patients. (C) Dendritic cells with long dendrites and dendritic cell bodies lacking dendrites in the central cornea 2 years after transplantation. (D) CD45$^+$ dendritic cells (red) in the rat corneal epithelium. (E) Inflammatory cells on the endothelial surface. (F) CD45$^+$ leukocytes (green) on the endothelial surface of a failed human transplant, with DAPI-counterstained nuclei (blue). (G) Mature macrophages in a patient with regressed neovessels. (H) KiM2R$^+$ mature rat macrophages (red) during neovessel regression (CD31$^+$ vessels, green). (I) Neutrophil-granulocytes invading the corneal epithelium in a patient. (J) CD11b$^+$ neutrophil-granulocytes (green) in the rat. (K) Massive influx of CD11b+ myeloid cells (green) from the conjunctiva into the corneal stroma in acute inflammation in the rat. (L) In vivo image of the myeloid cell influx during inflammation in the rat.

2.3. Feature recognition for diagnosis — Application to acanthamoeba keratitis

Acanthamoeba keratitis (AK) is a serious, sight-threatening infection of the cornea by the acanthamoeba parasite. [25] The parasite first establishes itself as an amoeboid body (trophozoite) on the ocular surface, where it gradually kills epithelial cells by phagocytosis or cell lysis. As it matures, the parasite forms characteristic double-walled cysts, varying in size from 5 - 200μm in diameter typically. If left untreated, the parasite will continue to migrate through to the stroma, where it induces inflammation, scarring, and eventually corneal ulceration and perforation, leading to blindness. AK can be difficult to diagnose owing to the relative insensitivity of current biosampling and culture methods; however, its timely diagnosis is paramount for administering the correct treatment and for optimal clinical outcomes.

In recent years, IVCM has become an important tool for screening and diagnosing patients suspected of having AK. The interested reader is referred to recent literature on the subject of AK diagnosis by IVCM. [26]- [28] Several unresolved issues remain, however, with this in vivo diagnostic technique. Sensitivity and specificity of diagnosis are instrument and operator-dependent, [26], [27] and the in vivo images are often difficult to interpret and no clear guidelines for interpretation exist. Guidelines for performing IVCM examination in suspected cases of AK are likewise lacking.

Generally, exams should include focusing on superficial epithelial layers and performing wide-area scans across the ocular surface, using the fixation target to access peripheral zones. Attention should especially be given to zones where the epithelium exhibits defects. Most cysts detected in the clinic are present in the superficial epithelial layers. Repeated examination sessions on the same day or on consecutive days are recommended, particularly in suspected cases where no positive confocal evidence is found initially.

Several in vivo images from patients with suspected AK are shown in Figure 6. It is apparent that the morphologic appearance in terms of size, reflectivity, density of cells, and presence of the double-wall in cysts can vary widely across patients. Upon comparing Figures 5 and 6, it can be seen that cysts can be easily misinterpreted as inflammatory cells and vice-versa. It is also of note that images obtained with white-light slit-scanning confocal systems [26] are markedly more difficult to interpret than those obtained with the laser-scanning system, due to the increased axial resolution of the latter.

2.4. Keratocyte apoptosis

Keratocytes normally appear in IVCM examinations as bright nuclei with transparent or semi-transparent cell bodies in the normal cornea (panels 6-8 in Figure 3). Keratocyte death by apoptosis occurs in the cornea in cases of stromal perturbation such as external trauma, or with aging in normal subjects. Comparing the stromal keratocytes in healthy younger and older subjects, a lower density of keratocytes is observed [29] and the presence of a large number of fine 'needle-like' structures is evident in older corrneas (Figure 7).

Needle-like structures are also observed in the transplanted cornea, concomitant with a sparse presence of keratocytes. The structures observed at various stages in the apoptotic process include a pronounced increase in reflectivity of the keratocyte body, elongation of the nucleus and cell body into a linear form, condensation of this linear form into a thin, needle-like structure, and the presence of condensed, punctate inclusions of smaller size and reduced reflectivity relative to the healthy keratocyte nucleus (Figure 8).

In keratoconus patients treated with UVA-riboflavin corneal collagen cross-linking (CXL), cells in the ultraviolet-irradiated region of the cornea undergo apoptosis *en masse,* resulting in a dramatic change in the corneal stromal appearance, with keratocytes initially absent in the treated region. In border regions of the stroma, needles are observed (Figure 8). The features observed

http://dx.doi.org/10.5772/55216

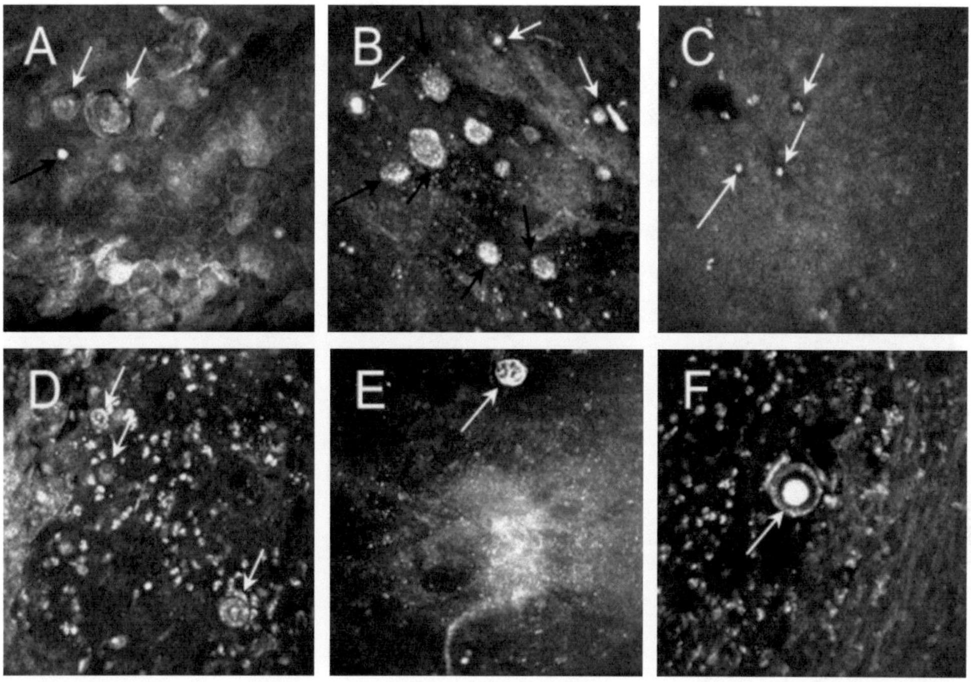

Figure 6. IVCM images in several suspected cases of AK. (A) Superficial epithelial double-walled cysts (white arrows) and suspected early-stage cyst (black arrow). (B) Acanthamoeba cysts of varying size (white arrows) and suspected trophozoites (black arrows). (C) Suspected early-stage cysts (arrows) appear similar in morphology to inflammatory cells, however, the latter generally appear in greater numbers. (D) Inflammatory cells and double-walled cysts (arrows). (E) a large cyst (arrow) in a region devoid of epithelial cells. (F) Large cyst with clear double-wall morphology (arrow).

in patients treated by CXL are mimicked in a rat model, where IVCM images indicate the same structures observed in humans, with immunohistochemical staining confirming apoptosis (Figure 8). The length of time apoptotic remnants persist in the corneal stroma is unknown; however, their continued appearance months or years after trauma suggests a slow turnover.

2.5. Detection of the limbal epithelial stem cell niche

An emerging clinical application for IVCM is in the assessment of the limbal epithelial stem cell niche in patients with limbal stem cell deficiency. A few studies have described the in vivo morphological characteristics of the limbal palisades region of the corneal epithelium, believed to harbor limbal epithelial stem cells. [30]- [33] The structures associated with the stem cell niche are the palisade ridges and focal stromal projections at the basal limbal epithelium, and cells positive for putative stem cell markers reside at the base of these structures. [31] In Figure 9, these structures are depicted in normal healthy subjects, where palisade ridge morphology and the distribution of focal stromal projections is seen to vary. The structures additionally have a morphologic appearance that varies with skin pigmentation of the subject [30], [33] which is an important consideration when assessing the palisade morphology.

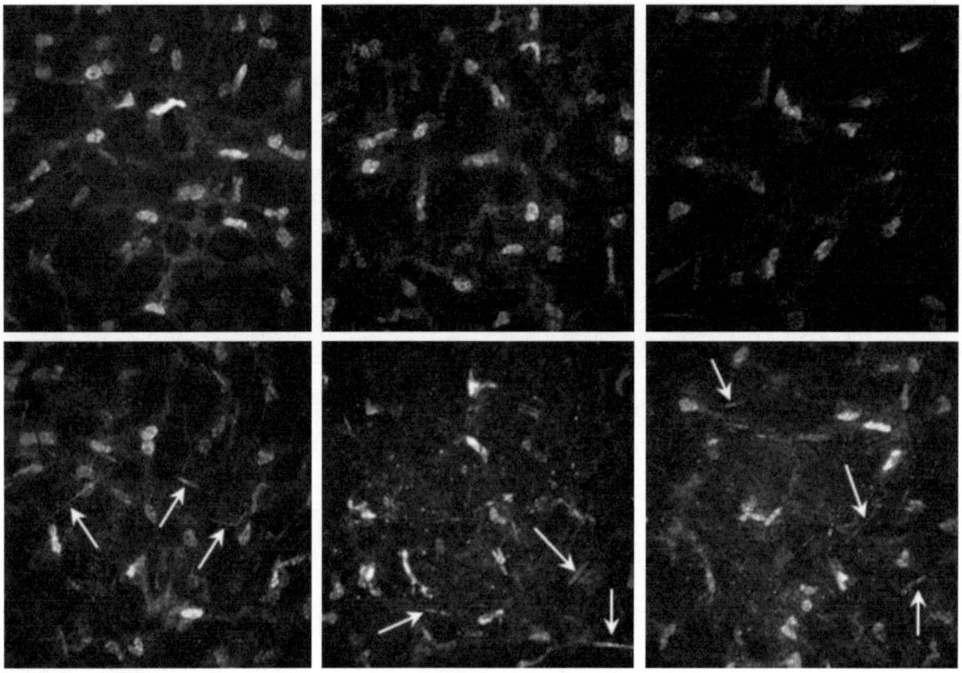

Figure 7. Normal healthy corneas at mid-stromal depth. Top row, young subjects (23, 21, 16 years of age, from left to right). Bottom row, older subjects (88, 77, 76 years) exhibit fine, reflective needle-like structures (arrows) indicating keratocyte apoptosis, not generally present in the younger corneas.

IVCM can also assist in assessing the degree of limbal stem cell deficiency in a patient. [34]- [36] In practice, the limbus of a patient is imaged at nasal, temporal, inferior and superior regions to document the limbal palisades. Most often, a superior and inferior examination suffices, as stem cells are predominantly located in these regions. In cases of limbal stem cell deficiency, the palisades may appear degraded or absent, replaced with blood vessels, inflammatory cells, and highly scattering conjunctival tissue (Figure 9).

With further detailed studies of normal and stem cell deficient corneas, IVCM could become an important tool in assessing degree of trauma, risk for surgical intervention, and prognosis in treated or stem cell transplanted patients.

3. Nerve analysis and quantification methods

3.1. Corneal nerve anatomy

The cornea is one of the most densely innervated tissues in the body. Nerves enter the cornea from a peripheral, mid-stromal depth as nerve trunks, then branch as they proceed anteriorly to eventually terminate at single nerve terminals between corneal epithelial cells at the ocular surface. For an in-depth discussion of corneal nerves and their detailed anatomy, the reader

http://dx.doi.org/10.5772/55216

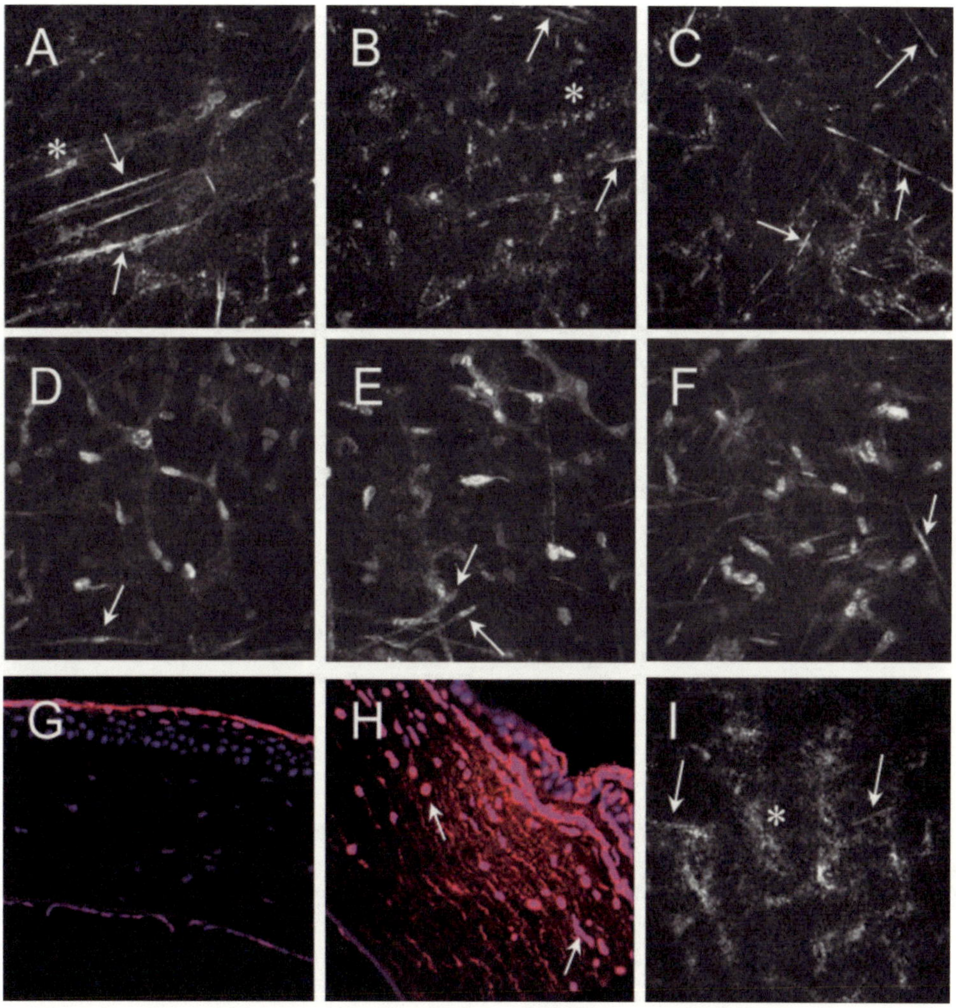

Figure 8. Morphologic appearance of apoptotic keratocytes. Top row: stroma of the transplanted cornea, middle row: stroma adjacent to a collagen-crosslinked region, bottom row, images from a rat cornea treated by the CXL procedure. (A) Prominent needles (arrows) and an elongating keratocyte (asterisk) in a patient 2 years after corneal transplantation. Note the scarcity of keratocytes. (B) A stromal region devoid of keratocytes, with needles (arrows) and condensed bodies (asterisk) visible in a patient 2 years after transplantation. (C) Needles (arrows) in a region of stroma devoid of keratocytes 4 years after transplantation. (D-F) Keratocyte elongation (arrows) during apoptosis in a 27-year old patient 4 months after receiving CXL treatment for keratoconus. (G) Immunofluorescent section from a normal, unaltered rat cornea marked with apoptotic marker BrdU (red) and nuclear counterstain 7-AAD (blue). Note the apoptosis in the superficial epithelium. (H) Section from a rat 1 day after CXL treatment, indicating significant anterior keratocyte apoptosis (arrows). (I) The same rat cornea at 1 day in vivo exhibited a loss of keratocytes, needles (arrows) and condensed bodies (asterisk).

is referred to several recent publications. [37]- [39] The subbasal nerve plexus in particular, located at the border between Bowman's layer and the basal epithelium, has become a landmark within the cornea owing to the concentration of nerves at this plane and the ease of imaging this layer with high contrast by confocal microscopy. The subbasal nerves gently

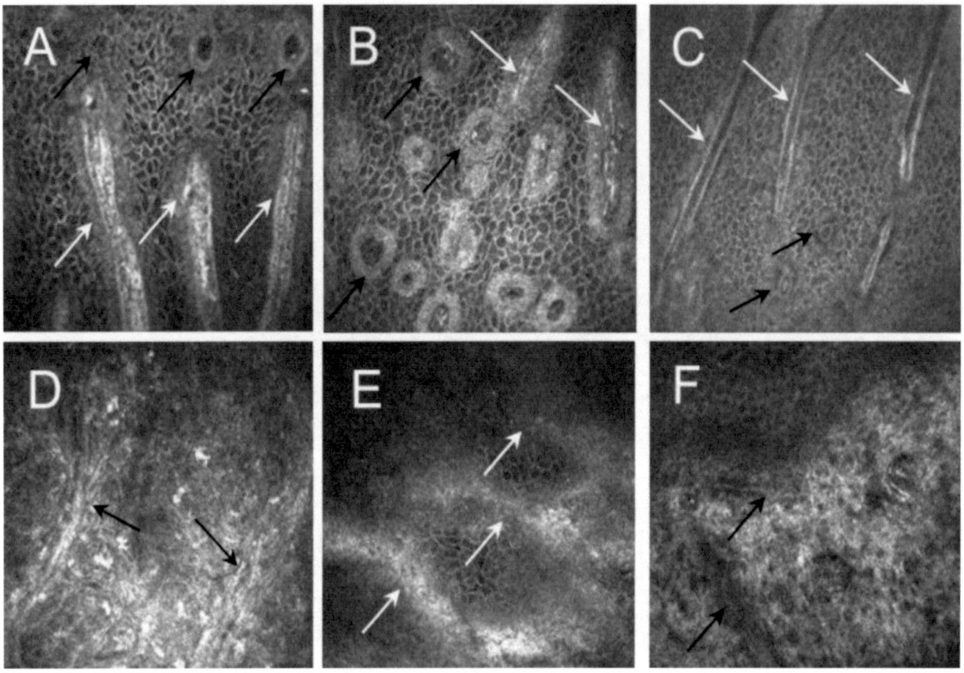

Figure 9. Limbal epithelial stem cell assessment in vivo. (A-C) appearance of the limbal palisades in normal healthy subjects. (D-F) appearance of the palisades in stem cell deficient patients with aniridia. (A-C) Palisade ridges (white arrows) and focal stromal projections (black arrows) exhibit varied diameter, cell appearance and distribution. (D,E) Degraded remnants of palisade ridges are visible (arrows). (F) Total absence of the stem cell niche in a conjunctivalized limbus, where blood vessels (arrows) are apparent amidst the highly scattering conjunctival tissue.

spiral inwards from the periphery to the corneal apex (Figure 10), and may follow the pattern of epithelial cells as they regenerate. [36], [40] Notably, the subbasal nerves have been shown to be perturbed in many pathologies and in cases of trauma. [38], [41] An example of a pathologic subbasal nerve architecture in keratoconus is depicted in Figure 10.

3.2. Nerve analysis techniques

Nerve analysis can be aided by qualitative and quantitative processing techniques. Nerve montages such as that depicted in Figure 10 can aid in qualitative assessment of nerve morphology. Recently, several authors have proposed various techniques to automate the process of creating nerve montages from the individual IVCM images obtained by adjusting fixation during patient examination. [42], [43] The quality of the resulting montage, however, is limited by the examination procedure. Care must be taken such that a minimum pressure is applied on the cornea by the plastic microscope objective cap during examination – otherwise, unwanted striations will occur and reduce the image quality. [44]

Quantitative nerve analysis has been performed by a number of investigators [36], [38], [39], [41] and in the case of the subbasal nerve plexus, reporting of nerve density has followed an accepted convention. Nerve density values determined from laser-scanning confocal instru-

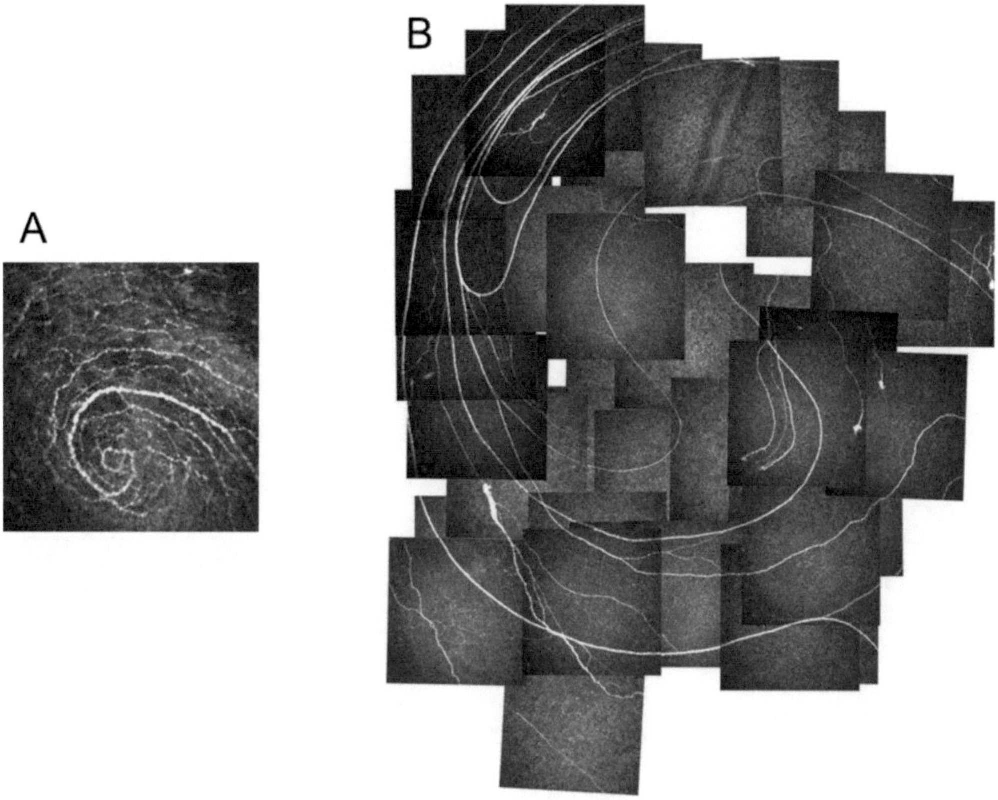

Figure 10. Human corneal subbasal nerve architecture. (A) A prominent whorl of subbasal nerves into the infero-central corneal apical region. (B) The whorl pattern is missing in this montage of the subbasal nerve plexus in a keratoconic cornea. Instead nerves, form characteristic looping structures.

ments can therefore be compared. Nerve density is reported in terms of total nerve length per unit area of the central corneal subbasal nerve plexus. In practice, a single 400 × 400μm field of view image obtained from the central cornea is chosen, typically with the greatest number of visible subbasal nerves. Nerves in the image are then semi-automatically traced by the aid of tracing software [12], [19], [36] to determine total nerve length, which is then divided by the area in the field of view, to determine nerve density in $\mu m/mm^2$. The subbasal nerve density in humans changes with age, [29] but in general the accepted value for the average density by IVCM is about 20,000 $\mu m/mm^2$ in the central cornea. [10], [29], [38], [39] Nerve density measurement according to this technique, however, is dependent upon where the single image is sampled. Images from the whorl region typically have the highest density while those in the paracentral region may have significantly reduced density. [39] Moreover, the clarity and contrast of the nerve image may render more nerves visible (including secondary connecting branches), while images with too much mechanical pressure placed on the cornea may disrupt the nerve pattern. Scattering from the epithelial basement membrane and/or anterior Bowman's layer can also influence visibility of the subbasal nerves (Figure 11). Therefore, wide-area montages such as that shown in Figure 10 are recommended for quantification of the average central subbasal nerve density, with the recognition that a qualitative picture of nerve architecture may be equally useful. While some parameters such as nerve branching and tortuosity have been quantified by some investigators, consensus on analysis techniques of the subbasal nerve architecture is still required. Moreover, the subbasal nerve density as observed by IVCM likely excludes most of the interconnecting nerve fibers between the larger subbasal nerve bundles. [45], [46] With these bundles included, the average subbasal nerve density is upwards of 50,000 μm/mm^2 in the central cornea. [45]

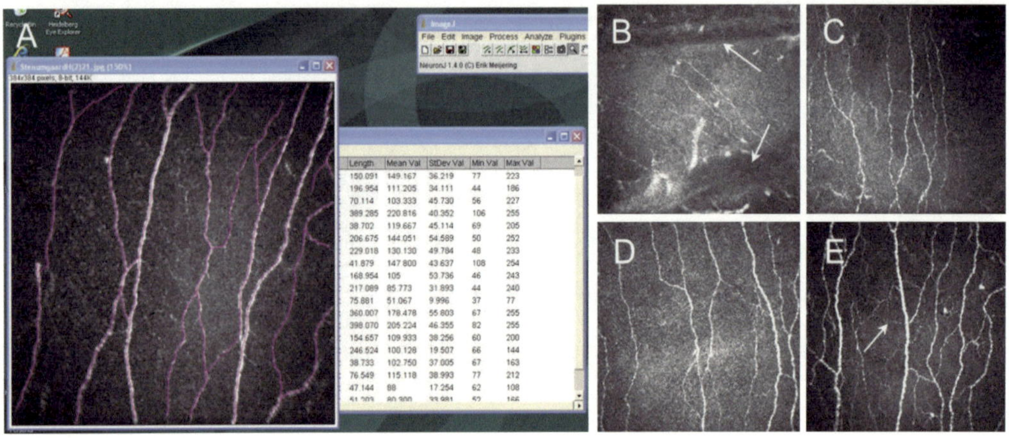

Figure 11. Subbasal nerve density analysis with IVCM. (A) An example of software-aided nerve tracing and quantification of nerve length. (B-E) Images taken from the central subbasal nerve plexus in healthy volunteer subjects. (B) Artifacts (arrows) caused by mechanical pressure on the cornea. These reduce the effective area in which nerves can be quantified, resulting in an artificially low nerve density. 80y old male subject, density 10,000 μm/mm^2. (C) A slightly oblique section with different corneal depths apparent in the same image, causing nerve density to be underestimated. 35 year old male, density 26,000 μm/mm^2. (D) Subbasal nerves with basal epithelial cells of increased reflectivity rendering few interconnecting nerve branches visible. 23 year old male, density 24,500 μm/mm^2. (E) Nerve plexus with good contrast and low basal cell reflectivity enables interconnecting nerve branches to be seen. 55 year old male, density 33,600 μm/mm^2.

3.3. Assessment of nerve density in corneal epithelial basement membrane dystrophy

In the cornea, epithelial basement membrane dystrophy (EBMD) is one of the most common diseases that can threaten vision. In EBMD, adhesion between the epithelium and the underlying Bowman's layer is defective. This disrupted epithelial barrier leads to abnormal epithelial cell and basement membrane physiology that can result in reduced vision and painful erosions of the epithelial cells. [19], [47]

As the disease affects the region where the subbasal nerve plexus is located, it is desirable to know whether the nerves are also affected. Central subbasal nerve density in corneas with EBMD (virgin eyes where no surgical treatment has been applied) can be determined by IVCM and nerve tracing techniques, and compared to the nerve density in a control group of healthy individuals. Comparing nerve density in 40 healthy control subjects with that of 24 corneas from 24 patients with EBMD, it is clear that the subbasal nerve density is reduced in EBMD (Figure 12).

The analysis was achieved by first selecting good quality subbasal nerve images without artifacts, then coding images randomly, such that in further analysis the analyzer is masked to the origin of each image. The coded images are then subjected to nerve tracing by two trained independent observers, ideally on the same computer monitor on separate occasions. The results are then unmasked and the average density per image is taken. In addition, a Bland-Altman analysis can be performed to indicate the overall level of agreement in nerve density between observers. [19]

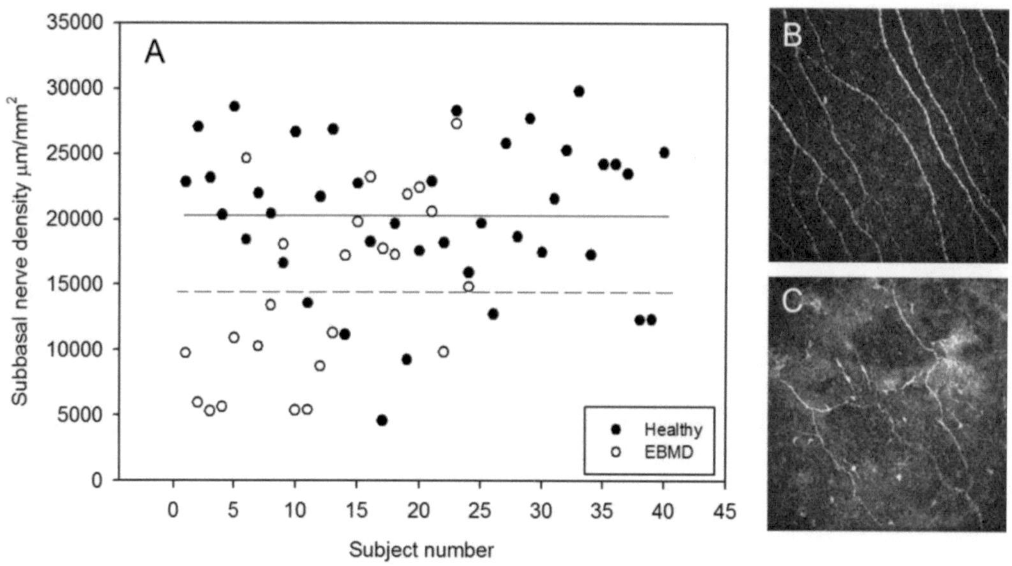

Figure 12. Assessment of nerve density in EBMD patients and healthy subjects. (A) Nerve density in EBMD patients (open circles) is on average 14,000 µm/mm² (dashed line), which is below the average level of 20,000 µm/mm² in healthy subjects (solid line). This difference is visible upon comparing images of a healthy plexus (B) with that obtained from an EBMD patient (C).

3.4. Longitudinal quantification of nerve regeneration in corneal transplants

Repeated, longitudinal quantification of nerve density can aid in the determination of nerve recovery after trauma, pharmacological treatment, or surgical intervention. Nerves can thereby serve as an indicator of wound healing and restoration of epithelium and ocular surface integrity. It is recommended, where possible, to correlate nerve presence with additional testing of nerve function by esthesiometry.

In the following example, the same group of 9 transplant patients was examined by IVCM yearly for 3 years after transplantation. Central subbasal nerve density was compared to a group of 20 healthy control subjects. With repeated examination of the transplant patients, care must be taken to scan approximately the same central corneal region to obtain an accurate representation of nerve regeneration.

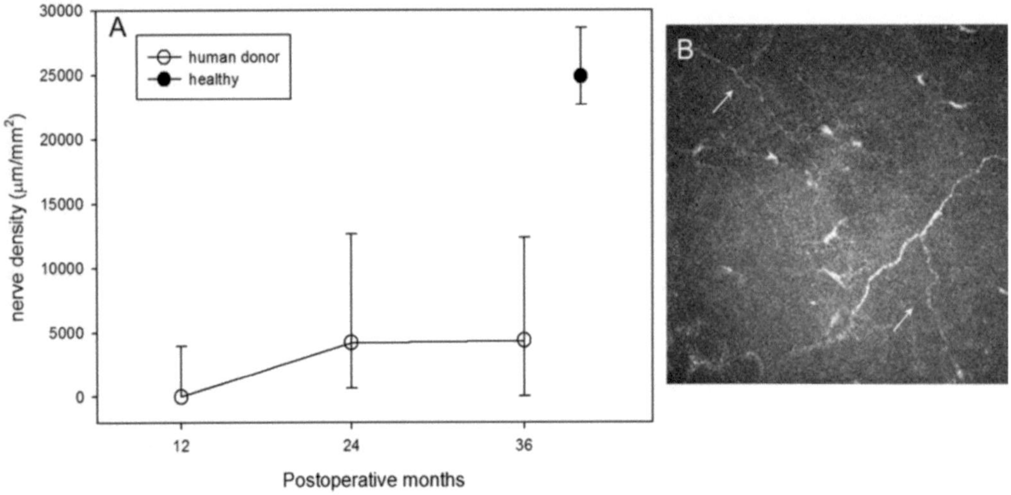

Figure 13. Recovery of central corneal subbasal nerve density after transplantation in a cohort of 9 patients. (A) A slow increase in subbasal nerve density is evident in the first postoperative years, however, density is significantly below the level in healthy corneas. (B) Image of regenerated nerves in the central cornea 3 years after transplantation, depicting thin, tortuous nerve branches (arrows) and a sparse presence of subbasal nerves.

In addition to longitudinal imaging of the central cornea, the peripheral cornea can provide insights into re-innervation of the transplanted cornea. The peripheral scar tissue that forms at the donor-to-recipient interface after surgery can serve as a clear landmark for repeated longitudinal imaging of the same microscopic field of view over time (Figure 14). Imaging the same microscopic field of view has been reported in both clinical [18], [20] and experimental animal studies [16], and provides a powerful demonstration of morphologic changes with time. In this example, imaging of an identical region in the cornea in 1 or 2 year intervals indicates changes in nerve presence and penetration into the implant, as well as longitudinal changes in the scar tissue itself.

4. Cell quantification techniques

In addition to nerves, corneal cells imaged by IVCM can also be subjected to quantitative analysis. The cells most often quantified are epithelial, stromal keratocytes, endothelium, and inflammatory. [10], [23], [24]

4.1. Quantification of epithelial cell and keratocyte density

A review of cell density in the various corneal layers reported by investigators using different types of confocal instruments is presented in Reference 10. With laser-scanning

http://dx.doi.org/10.5772/55216

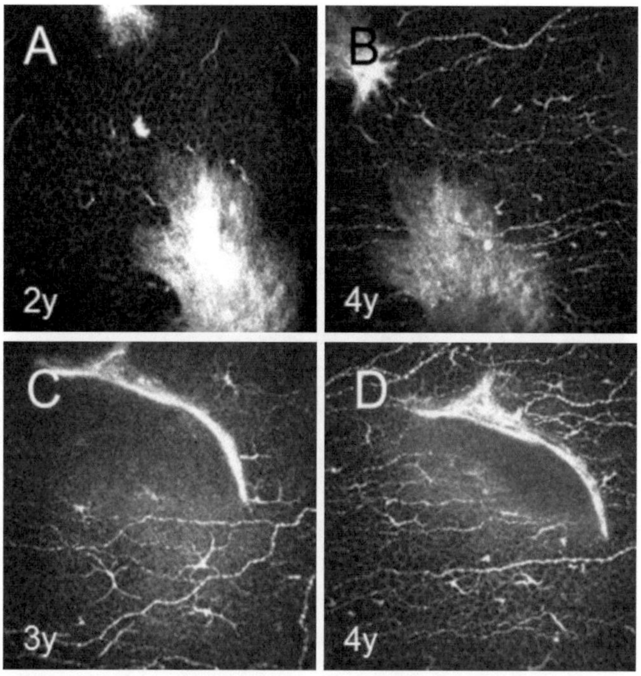

Figure 14. Longitudinal images of an identical region of peripheral cornea with scar tissue present between recipient and donor tissue. (A, B) In one patient, no subbasal nerves were observed in this peripheral region 2 years after transplantation. 4 years after transplantation, however, the identical region is infiltrated with a high density of nerves that traverse the scar tissue. In addition, light scatter from scar tissue has diminished significantly. (C, D) In another patient, the density of nerves in the immediate region of the scar increased significantly from 3 to 4 years postoperative. Note that the subbasal nerve paths are not constant but are constantly changing with time, a phenomenon that has been reported. [40]

confocal microscopy of the cornea, the fine axial resolution requires specification of the exact depth of the layer to be quantified. For example, in one study, epithelial cell and anterior stromal keratocyte density were determined longitudinally in patients with recurrent corneal erosions treated by laser ablation. [19] Epithelial cells were quantified at the superficial, wing, and basal cell layers, with wing cell images taken 20-25 μm below the corneal surface. Similarly, keratocytes were quantified at a distance of 10-15 μm below Bowman's layer. The resulting densities are specified as cells/mm^2. While some investigators prefer to report volumetric density (i.e., cells/mm^3), this notation is dependent upon the axial depth of field which varies between laser and non-laser confocal systems. Comparison between results obtained with laser confocal microscopes is possible by reporting density in cells per unit area.

Once depth is specified, the desired region of interest must be chosen, as it is often impractical to quantify all cells in an image, and desirable to exclude edge artifacts. Image processing techniques are then applied to the region of interest to enhance cell boundaries and facilitate manual cell counting. In Figure 15, a technique for image enhancement and manual cell density quantification is described. [19]

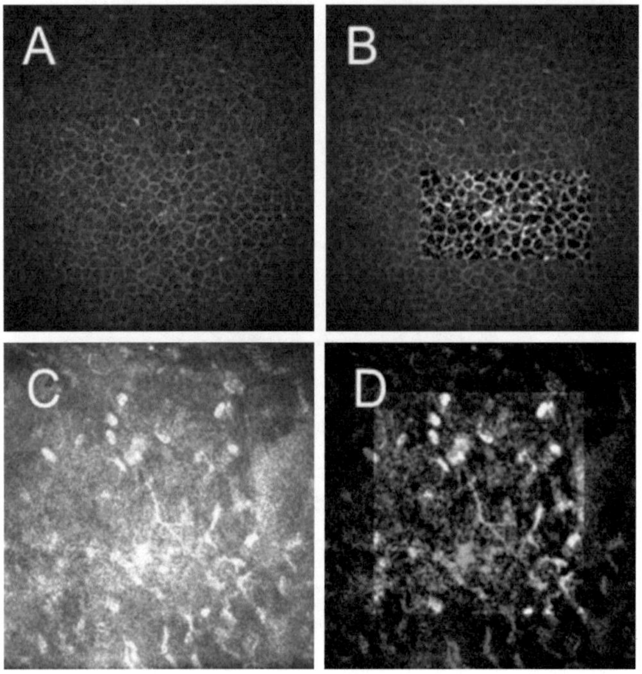

Figure 15. Example of cell quantification by IVCM in corneas affected by recurrent corneal erosions. (A) raw epithelial cell image. (B) image after selection of a 100 × 200 μm region of interest, which has been enhanced by bandpass filtering. Cells are then counted using point-and-click software, but excluding cells touching the bottom or left edge of the region, by convention (to avoid overestimation of density). Two observers perform the counting, and average cell count is converted to density by dividing by the area of the region of interest. The full technique is described in detail elsewhere. [19] (C) Image of keratocytes in the anterior stroma in a patient. (D) images after selection of a 250 × 250 μm region of interest, which has been enhanced by bandpass filtering. Cell quantification then proceeds in a similar manner as for epithelial cells.

4.2. Quantification of dendritic cell density

Inflammatory cells, in particular dendritic cells with long dendrites, migrate into the central cornea at the level of the subbasal nerve plexus in cases of inflammation. Several investigators have quantified the dendritic cell density in the normal and inflamed human cornea. [23], [24], [36] The threshold dendritic cell density in the normal central cornea is 34 ± 3 cells/mm^2, above which dendritic cell densities are pathologically elevated. [23] Contact lens wearers, for example, do not have inflamed corneas, but have an elevated central corneal dendritic cell density reported to be 78 ± 25 cells/mm^2. [48] Active inflammation was shown to correspond to central dendritic cell densities greater than 137 ± 27 cells/mm^2. [24] Quantification of dendritic cell density is achieved by manual counting of dendritic cells with dendrites in an IVCM field of view with the greatest number of visible dendritic cells in the central cornea. [36] The number of cells is then divided by the field of view (400 × 400 μm, or 0.16 mm^2) to yield density. An example of dendritic cell density quantification in patients with aniridia is given in Figure 16. Aniridia patients clearly have pathologically high dendritic cell density, however, only 2 of 11 subjects meet the criteria for having a clinically inflamed cornea.

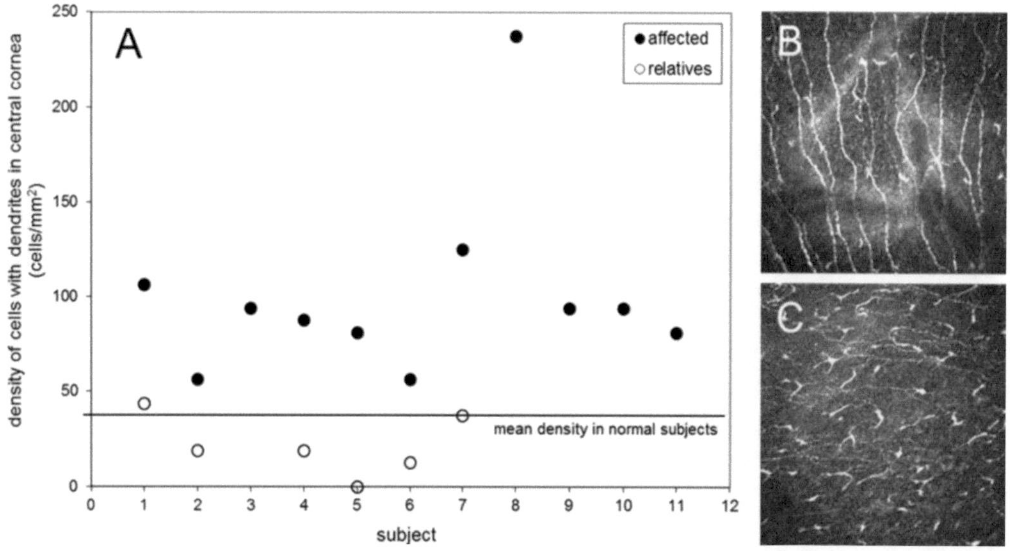

Figure 16. Quantification of mature dendritic cells in patients with aniridia. (A) dendritic cell density in aniridia patients (solid circles) is significantly elevated compared to unaffected relatives (open circles), who have a density considered to be normal (average normal density indicated by solid line, according to [23]). 2 of 11 aniridia subjects had cell density high enough to be considered clinically inflamed. (B) Image of the central subbasal plexus in an unaffected relative, with sparsely distributed mature dendritic cells, density 38 cells/mm^2. (C) Corresponding subbasal plexus in an aniridia patient, density 238 cells/mm^2.

4.3. Fully automated cell counting technique

Manual counting of epithelial cells and keratocytes can be time consuming and subject to observer interpretation and error. It is therefore desirable to automate the cell counting method. We have recently achieved fully-automated cell counting of epithelial wing cells and keratocytes from IVCM images taken from normal, healthy corneas. The automated technique is easy to implement and does not require expensive specialized software, but instead uses the free ImageJ platform. [49] In this method, one starts with the processed images depicted in Figure 15. Images are cropped to the region of interest, and thresholding is applied with an automatic (default) setting used to determine the threshold level. Alternatively, a fixed threshold level can be utilized for all images of a given cell type, based on prior experience of the optimum grayscale level for the particular type of images to be analyzed. Next, images should be inverted as required, such that cells appear as dark objects on a bright background. This is necessary for keratocytes, but not for wing cells as imaged in darkfield microscopy. The image is then converted into a binary representation (i.e., grayscale values are converted to either black or white), and a watershed function is used to automatically add boundaries where required, to depict a full separation between adjacent cells. Wing cells or keratocytes can then be automatically counted using the 'analyze particles' function in ImageJ. Because the function does not allow inclusion of cells touching only two of the four boundaries, the function is applied twice (once including and once excluding all four boundaries) and the cell count values are averaged. An example of this procedure is given in Figure 17.

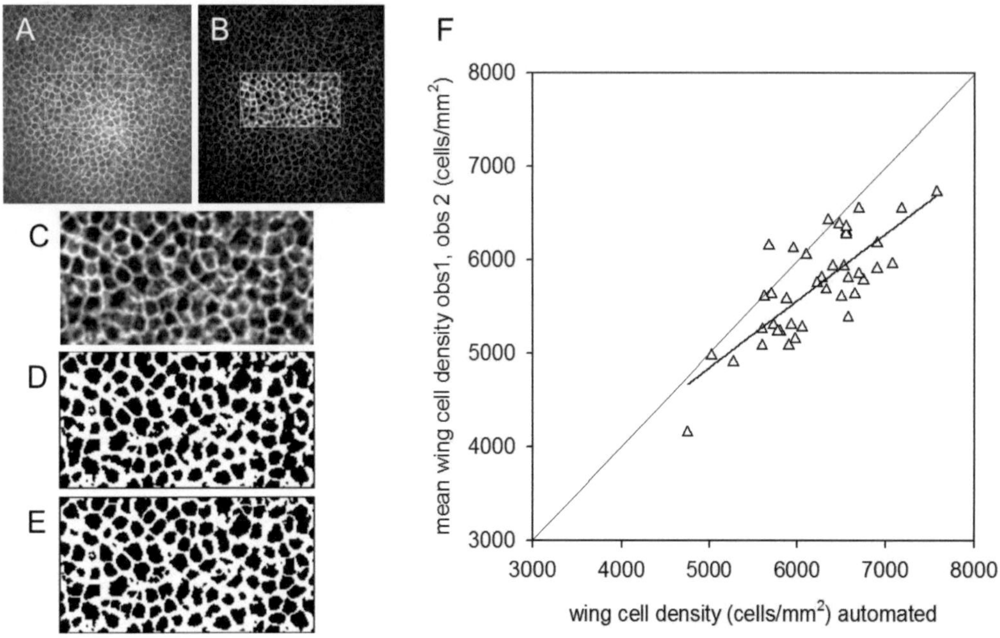

Figure 17. Fully automated analysis of wing cell density from IVCM images. (A) Wing cell layer from a healthy human cornea. (B) Bandpass-filtered region of interest. (C) Cropped, magnified view of region of intrest with automatic thresholding applied. (D) Binary-converted image. (E) Watershed function applied to the binary image to delineate separate cells. Note that the subsequent automatic cell counting is configured to ignore dark areas smaller than a given size in square pixels. This avoids the small point-like artifacts in the image from being included as cells. (F) It is always important to compare the agreement of a new automated method with the original manual cell counting method, to determine the degree of agreement between methods and isolate any source of bias. Here, the automated technique appears to yield slightly larger values for cell density than the manual method (averaged value of two observers).

5. Improved axial resolution and 3D rendering

5.1. Technical requirements

Although the axial resolution of IVCM is approximately 4μm, situations arise where a finer resolution or depth spacing of images is required. This is of particular importance when analyzing small structures or attempting to create 3D models or high-resolution cross-sections for in vivo histology-type analysis. These requirements can be addressed by combined hardware and software-analysis approaches. In terms of hardware, the HRT3-RCM system gives a minimum axial (depth) spacing of 1-2μm between adjacent images. Accurate characterization of structures as thin as Bowman's layer (approximately 10 μm thick in humans) is therefore difficult. [50] An attachment to the HRT3-RCM system, however, can be used to achieve a finer depth spacing of images. The attachment enables fine motorized adjustment of

http://dx.doi.org/10.5772/55216

the microscope objective in the axial direction, to achieve images with axial spacing as fine as 0.5 μm.[4] Analysis of images can be manual or software-assisted. In the following example, software-assisted analysis is used to produce high-resolution histologic-type sections from live tissue and 3D views of lymphatic vessels trafficking inflammatory cells in the rat cornea.

5.2. High-resolution 3D in vivo imaging of corneal lymphatics

Corneal lymphatics are not normally present in the cornea, however, they can invade the corneal stroma during inflammation. [17], [51] In a rat model of inflammatory angiogenesis, the first corneal lymph vessels are observed from 7 to 14 days after the induction of inflammation by placement of a nylon suture through the corneal stroma. [17] Lymphatic vessels are normally analyzed ex-vivo by histologic staining in tissue sections to reveal details of the lymphatic vessel lumen in cross-section [52], however, IVCM enables the histologic cross-section to be analyzed in vivo without removing tissue from the animal. To achieve this, confocal image stacks containing lymph vessels are first obtained using the motor-driven axial positioning attachment. [4] Setting the acquisition rate to 30 frames/sec, a stack of images is obtained during a depth scan of 30 μm in the axial direction. The stack consists of 60 confocal image frames, with an axial spacing of 0.5 μm between images. Using 3D tools from ImageJ, any desired cross-section in the xz or yz directions (z is the axial direction) can be obtained for analysis (Figure 18). An image stack can also serve as input for 3D rendering software, for example the Volocity Visualization software (Volocity 6.0, Perkin Elmer Inc., Waltham, MA, USA). Motion artifacts in the stack can be auto-corrected using an image alignment tool in Volocity, resulting in smoother boundaries when creating a 3D model. With the model, structures can be studied from any desired viewpoint (Figure 18).

6. Future perspective — Correlative microscopy

Although IVCM provides unparalleled possibilities for microscopic in vivo investigation of the cornea, one of the principal drawbacks of the technique is that it provides only morphologic information. The composition and identity of structures and cells viewed by IVCM can only be determined by tissue excision and immunohistological methods. In many cases, however, tissue sampling or excision is impractical or unethical, particularly in human subjects where non-destructive imaging is required.

In the future, correlative microscopy techniques will be required to overcome this limit. With this approach, the same tissue is first examined in vivo, then after tissue excision (in animal models, or after transplantation in humans, for example) in stained sections or by light, fluorescence, or electron microscopy. Comparison of the same structures in vivo and ex vivo by this approach yields insights into the correlation between in vivo morphology and identity and composition of the structures. An example of corneal tissue examined by IVCM prior to transplantation and further analysis by light microscopy and transmission electron microscopy is given in Figure 19. Such analysis improves the ability to interpret IVCM images and recognize pathologic features in vivo.

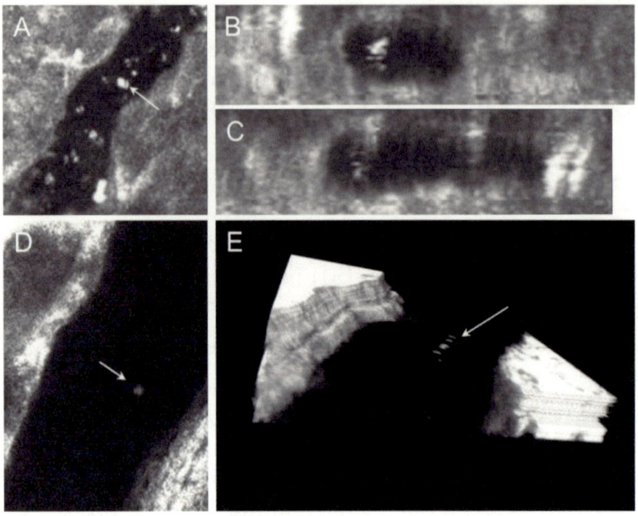

Figure 18. High-resolution 3D imaging of corneal lymphatics in the rat, using IVCM. (A) A standard xy confocal section depicting a lymphatic in the rat cornea during sustained inflammation, with leukocyte traffic visible within the lymph vessel (arrow). (B) An image stack of the same lymph vessel viewed in the xz-plane provides a live histological-type view of the lymph vessel cross-section, with a cell visible in the lymph fluid. (C) yz-plane of the same lymph vessel. (D) A larger, limbal lymph vessel in the rat contains a flowing leukocyte. (E) A confocal image stack obtained during the flow of the leukocyte is rendered into a 3D model, enabling the cell trajectory within the lymphatic lumen to be visualized (arrow).

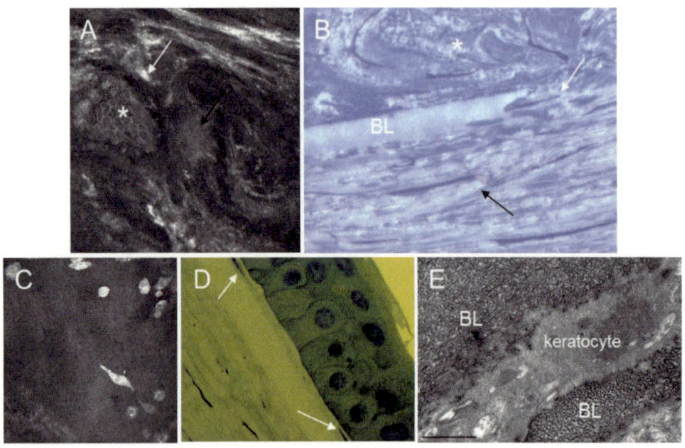

Figure 19. Example of correlating confocal microscopy findings with light and electron microscopy in the same tissue sample. (A) Pre-transplantation IVCM image from a 35y old female with lattice dystrophy, depicting disrupted Bowman's layer (black arrow) along with a presumed keratocyte (white arrow) and basal epithelial cells (asterisk) present at the same corneal depth. Note the uneven light scattering from the stroma, indicating disrupted collagen organization. (B) Light microscope image of a section of the same cornea as in (A) after resection, indicating the histologic appearance of a truncated Bowman's layer (BL), presumed keratocyte at the same depth as BL (white arrow), and disruptions in the stromal collagen (black arrow). (C) Pre-transplantation IVCM image from a 61y old female with endothelial dystrophy, with keratocytes present in the normally acellular BL.(D) Light microscopy section of the same cornea after resection, with keratocytes clearly visible in or anterior to BL. (E) Confirmation by electron microscopy that a keratocyte is present within the non-lamellar collagen of BL.

http://dx.doi.org/10.5772/55216

An example of the use of multiple imaging modalities to confirm diagnosis is given in Figure 20. In this case, upon routine examination at the optometrist an asymptomatic 35y old female was found to have fine, refractile structures distributed throughout both corneas. Slit-lamp biomicroscopy revealed golden-brown colored deposits, which by optical coherence tomography were highly light-scattering and determined to be confined to the anterior one-third of the stroma, which appeared otherwise normal. IVCM confirmed the anterior stromal location of a dense population of light-scattering crystalline structures which were not co-located with keratocyte nuclei. IVCM confirmed the presence of the structures peripherally, and a peripheral biopsy was taken. The biopsy sample was analyzed with transmission electron microscopy, where the crystalline structures could be localized to the keratocyte cytoplasm, confirming the diagnosis of non-nephropathic adult cystinosis.

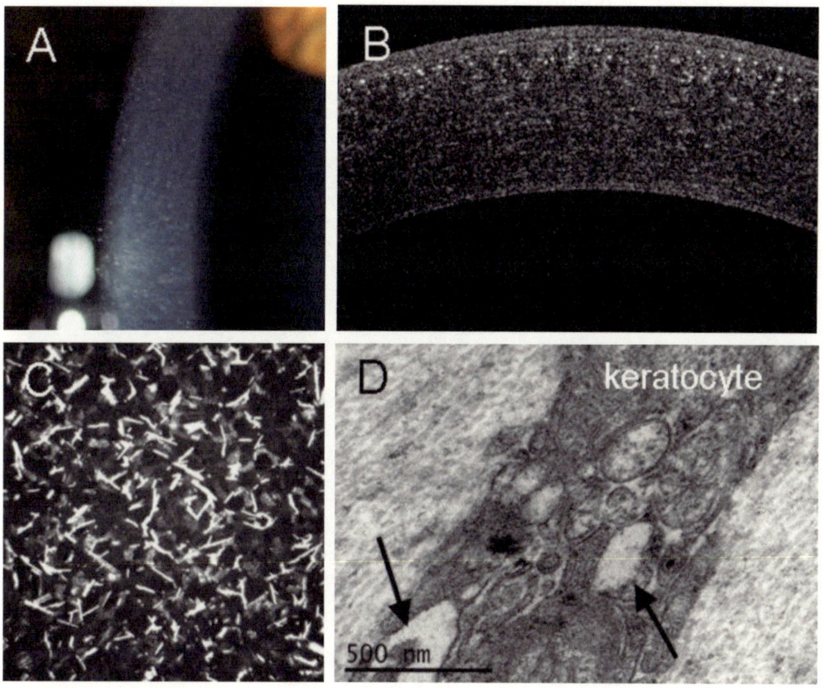

Figure 20. Example of multiple imaging techniques used to characterize and diagnose non-nephropathic adult cystinosis in the cornea of a 35y old female. (A) Slit lamp image of fine, golden refractile particles distributed throughout the cornea. (B) Fourier-domain optical coherence tomography image of the cornea indicating distribution of light-scattering objects in the anterior one-third of the cornea. (C) In vivo confocal microscopy in the anterior cornea indicates linear, crystalline, highly scattering structures in the anterior stroma, located outside the keratocyte nucleus. (D) Transmission electron microscopy image from a biopsy sample taken from the same patient, indicating the presence of cysteine crystals within the cytoplasm of the keratocytes (arrows).

An example of exact correlation between in vivo and immunologically stained tissue is given in Figure 21. In a rat model of angiogenesis, blood vessels and inflammatory cells were observed first by IVCM, by the technique depicted in Figure 2B. Following live imaging, the cornea was excised at specific time points (2 and 7 days after induction of inflammation in this example). The tissue was then prepared as a whole-mount corneal sample, immunologically labeled with primary antibodies against specific cellular and vascular structures, and imaged in the same en face view as the IVCM provides, but instead using a high-resolution laser-scanning confocal fluorescence microscope to detect fluorescent signatures of the secondary antibodies. The whole-mount sample is then manually translated during scanning to locate the same vessel structures observed in vivo. This technique requires access to the pre-saved IVCM images while performing the fluorescence microscopy, and additionally requires distinct vessel configurations that serve as landmarks to aid in the localization of the same structures in fluorescence. When the exact region is found in fluorescence, comparison with the in vivo images can yield a wealth of information about the structures viewed in vivo. For example, as shown in Figure 21, the technique has confirmed pericyte coverage of certain pathologic blood vessels, and has resulted in the first observations of corneal lymphatics in a live animal cornea without labeling. [17]

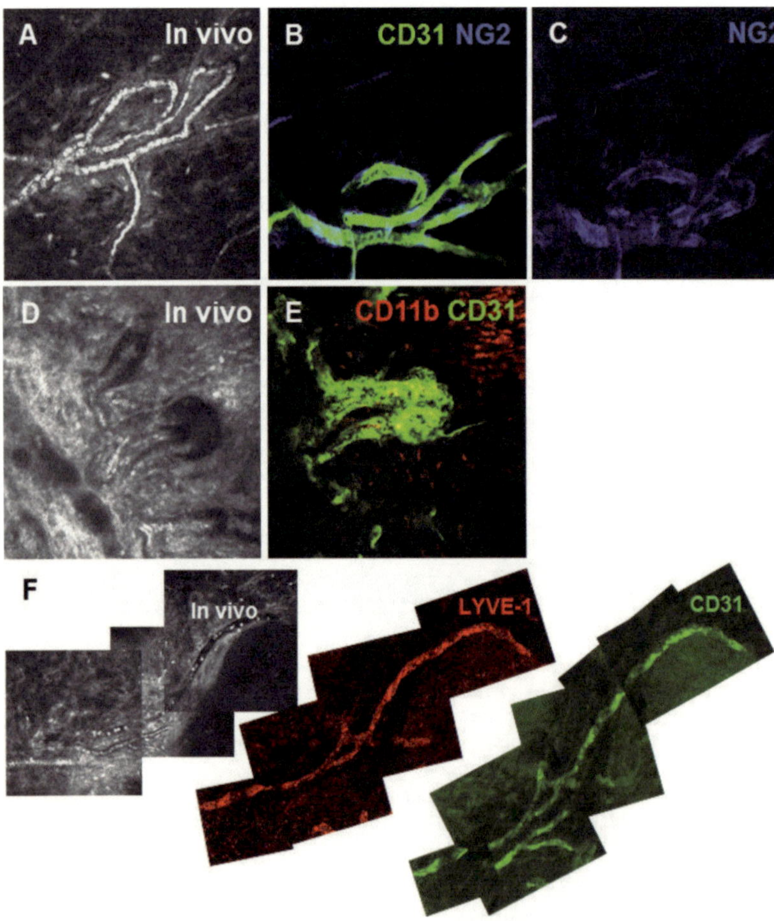

Figure 21. Examples of exact correlative in vivo/ex vivo microscopy. (A-C) The same pathologic corneal blood vessels viewed in vivo and ex vivo in an *en face* orientation, after immunofluorescent staining for blood vessel vascular endo-

thelium (CD31) and pericytes/mural cells (NG2). The images indicate the appearance of mature, pericyte-covered blood vessels in the cornea that are more resistant to regression. (D, E) In vivo and immunofluorescent view of a highly dilated limbal blood vessel prior to angiogenic sprouting. The blood vessel is $CD31^{+}$ while cells inside and around the vessel are $CD11b^{+}$ inflammatory myeloid cells. The images indicate the inflammatory cells do not physically incorporate into the vascular endothelium during the budding process, but likely serve a paracrine role. (F) Confirmation of the in vivo appearance of lymphatic vessels in the rat cornea, which have a different morphology than that of blood vessels in vivo. Localization of the same lymph vessel after tissue whole mounting and immunostaining for lymphatic endothelial marker LYVE-1 (which blood vessels do not express) served to confirm the lymphatic nature of the in vivo vessel structure.

7. Conclusion

Besides the direct clinical and preclinical application of IVCM to the cornea – which in itself yields a wealth of information pertaining to corneal anatomy, physiology, and pathology – the technique can serve as an instructional tool in the use of laser-scanning confocal microscopy. Many of the acquisition, processing, and analysis techniques described in this chapter are applicable to other confocal microscopes, different tissue types, and other fields of research and practice. Moreover, IVCM of the cornea is a constantly growing and evolving field, and it is envisaged that the technique will extend our knowledge of not only the eye, but of basic biologic and physiologic processes occurring in other parts of the body. Finally it is hoped that other fields of scientific endeavor will not only benefit from knowledge gained from IVCM of the cornea, but, conversely, that knowledge obtained from laser-scanning confocal microscopy in other fields can be applied to the cornea.

Acknowledgements

The authors wish to acknowledge kind contributions to the work presented from the following individuals: Tor Paaske Utheim and Xiangjun Chen, Oslo University Department of Ophthalmology, Catharina Traneus-Röckert, Department of Pathology, Linköping University, Thu Ba Wihlmark and Marina Koulikovska, Department of Ophthalmology, Linköping University Hospital, Joachim Stave, Department of Ophthalmology, University of Rostock, and Ulf Stenevi and Charles Hanson, Sahlgrenska University Hospital, Gothenburg. The authors also wish to acknowledge financial support from the King Gustav V and Queen Margaretas Freemasons Foundation and the Carmen and Bertil Regnérs Foundation.

Author details

Neil Lagali*, Beatrice Bourghardt Peebo, Johan Germundsson, Ulla Edén, Reza Danyali, Marcus Rinaldo and Per Fagerholm

Department of Clinical and Experimental Medicine, Faculty of Health Sciences, Linköping University, Linköping, Sweden

References

[1] C. Stephen Foster, Dimitri T Azar, Claes H Dohlman, Eds. Smolin and Thoft's The Cornea: Scientific Foundations and Clinical Practice, 4th Edition, Lippincott Williams Wilkins 2004.

[2] Oliveira-soto, L, & Efron, N. Assessing the cornea by in vivo confocal microscopy. Clin Experiment Ophthalmol (2003). , 31, 83-84.

[3] Atlas of Confocal Laser Scanning In-vivo Microscopy in OphthalmologyBy R.F. Guthoff, C. Baudouin, J. Stave,. Springer, Berlin (2006).

[4] Rostock Cornea Module Z-motor drive Unit. Joachim Stave (Rostock). Personal communication.

[5] Niederer, R. L, & Mcghee, C. N. Clinical in vivo confocal microscopy of the human cornea in health and disease. Prog Retin Eye Res. (2010). , 29, 30-58.

[6] Mclaren, J. W, Nau, C. B, Kitzmann, A. S, & Bourne, W. M. Keratocyte density: comparison of two confocal microscopes. Eye Contact Lens. (2005). , 31, 28-33.

[7] Lagali, N, Griffith, M, Fagerholm, P, Merrett, K, Huynh, M, & Munger, R. Innervation of tissue-engineered recombinant human collagen-based corneal substitutes: a comparative in-vivo confocal microscopy study. Invest Ophthalmol Vis Sci, (2008). , 49, 3895-3902.

[8] Guthoff, R. F, Zhivov, A, & Stachs, O. In vivo confocal microscopy, an inner vision of the cornea- a major review. Clin Experiment Ophthalmol. (2009). Jan;, 37(1), 100-17.

[9] Erie, J. C, Mclaren, J. W, & Patel, S. V. Confocal microscopy in ophthalmology. Am J Ophthalmol. (2009). , 148, 639-46.

[10] Patel, D. V, & Mcghee, C. N. Contemporary in vivo confocal microscopy of the living human cornea using white light and laser scanning techniques: a major review. Clin Experiment Ophthalmol. (2007). , 35, 71-88.

[11] Labbé, A, Liang, H, Martin, C, Brignole-baudouin, F, Warnet, J. M, & Baudouin, C. Comparative anatomy of laboratory animal corneas with a new-generation high-resolution in vivo confocal microscope. Curr Eye Res. (2006). , 31, 501-9.

[12] Lagali, N, Griffith, M, Shinozaki, N, Fagerholm, P, & Munger, R. Innervation of tissue-engineered corneal implants in a porcine model: a 1-year in-vivo confocal microscopy study. Invest Ophthalmol Vis Sci (2007). , 48, 3537-3544.

[13] Hackett, J, Lagali, N, Merrett, K, Edelhauser, H, Sun, Y, Gan, L, Griffith, M, & Fagerholm, P. Biosynthetic corneal implants for replacement of pathologic corneal tissue: performance in a controlled rabbit alkali burn model. Invest Ophthalmol Vis Sci (2011). , 52, 651-657.

[14] Hovakimyan, M, Guthoff, R, Reichard, M, Wree, A, Nolte, I, & Stachs, O. In vivo confocal laser-scanning microscopy to characterize wound repair in rabbit corneas after collagen cross-linking. Clin Experiment Ophthalmol. (2011)., 39, 899-909.

[15] Liang, H, Brignole-baudouin, F, Labbé, A, Pauly, A, Warnet, J. M, & Baudouin, C. LPS-stimulated inflammation and apoptosis in corneal injury models. Mol Vis. (2007)., 13, 1169-80.

[16] Peebo, B. B, Fagerholm, P, Traneus-röckert, C, & Lagali, N. Time-lapse in vivo imaging of corneal angiogenesis: the role of inflammatory cells in capillary sprouting. Invest Ophthalmol Vis Sci (2011)., 52, 3060-8.

[17] Peebo, B. B, Fagerholm, P, Traneus-röckert, C, & Lagali, N. Cellular-level characterization of lymph vessels in live, unlabelled corneas by in-vivo confocal microscopy. Invest Ophthalmol Vis Sci (2010)., 51, 830-835.

[18] Bourghardt Peebo BFagerholm P, Lagali N. Transient anterior corneal deposits in a human immunodeficiency virus-positive patient. Cornea (2010)., 29, 1323-1327.

[19] Lagali, N, Germundsson, J, & Fagerholm, P. The role of Bowman's layer in anterior corneal regeneration after shallow-depth phototherapeutic keratectomy: a prospective, morphological study using in-vivo confocal microscopy. Invest Ophthalmol Vis Sci (2009)., 50, 4192-4198.

[20] Peebo, B. B, Fagerholm, P, Traneus-röckert, C, & Lagali, N. Cellular level characterization of capillary regression in inflammatory angiogenesis using an in vivo corneal model. Angiogenesis (2011)., 14, 393-405.

[21] Knappe, S, Stachs, O, Zhivov, A, Hovakimyan, M, & Guthoff, R. Results of confocal microscopy examinations after collagen cross-linking with riboflavin and UVA light in patients with progressive keratoconus. Ophthalmologica. (2011)., 225, 95-104.

[22] Erie, J. C, Hodge, D. O, & Bourne, W. M. Confocal microscopy evaluation of stromal ablation depth after myopic laser in situ keratomileusis and photorefractive keratectomy. J Cataract Refract Surg. (2004)., 30, 321-5.

[23] Zhivov, A, Stave, J, Vollmar, B, & Guthoff, R. In vivo confocal microscopic evaluation of Langerhans cell density and distribution in the normal human corneal epithelium. Graefes Arch Clin Exp Ophthalmol. (2005)., 243, 1056-61.

[24] Mastropasqua, L, Nubile, M, Lanzini, M, Carpineto, P, Ciancaglini, M, & Pannellini, T. Di Nicola M, Dua HS. Epithelial dendritic cell distribution in normal and inflamed human cornea: in vivo confocal microscopy study. Am J Ophthalmol. (2006)., 142, 736-44.

[25] Patel, D. V. McGhee CNJ. Acanthamoeba keratitis: a comprehensive photographic reference of common and uncommon signs. Clin Experiment Ophthalmol (2009)., 37, 232-238.

[26] Vaddavalli, P. K, Garg, P, Sharma, S, Sangwan, V. S, Rao, G. N, & Thomas, R. Role of confocal microscopy in the diagnosis of fungal and acanthamoeba keratitis. Ophthalmology. (2011)., 118, 29-35.

[27] Hau, S. C, Dart, J. K, Vesaluoma, M, Parmar, D. N, Claerhout, I, Bibi, K, & Larkin, D. F. Diagnostic accuracy of microbial keratitis with in vivo scanning laser confocal microscopy. Br J Ophthalmol (2010)., 94, 982-7.

[28] Shiraishi, A, Uno, T, Oka, N, Hara, Y, Yamaguchi, M, & Ohashi, Y. In vivo and in vitro laser confocal microscopy to diagnose acanthamoeba keratitis. Cornea (2010)., 29, 861-5.

[29] Niederer, R. L, Perumal, D, Sherwin, T, & Mcghee, C. N. Age-related differences in the normal human cornea: a laser scanning in vivo confocal microscopy study. Br J Ophthalmol (2007)., 91, 1165-9.

[30] Patel, D. V, Sherwin, T, & Mcghee, C. N. Laser scanning in vivo confocal microscopy of the normal human corneoscleral limbus. Invest Ophthalmol Vis Sci. (2006)., 47, 2823-7.

[31] Shortt, A. J, Secker, G. A, Munro, P. M, Khaw, P. T, Tuft, S. J, & Daniels, J. T. Characterization of the limbal epithelial stem cell niche: novel imaging techniques permit in vivo observation and targeted biopsy of limbal epithelial stem cells. Stem Cells. (2007)., 25, 1402-9.

[32] Takahashi, N, Chikama, T, Yanai, R, & Nishida, T. Structures of the corneal limbus detected by laser-scanning confocal biomicroscopy as related to the palisades of Vogt detected by slit-lamp microscopy. Jpn J Ophthalmol. (2009)., 53, 199-203.

[33] Miri, A, Al-aqaba, M, Otri, A. M, Fares, U, Said, D. G, Faraj, L. A, & Dua, H. S. In vivo confocal microscopic features of normal limbus. Br J Ophthalmol. (2012)., 96, 530-6.

[34] Deng, S. X, Sejpal, K. D, Tang, Q, Aldave, A. J, Lee, O. L, & Yu, F. Characterization of limbal stem cell deficiency by in vivo laser scanning confocal microscopy: a microstructural approach. Arch Ophthalmol. (2012)., 130, 440-5.

[35] Hong, J, Zheng, T, Xu, J, Deng, S. X, Chen, L, Sun, X, Le, Q, & Li, Y. Assessment of limbus and central cornea in patients with keratolimbal allograft transplantation using in vivo laser scanning confocal microscopy: an observational study. Graefes Arch Clin Exp Ophthalmol. (2011)., 249, 701-8.

[36] Edén, U, Fagerholm, P, Danyali, R, & Lagali, N. Pathologic epithelial and anterior corneal nerve morphology in early-stage congenital aniridic keratopathy. Ophthalmology. (2012)., 119, 1803-10.

[37] Müller, L. J, Marfurt, C. F, Kruse, F, & Tervo, T. M. Corneal nerves: structure, contents and function. Exp Eye Res. (2003)., 76, 521-42.

[38] Patel, D. V, & Mcghee, C. N. In vivo confocal microscopy of human corneal nerves in health, in ocular and systemic disease, and following corneal surgery: a review. Br J Ophthalmol. (2009). , 93, 853-60.

[39] Patel, D. V, & Mcghee, C. N. Mapping of the normal human corneal sub-Basal nerve plexus by in vivo laser scanning confocal microscopy. Invest Ophthalmol Vis Sci. (2005). , 46, 4485-8.

[40] Patel, D. V, & Mcghee, C. N. In vivo laser scanning confocal microscopy confirms that the human corneal sub-basal nerve plexus is a highly dynamic structure. Invest Ophthalmol Vis Sci. (2008). , 49, 3409-12.

[41] Patel, D. V, & Mcghee, C. N. Mapping the corneal sub-basal nerve plexus in keratoconus by in vivo laser scanning confocal microscopy. Invest Ophthalmol Vis Sci. (2006). , 47, 1348-51.

[42] Zhivov, A, Blum, M, Guthoff, R, & Stachs, O. Real-time mapping of the subepithelial nerve plexus by in vivo confocal laser scanning microscopy. Br J Ophthalmol. (2010). , 94, 1133-5.

[43] Turuwhenua, J. T, Patel, D. V, & Mcghee, C. N. Fully automated montaging of laser scanning in vivo confocal microscopy images of the human corneal subbasal nerve plexus. Invest Ophthalmol Vis Sci. (2012). , 53, 2235-42.

[44] Szaflik, J. P. Comparison of in vivo confocal microscopy of human cornea by white light scanning slit and laser scanning systems. Cornea. (2007). , 26, 438-45.

[45] Marfurt, C. F, Cox, J, Deek, S, & Dvorscak, L. Anatomy of the human corneal innervation. Exp Eye Res. (2010). , 90, 478-92.

[46] Al-aqaba, M. A, Fares, U, Suleman, H, Lowe, J, & Dua, H. S. Architecture and distribution of human corneal nerves. Br J Ophthalmol. (2010). , 94, 784-9.

[47] Waring, G. O. rd,. Rodrigues MM, Laibson PR.Corneal dystrophies. I. Dystrophies of the epithelium, Bowman's layer and stroma. Surv Ophthalmol. (1978). , 23, 71-122.

[48] Zhivov, A, Stave, J, Vollmar, B, & Guthoff, R. In vivo confocal microscopic evaluation of langerhans cell density and distribution in the corneal epithelium of healthy volunteers and contact lens wearers. Cornea. (2007). , 26, 47-54.

[49] Rasband, W.S, & Image, . , U. S. National Institutes of Health, Bethesda, Maryland, USA, http://imagej.nih.gov/ij/, 1997-2012.

[50] Germundsson, J, Fagerholm, P, Koulikovska, M, & Lagali, N. S. An accurate method to determine Bowman's layer thickness in vivo in the human cornea. Invest Ophthalmol Vis Sci. (2012). , 53, 2354-9.

[51] Regenfuss, B, Bock, F, Parthasarathy, A, & Cursiefen, C. Corneal (lymph)angiogenesis--from bedside to bench and back: a tribute to Judah Folkman. Lymphat Res Biol. (2008).

[52] Cursiefen, C, Schlötzer-schrehardt, U, Küchle, M, Sorokin, L, Breiteneder-geleff, S, Alitalo, K, & Jackson, D. Lymphatic vessels in vascularized human corneas: immunohistochemical investigation using LYVE-1 and podoplanin. Invest Ophthalmol Vis Sci. (2002). , 43, 2127-35.

Chapter 5

Laser Scanning Confocal Microscopy: Application in Manufacturing and Research of Corneal Stem Cells

Vanessa Barbaro**, Stefano Ferrari**, Mohit Parekh,
Diego Ponzin, Cristina Parolin and Enzo Di Iorio

Additional information is available at the end of the chapter

http://dx.doi.org/10.5772/55809

1. Introduction

Laser scanning confocal microscopes (LSCMs) are powerful devices used to acquire high definition optical images by choosing the required depth selectively. The presence of specific laser beams and features such as fluorescence recovery after photobleaching (FRAP), fluorescence lifetime imaging microscopy (FLIM), and fluorescence resonance energy transfer (FRET) allow to:

i. increase the quality of the image;

ii. observe and analyze subcellular organelles;

iii. track the localization of any given labeled molecule within the cell;

iv. identify specific areas within a tissue/organ (Figure 1).

In parallel, the development and manufacturing of fluorescent probes (=fluorophores) characterized by low toxicity profiles are allowing to perform the above mentioned studies using living cell cultures or tissues that are not fixed. Furthermore, fluorescent proteins such as the Green Fluorescent Protein (GFP) and its derivatives allow to detect how the biosynthetic machinery of the cell works or a transgene (driven by a plasmid or a genetically engineered virus) is expressed (Figure 2) or a chimeric protein interacts with other cellular components.

The aim of this chapter is therefore to describe how LSCM functions and features have helped vision sciences and regenerative medicine applications in the field of ophthalmology. The next sections will analyze how LSCM-based analyses have helped to:

1. evaluate how the ocular surface is formed;
2. define the role of p63 as stem cell marker;
3. set up quality control assays required for clinical applications of limbal stem cells in patients with limbal stem cell deficiency (LSCD);
4. validate the use of impression citology as a diagnostic tool for LSCD;
5. study gene therapy-based potential ways to treat rare genetic disorders of the ocular surface.

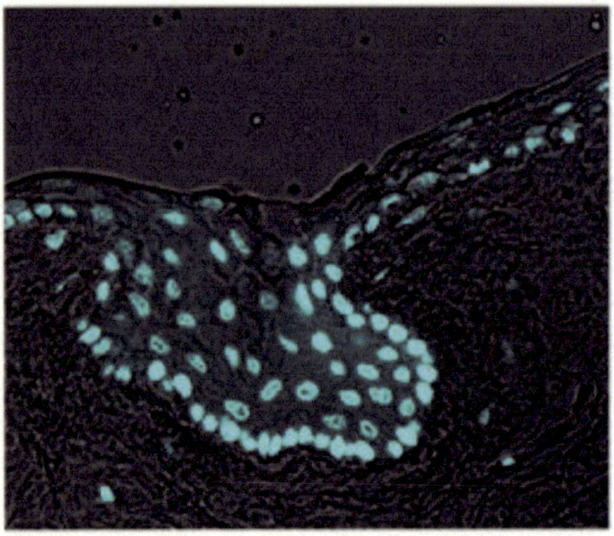

Figure 1. DAPI staining of the Palisades of Vogt in the limbus of human ocular surfaces.

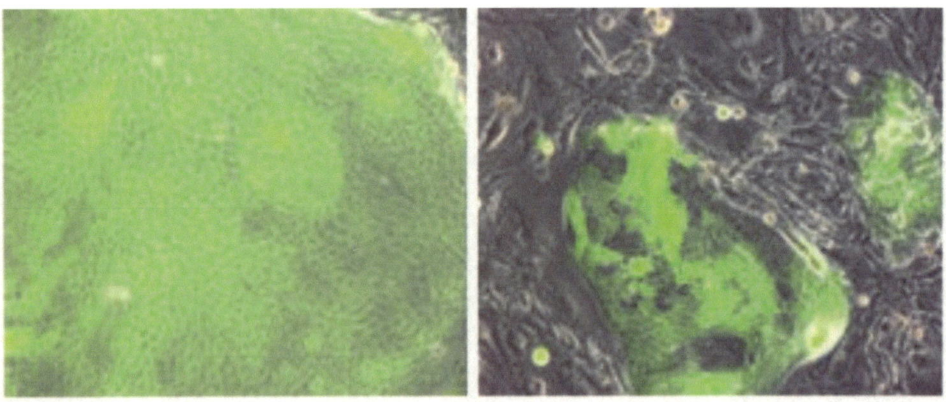

Figure 2. Human primary corneal epithelial stem cells showing a) high GFP expression in normal cells and b) low GFP expression in partially deceased or suffering cells.

2. LSCM as a mean to study the human ocular surface

2.1. The human ocular surface and limbal stem cell deficiency

The human ocular surface is made up of cornea, conjunctiva and limbus (Figure 3). The cornea is the anterior part of the eye which helps to transmit the light through the lens to the retina. Any alterations to the cornea may lead to poor visual outcome. The limbus is the intermediate layer between cornea and conjunctiva (Figure 4). It is a reservoir of limbal epithelial stem cells, which are essential for the renewal of the epithelium and the integrity of the corneal stroma. Pathologies/injuries affecting the limbus lead to LSCD, which can be caused either by inherited pathologies or, more commonly, are the result of acquired factors, such as chemical/thermal injuries, ultraviolet and ionizing radiations, contact lens keratopathy, limbal surgery and conditions like Stevens-Johnson syndrome. When LSCD occurs, the neighbouring conjunctival epithelium, which is normally prevented from encroaching on the corneal surface by LSCs, migrates over the stroma [1]. This process is known as conjunctivalization and usually is accompanied by neovascularization and abnormal fibrovascular tissue covering the corneal surface (pannus). This eventually leads to chronic inflammation, corneal opacity and vision impairment (Figure 5). Conventional corneal transplantation is not feasible as, in order to succeed, it requires the gradual replacement of the donor's corneal epithelium with the recipient's. LSCD, instead, allows/stimulates conjunctival cell ingrowth with accompanying neovascularization and inflammation, resulting in cornea graft failure. Patients with total LSCD are therefore poor candidates for corneal transplantation.

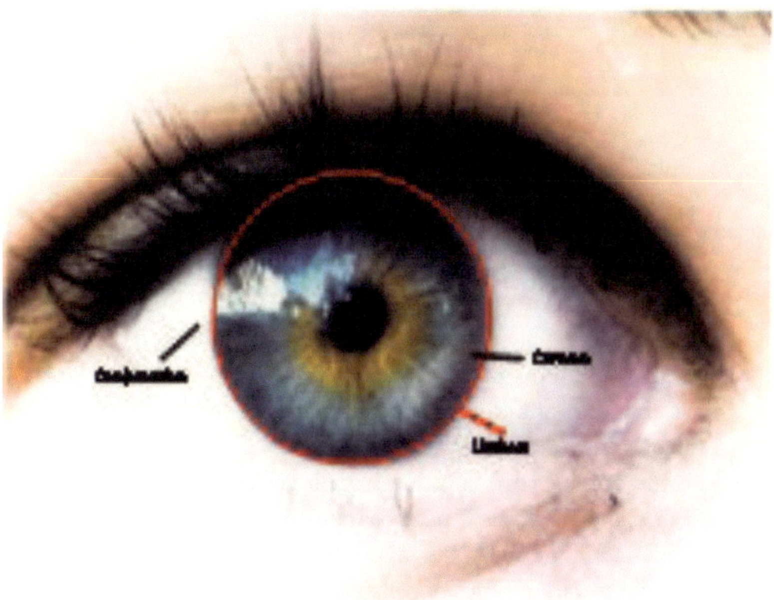

Figure 3. Anatomy of the human ocular surface

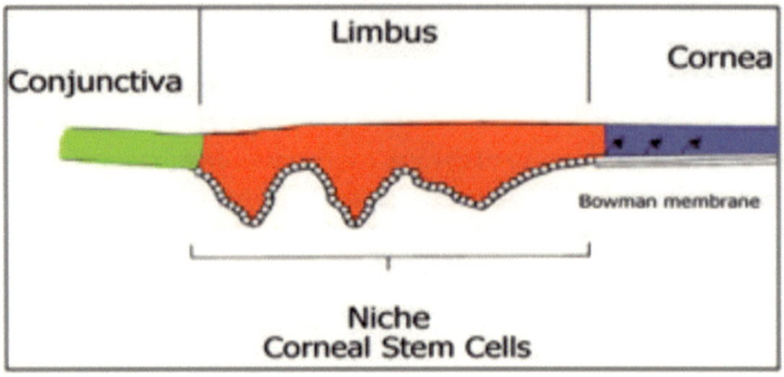

Figure 4. The corneal stem cell niche in the limbus

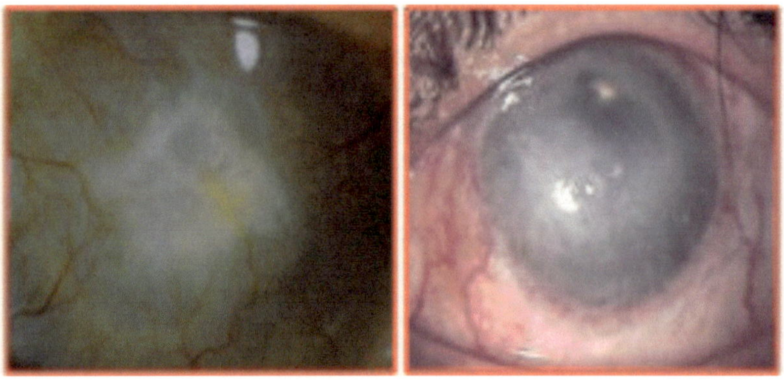

Figure 5. Corneal opacification due to conjunctivalization and vascularization, as seen in LSCD patients.

2.2. Limbal Stem Cells (LSCs) and clonal analysis

The corneal epithelium provides an ideal model to distinguish between three major types of cells such as, a) corneal keratinocyte stem cells (KSC) which governs the renewal of corneal squamous epithelium, b) transient amplifying cells (TA) which migrate from the limbus to form the corneal epithelium and c) post-mitotic (PM) or differentiated cells which terminally differentiate after a limited number of cell divisions [2-4], as shown in figure 6. The determination of the LSCs is an important criterion to anticipate the positive/negative consequences of ocular surface reconstruction during cell therapy-based treatments carried out to treat LSCD. Evaluation of the number of LSCs does require assays that allow to determine their number or percentage. Clonal analysis is used to investigate the properties of individual cells and is essential to understand the self-renewal potency of each cell. Clonal analysis of limbal epithelial cells has shown different types of cells such as:

a. Holoclones (Figure 7 a, d): Putative stem cells with a diameter of 6-10 μm. These cells have a high proliferating capability with ≤5% aborted colonies and ≥100 cell doublings;

http://dx.doi.org/10.5772/55809

b. Meroclones (Figure 7 b, e): Young transient amplifying cells with intermediate proliferating capacity having a diameter of 10-18 μm. These cells usually have 5-95% aborted colonies;

c. Paraclones (Figure 7 c, f): Terminally differentiated cells with 15-20 cell doublings and very low proliferative capability. These cells are 18-36 μm long in diameter.

While the evaluation of the colonies generated by the three clonal types can be carried out through traditional microscopy techniques, it was only with the advent of LSCM-based techniques that the size of the cells was determined, thus allowing to identify the cell types earlier during the culturing process.

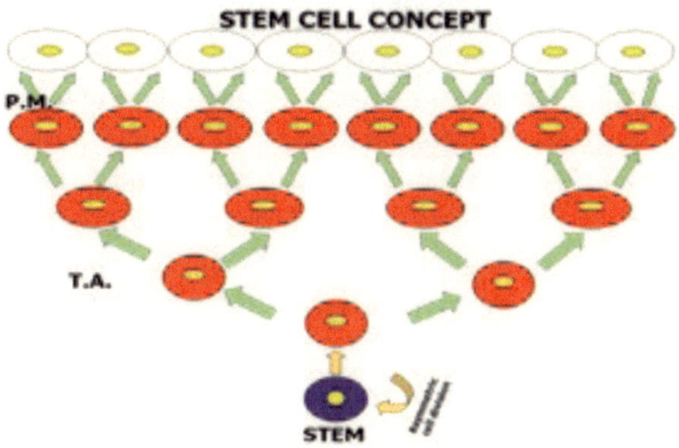

Figure 6. Proliferation of stem cells

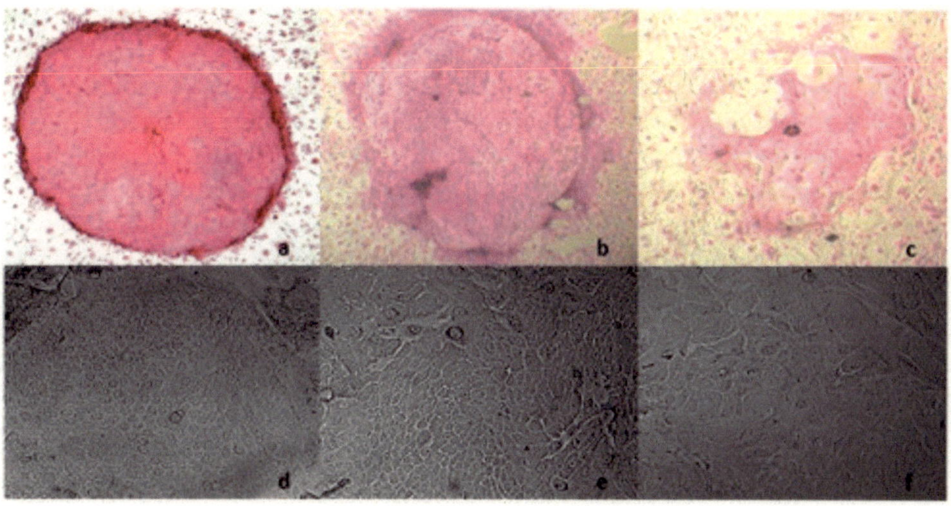

Figure 7. Clonal analysis of limbal epithelial stem cells, holoclones (a, d), meroclones (b, e), paraclones (c, f).

3. The role of p63 as a stem cell marker

3.1. p63 expression in the cornea

"p63" is a transcription factor belonging to the same family that includes p53 and p73. Whereas p53 plays a well-established role in tumor suppression, p63 and p73 play unique roles in morphogenesis [5-8]. In particular, p63-/- mice show major defects in limb and craniofacial development, as well as a striking absence of stratified epithelia. p63 is essential for regenerative proliferation in epithelial development, distinguishes human keratinocyte stem cells from their TA progeny, is expressed by the basal cells of the limbal epithelium (but not by TA cells covering the corneal surface), and is abundantly expressed by epidermal and limbal holoclones, but undetectable in paraclones. The p63 gene generates six isoforms, the transactivating (TA) and the ΔN isoforms. In both cases, alternative splicing gives rise to 3 different C termini, designated α, β and γ. In human corneal epithelia, ΔNp63α is the major p63 isoform expressed and it is necessary for the maintenance of the proliferative potential of limbal stem cells and essential for regenerative proliferation in the ocular surface [9]. Limbal-corneal keratinocytes express not only ΔNp63α but also the ΔNp63β and ΔNp63γ isoforms. However, while expression of ΔNp63α is restricted to the limbal stem cell compartment, the expression of ΔNp63β and ΔNp63γ correlates with limbal cell migration, corneal wound healing and corneal differentiation. ΔNp63α is expressed in a small amount of undifferentiated and small cells (stem cells). The percentage of these cells in primary cultures ranges between 3% and 8% and decreases progressively both during clonal conversion (the transition from holoclones to meroclones and paraclones) and serial propagation of stem cells in vitro (life span).

3.2. Immunofluorescence for p63 in corneal tissues

A series of experiments were performed by using the 4A4 antibody, able to recognize all p63 isoforms, and LSCM-based assays. The p63 staining was performed on cryosections of various human corneal tissues, classified in two groups: normal unperturbated corneas (referred to as *resting corneas*) and wounded corneas (referred to as *activated corneas*) [9]. It was observed that in the resting ocular epithelium, the α isoform of ΔNp63 is present only in the basal layer of the limbus, thus meaning that ΔNp63α is likely to identify the stem cell population of the human limbus and supporting the concept that α is the isoform of ΔNp63 essential for regenerative proliferation. The number of limbal cells positive for α isoform was significantly higher in wounded corneas, thus suggesting that human limbal stem cells divide upon corneal injury. Neither β nor γ isoforms are present in substantial amounts in resting corneas, but both become abundant in activated limbal and corneal epithelia. The presence of p63+ cells in the activated central corneal epithelium is due to migration of p63+cells from the limbus. This explains why corneal cells cultivated from a resting cornea proliferate very little and do not express p63.

3.3. In situ hybridization of p63 isoforms

In order to strengthen the knowledge about p63 and its isoforms, an *in situ* hybridization assay has been performed and LSCM used to analyze the results. With respect to the ΔNp63 isoforms,

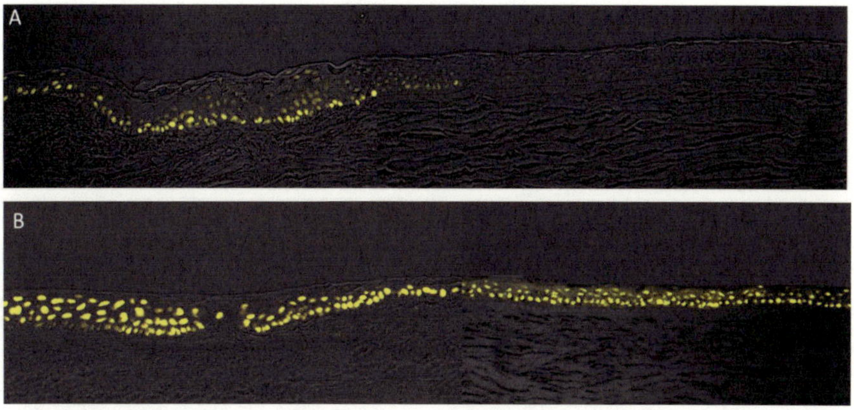

Figure 8. Expression of p63 in resting (A) and activated (B) corneas.

α mRNA was present in patches of basal cells of the resting limbus β (Figure 9a) but was indetectable in the entire resting corneal epithelia (Figure 9b). β mRNA was indetectable in resting limbal and corneal epithelia (Figure 9 c,d) whereas γ mRNA was barely detectable in the uppermost layers of both epithelia (Figure 9 e,f). In sharp contrast, α, β and γ mRNAs were detected in both limbal and corneal epithelia from activated corneas (Figure 9 g-l).

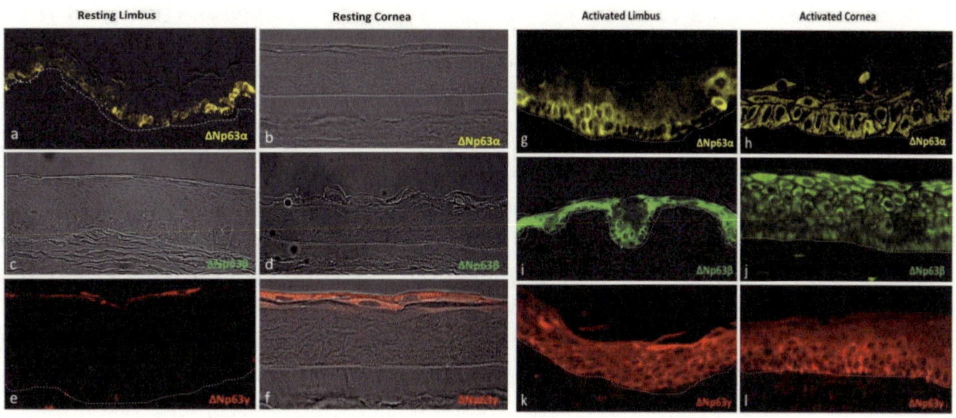

Figure 9. ΔNp63α, β, and γ transcripts in resting and activated limbal and corneal epithelia. The dotted line indicates the basal layer of activated limbal and corneal epithelia.

3.4. Double Immunofluorescence for p634A4 and ΔNp63α in limbal clonal types

Double immunofluorescence (DI) is used in order to examine the co-distribution of expression of two or more different markers in the same sample and results can only be achieved by using the features of a LSCM. DI was performed on the different types of clones isolated from primary limbal cultures by using 4A4 and ΔNp63α antibodies. It was found that all cells of colonies produced by holoclones expressed ΔNp63α while cells of colonies formed by paraclones lacked this isoform. Because the cells of paraclones are stained by the 4A4 mAb, this staining must be due to β and γ. Cells of colonies formed by meroclones were well stained by 4A4, but very much less by the antibody to α. So we can conclude that the isoform of ΔNp63 that most precisely characterizes clonal types is the α isoform.

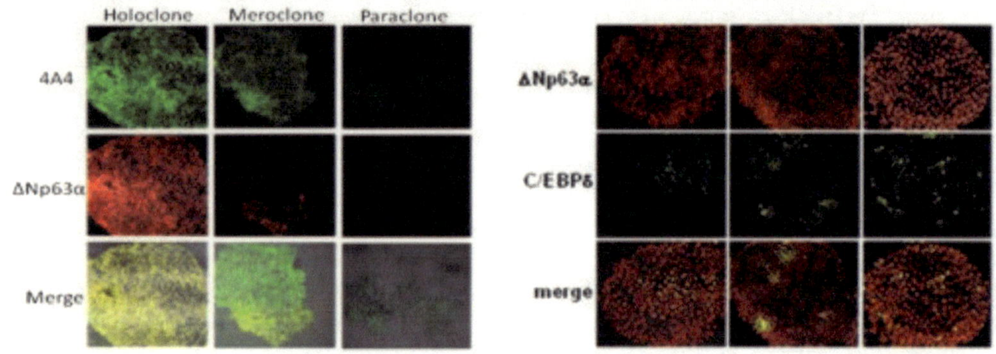

Figure 10. DI of Holoclone, Meroclone and Paraclone cell types by using 4A4 and ΔNp63α antibodies

4. LSCM features as a way to set up quality control assays required for clinical applications of limbal stem cells in patients with limbal stem cell deficiency

The clinical success of keratinocyte-mediated cell therapy for LSCD patients depends primarily on the quality of the cultures used to prepare the corneal grafts [10-12]. These must contain a sufficient number of stem cells in order to guarantee long-term epithelial renewal. Corneal epithelial stem cells mainly express the ΔNp63α isoform, essential for the maintenance of the proliferative potential of limbal stem cells. In order to obtain a more accurate evaluation of the stem cell content within corneal cell grafts, a quantitative evaluation of ΔNp63α content has been performed by means of Q-FIHC assay, a tool based on the use of LSCM for the detection and quantification of fluorescent intensity (FI) in human corneal cells and tissues [13].

Primary cultured corneal epithelial cells (ranging from 500 to 15,000 per slide) were trypsnised, cytospinned onto ThermoShandon glass slides and fixed in 3% paraformaldehyde for 10 min. The slides were incubated with antibody against ΔNp63α for 1 h at 37°C. Fluorescence-conjugated secondary antibodies were incubated for 1 h at room temperature. Sections were drained and coverslipped with glass slides using Vectashield mounting medium with DAPI [15].

A cohort of almost 200 patients was analyzed and quantified for expression of ΔNp63α (highly p63+ cells ranged from 2-8% with mean values of 5.6 ± 0.2, n=180) as shown in Figure 11.

http://dx.doi.org/10.5772/55809

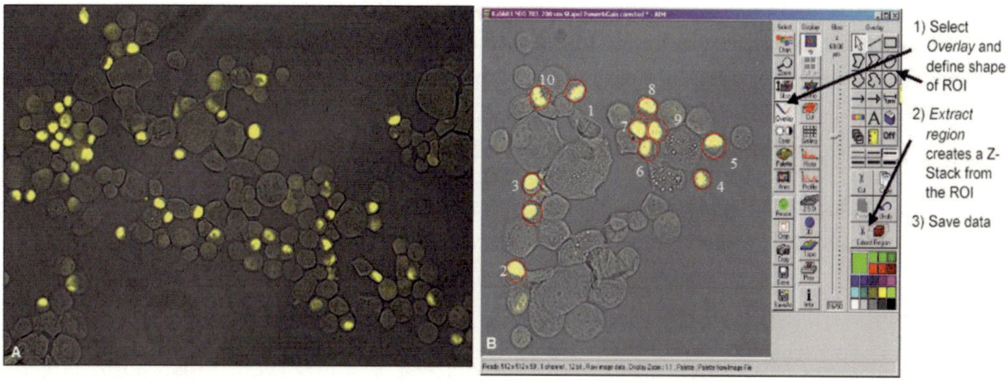

Figure 11. ΔNp63α expression in keratinocytes primary culture (A); cells with different sizes and ΔNp63α level were chosen (B): the software allows the interactive definition of area for size and intensity measurements (ROI analysis).

In addition to ΔNp63*α*, other specific markers have been used to check the quality of corneal graft, including K12 or K3 (to evaluate the amount of corneal cells), ERTR7 (to determine the percentage of murine cells that might be present on the limbal stem cell graft), K19 and MUC-1 (to evaluate the amount of conjunctival cells in the graft). (Figure 12):

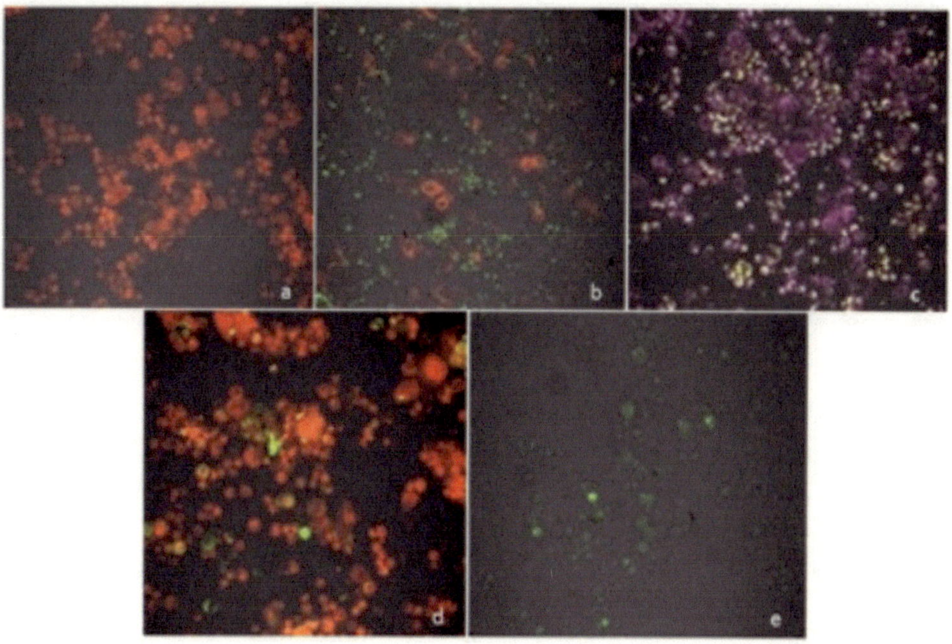

Figure 12. Expression analysis of a) K12; b) ΔNp63α- ERTR7; c) K3,-K19; d) MUC-1, K12; e) MUC-1

However, LSCM is not only useful for quality checks of the limbal stem cell grafts before transplantation, but also for post-transplantation quality checks, performed on excised corneal buttons from patients undergoing penetrating keratoplasty after limbal stem cell grafting. This assay allows to predict the outcome of limbal stem cell transplantation. Figure 13 shows post-transplantation quality control checks performed on 2 patients:

Patient 1: staining for K12 (blue) and K3 (green) was seen throughout the thickness of the epithelium, thus confirming the corneal phenotype and the success of limbal stem cell grafting. Markers of the conjunctiva, K19 (yellow) and MUC-1 were barely detectable and negative, respectively.

Patient 2: staining for K19 (yellow) and MUC-1 (red) was observed, thus confirming the conjunctival phenotype epithelium and the failure of limbal stem cell grafting. K12 was detected as negative whereas K3 (green), was found weakly positive.

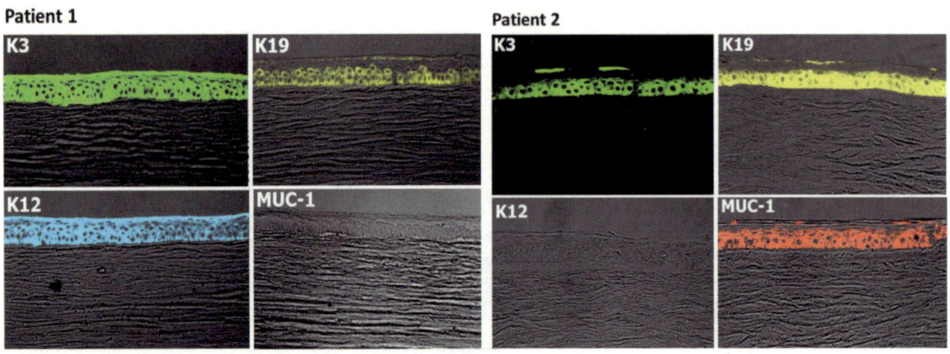

Figure 13. Post transplantation quality analyses on corneal buttons from patients undergoing penetrating keratoplasty after limbal stem cell grafting: staining for Keratin 3 (green). Keratin 19 (yellow), Keratin 12 (blue) and Mucin 1 (red).

5. Impression cytology (IC) as a diagnostic tool to evaluate the grading of limbal stem cell deficiency (LSCD)

5.1. Criteria for selection of markers to be used for analysis of impression cytology specimens

Diagnosis of LSCD relies on the confirmation of cornea conjunctivalisation, either through the presence of goblet cells or the altered expression of keratins in specimens obtained by impression cytology (IC).

IC is a minimally invasive technique, allowing ophthalmologists to evaluate rapidly the 'health status' of the ocular surface. It requires (1) specific markers of the ocular surface epithelia and (2) the expression of these markers in the uppermost layers of the ocular surfaces. In fact, Z-stack analyses have shown that the thickness of the specimens obtained through impression cytology corresponds to that of just one cell layer (the apical one) (Figure 14) and only occasionally includes the underneath sub-apical flattened cell layers. Only the most superficial cells of the ocular surface are therefore collected onto the IC membranes and analysed.

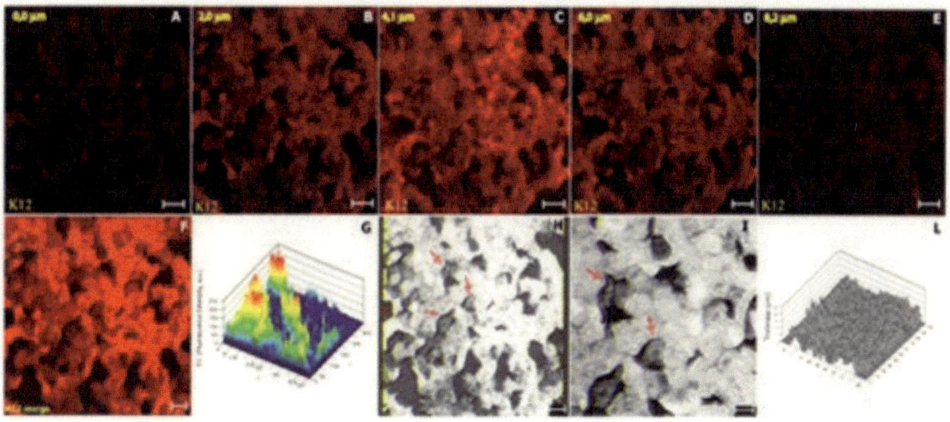

Figure 14. Three-dimensional (3D) information and analysis of impression cytology specimens after reconstruction of Z-stack data. Example of specimens stained with an antibody against K12 is shown. Z-stack of confocal microscopy images taken from impression cytology samples is shown at A) 0, B) 2, C) 4.1, D) 6 and E) 8.2 μm. F) Merge of Z-stack gallery of images. G) Distribution of fluorescence intensity (FI) in the specimen. 3D-reconstruction of the epithelial cells impressed onto the cytology membrane (magnification in I is twice than in H) showed that the thickness of the samples is lower than 20μm, thus corresponding to just 1-2 cell layers from the apical part of the ocular surface (L). Arrows indicate sub-apical corneal epithelial cells.

In order to select more reliable markers of the cornea and conjunctiva to use in IC specimens, the expression of K12, MUC-1, K3 and K19 was evaluated in sections obtained from corneoscleral buttons, comprising corneal, limbal and conjunctival epithelia, thus allowing to elucidate their expression pattern [14].

As shown in Figure 15, K12 expression was restricted to corneal epithelium and the suprabasl layers of the limbus. As opposed to K3, K12 was never observed in conjunctiva, confirming that K12 is a specific marker of the cornea. K19 was found expressed in the basal and suprabasal layers of all three epithelia with a higher expression level in conjunctiva. In contrast, the expression of MUC-1was restricted to the superficial layers of the conjunctival epithelium.

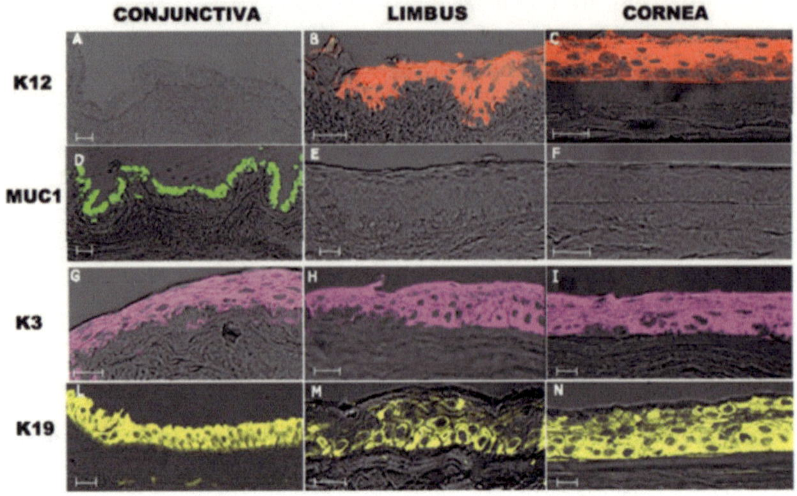

Figure 15. Analysis of various markers on sections from corneoscleral buttons. Keratin 12 (K12-red) is absent in the conjunctiva (A) but expressed specifically in the limbal (B) and corneal region (C). Mucin 1 (MUC 1-green) is present in the conjunctiva (D) but not in the limbus (E) or cornea (F). Keratin 3 (K3-purple) and keratin 19 (K19-yellow) are expressed in all three districts of the ocular surface (G-I and L-N).

Whole corneoscleral button stained with antibodies against MUC-1 and K12 allows to appreciate better the specificity of the two markers (Figure 16):

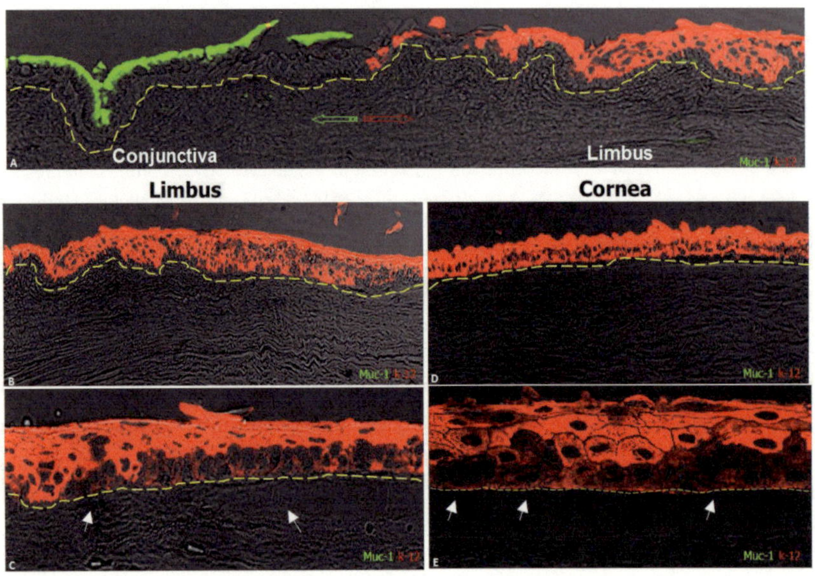

Figure 16. Analysis of MUC-1 (green) and K12 (red) in whole corneoscleral button. The expression of MUC1 was restricted to the superficial layers of the conjunctival epithelium (A) and no, or below threshold levels, staining was observed in the limbus (B, C) and cornea (D, E).

5.2. Expression of K12 and Muc1 on IC specimens from healthy donors

IC specimens obtained from the ocular surfaces of healthy donors (limbus area), were evaluated using the pair of markers MUC-1/K12 (double immune-staining). As shown in Figure 17, distinct expression patterns were observed, with no overlapping signals between MUC1 and K12 staining.

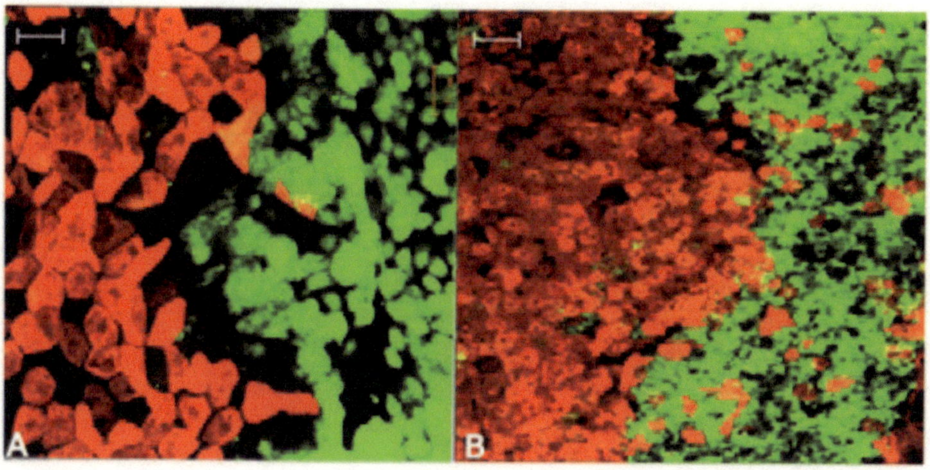

Figure 17. Double immunostaining on IC specimen using K12 (red) and MUC-1 (green) in the limbus area.

5.3. Expression of K12 and MUC-1 on IC specimens from patients affected by LSCD

IC specimens obtained from 3 patients (left panels) with ocular surface disorders and stained with K12/MUC-1 showed the following results (Figure 18):

Patient A (top panel) and C (middle panel) : completely conjunctivalised corneas, presence of MUC-1 (green staining) and disappearance of K12 (red staining);

Patient E (bottom panel): after limbal stem cell grafting, cells are positive for K12, with absence of MUC-1.

5.4. Evaluation of marker co-expression in impression cytology samples after Q-FICH

Fluorescence intensity (FI) values (expressed as pixel intensity) from impression cytology samples stained for K3/K19 or K12/MUC-1 markers were plotted onto scatter plots [13]. For K3/K19, all pixels were found in scatter region 3 which concluding that the markers were co-expressed within the same cell, and signals were overlapping, as clearly visible when signals for K3 and K19 are merged (Figure 19A). For K12/MUC-1, FI values were shifted towards scatter regions 1 and 2, thus meaning no co-expression and higher specificity of the two markers (Figure 19E), which is clearly visible when signals for K12 and MUC-1 are merged. Thus, in general, when there is an overlapping, the scatter merges (especially with non-cell

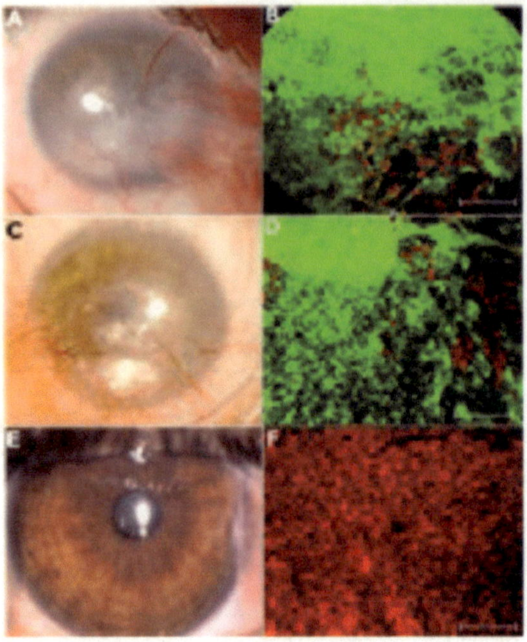

Figure 18. The physical characteristics of 2 patients affected by LSCD (bottom and middle panels) and of a patient after successful treatment (bottom panel) and the representative impression cytology analyses using K12 and MUC-1 markers.

specific marker) whereas when there is no overlapping and the markers show clear difference in immunofluorescence then the pixels scatter in different regions (for cell specific marker). All these information were obtained using the features and softwares of LSCM.

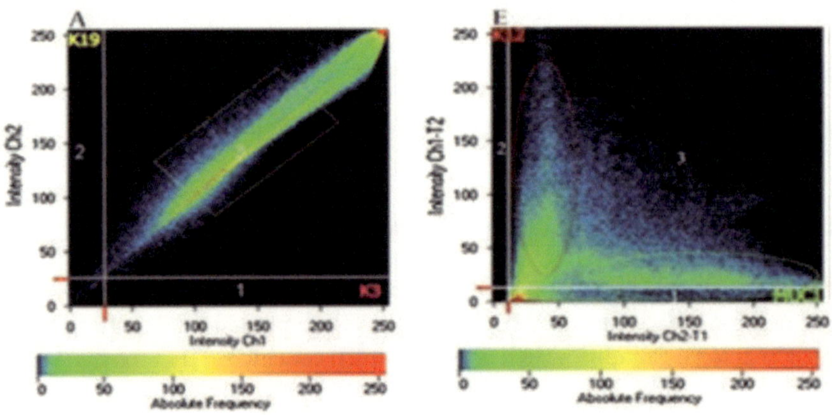

Figure 19. Scatter plot to determine cell specificity. The figure show the analysis performed by the software using CM. Figure A, shows merge of two different markers and E shows the separation using cell specific markers for conjunctiva and cornea.

6. LSCM to study gene therapy-based potential ways to treat rare genetic disorders of the ocular surface

6.1. EEC syndrome

Ectrodactyly Ectodermal dysplasia Clefting (EEC) syndrome is a rare autosomal dominant inherited disease characterized by ectrodactyly (split-hand-food malformation), ectodermal dysplasia and cleft lip and palate. It affects the skin, nails, hair, teeth, sweat glands and the ocular ectodermal derivatives. These patients are generally characterized by dense vascularized corneal pannus, leading to progressive corneal clouding and eventually severe visual impairment. It has been found that p63 mutational analysis in 11 heterozygous missense mutations have resulted in EEC phenotype, the most common being R304Q and R279H. These patients have ocular involvement and the major cause of visual morbidity was found to be LSCD with a progressive degeneration of corneal epithelial tissues [15,16].

LSCM-based techniques and assays have been fundamental in determining the causes leading to LSCD in EEC syndrome. When the pannus removed from the ocular surfaces of patients with EEC syndrome was analyzed, the phenotype of the cells was of conjunctival type, as shown by negative cornea-specific K12 staining and strong MUC-1 expression. This confirmed that the corneal epithelium was replaced by conjunctival overgrowth (conjunctivalisation). LSCM also helped to evaluate that EEC epithelial stem cells have defects in stratification and differentiation. In fact, when grown onto human keratoplasty lenticules, the epithelia generated by mutant cells were thinner, with only 1-2 layers, some devoid of cells and with flat irregular cells. Severe tissue hypoplasia was also observed and the defects were prevalent in both stratification and differentiation. The epithelial thickness significantly differed between the mutant tissues and the WT-p63 because of stem cell incapacity to give rise to a full thickness stratified and differentiated corneal tissue (unpublished data).

6.2. Potential treatment options for EEC syndrome

Small interfering RNA (siRNA) is a class of double stranded RNA molecules. They play a major role in RNA interference (RNAi) where they interfere with the expression of specific genes by means of complementary nucleotide sequences. This is a new potential therapeutic measure that is believed to be suitable for treating the ocular surface disorders of patients with EEC syndrome as it would silence the expression of mutant alleles differing from wild-type ones. In EEC patients, the most common mutation is found in p63 gene on R(arg)279H(his). The prospective therapy includes a single nucleotide difference between two alleles that may not be sufficient to confer allele specificity siRNA, but by introducing mismatches, only the mutant allele would be inactivated, without altering the expression of wild-type mRNA. Binding of siRNA to the mutant mRNA causes the formation of a double stranded RNA which is cleaved along with the mutant p63 mRNA degradation, while the normal mRNA is not recognized as specific target (unpublished data).

However before such a strategy is used, fluorescent-labelled siRNAs need to be delivered to the cells in order to see the optimal dose, the transfection efficiency, any potential toxic effect and where the siRNAs are delivered (to the nucleus? to the cytoplasm?). As shown in Figure 20, LSCM was instrumental to define all these parameters and understand which was the optimal, non-toxic dose of siRNAs that needs to be delivered to obtain a potential therapeutic effect.

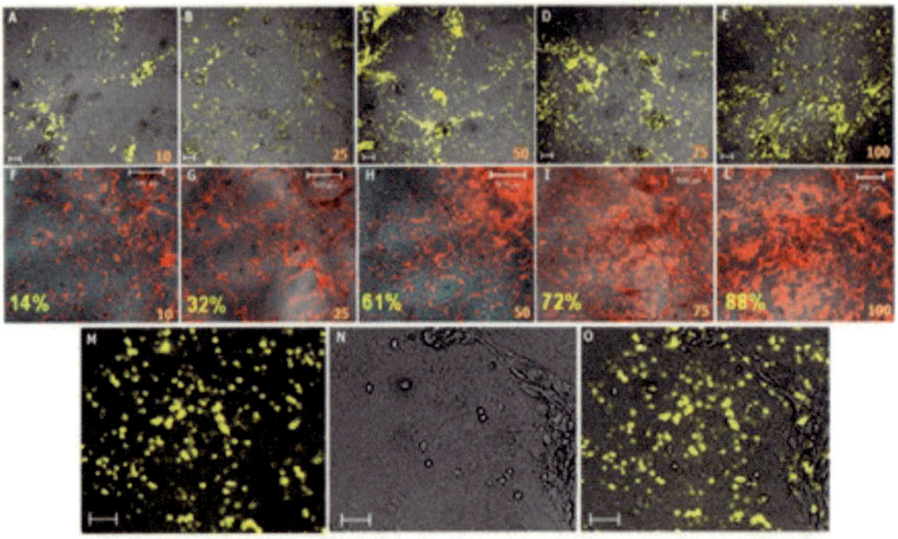

Figure 20. Fluorescently labeled siRNAs used as control to estimate the transfection efficiency of siRNAs, allowing direct observation of their cellular uptake, distribution and localization. Optimization of transfection conditions using an extensive concentration range from 1 to 100 nM using two different batches of siRNAs (A-L) in human keratinocytes. Confocal microscope grid (yellow, M), transmitted light (grey, N) and merge (O).

7. Conclusions and future options

The possibility to get cellular images of the zone of interest in real time and with different depth in the various layers of the specimens (through Z-stack analysis) are all features that can be achieved by using a LSCM and its options. In the previous sections we have shown how the characteristics of LSCM can help to study the ocular surface and evaluate potential pathologies. In addition, LSCM and high resolution image analysis can help evaluating whether stem cell-based clinical applications are successful. The techniques described can, in fact, be highly reliable

- **i.** for quality control of the finished products (stem cell grafts) using cell specific markers (such as $\Delta Np63\alpha$, K12, MUC-1);
- **ii.** to predict the stem cell content and potency, identity and impurity of the graft;
- **iii.** for post-transplantation follow-up studies;
- **iv.** to evaluate the results of gene therapy-base therapies for the treatment of patients affected by genetic disorders of the ocular surface or other similar disorders.

In the future, the techniques described in this chapter might help setting up procedures and solutions for other clinical applications.

Acknowledgements

The work described in this book chapter was partly supported through grants from the Veneto Region (Ricerca Sanitaria Finalizzata 2009, project 306/09, CUP n.: J71C09000050002), the Association Francaise contre les Myopathies (AFM2012/Project 15651) and the Italian Ministry of Health (GR-2009-1555694, CUP n.: H31J11000260001).

Author details

Vanessa Barbaro**[1], Stefano Ferrari**[1], Mohit Parekh[1], Diego Ponzin[1], Cristina Parolin[2] and Enzo Di Iorio[3*]

*Address all correspondence to: enzo.diiorio@fbov.it, vincenzo.diiorio@unipd.it

1 Fondazione Banca Degli Occhi Del Veneto Onlus, Zelarino, Venezia, Italy

2 Department of Biology, University of Padova, Padova, Italy

3 Department of Molecular Medicine, University of Padova, Padova, Italy and Fondazione Banca Degli Occhi Del Veneto Onlus, Zelarino, Venezia, Italy

**The authors equally contributed to this work

References

[1] Ahmad, S. Concise Review: Limbal Stem Cell Deficiency, Dysfunction, and Distress. Stem Cells Trans Med (2012). , 1(2), 110-115.

[2] Barrandon, Y, & Green, H. Three clonal types of keratinocyte with different capacities for multiplication. Proc. Natl. Acad. Sci. (1987). , 84-2302.

[3] Rochat, A, Kobayashi, K, & Barrandon, Y. Location of stem cells of human hair follicles by clonal analysis. Cell (1994). , 76-1063.

[4] Pellegrini, G, Golisano, O, Paterna, P, Lambiase, A, Bonini, S, Rama, P, & De Luca, M. Location and clonal analysis of stem cells and their differentiated progeny in the human ocular surface. J. Cell Biol. (1999). , 145-769.

[5] Yang, A, Kaghad, M, Wang, Y, Gillett, E, Fleming, M. D, Dötsch, V, Andrews, N. C, Caput, D, Mckeon, F. "p, & Homolog, a p. at 3q27-29, encodes multiple products with transactivating, death-inducing, and dominant-negative activities". Mol. Cell (1998)., 2(3), 305-16.

[6] Osada, M, Ohba, M, Kawahara, C, Ishioka, C, Kanamaru, R, Katoh, I, Ikawa, Y, Nimura, Y, Nakagawara, A, Obinata, M, & Ikawa, S. Cloning and functional analysis of human which structurally and functionally resembles p53". Nat. Med. (1998)., 51.

[7] Zeng, X, Zhu, Y, & Lu, H. NBP is the homolog p63". Carcinogenesis (2001)., 53.

[8] Tan, M, Bian, J, & Guan, K. Sun Y (February 2001). "53CPis p51/p63, the third member of the p53 gene family: partial purification and characterization". Carcinogenesis (2001).

[9] Di Iorio EBarbaro V, Ruzza A, Ponzin D, Pellegrini G, and De Luca M. Isoforms of ΔNp63 and the migration of ocular limbal cells in human corneal regeneration. PNAS (2005)., 102(27), 9523-9528.

[10] Rama, P, Bonini, S, Lambiase, A, Golisano, O, Paterna, P, De Luca, M, & Pellegrini, G. Autologous fibrin-cultured limbal stem cells permanently restore the corneal surface of patients with total limbal stem cell deficiency. Transplantation. (2001). Nov 15;, 72(9), 1478-85.

[11] Ronfard, V, Rives, J. M, Neveux, Y, Carsin, H, & Barrandon, Y. Long-term regeneration of human epidermis on third degree burns transplanted with autologous cultured epithelium grown on a fibrin matrix. Transplantation. (2000). Dec 15;, 70(11), 1588-98.

[12] De Luca, M, Pellegrini, G, & Green, H. Regeneration of squamous epithelia from stem cells of cultured grafts. Regen Med. (2006). Jan;Review., 1(1), 45-57.

[13] Di Iorio E Barbaro V, Ferrari S, Ortolani C, De Luca M and Pellegrini G. Q-FIHC: Quantification of fluorescence immunohistochemistry to analyse isoforms and cell cycle phases in human limbal stem cells. Micr Res Tech (2006)., 63.

[14] Barbaro, V, Ferrari, S, Fasolo, A, Pedrotti, E, Marchini, G, Sbabo, A, Nettis, N, & Ponzin, D. Di Iorio E. Evaluation of ocular surface disorders: a new diagnostic tool based on impression cytology and confocal laser scanning microscopy. Br J Ophthalmol (2010)., 94-926.

[15] Barbaro, V, Confalonieri, L, Vallini, I, Ferrari, S, Ponzin, D, Mantero, G, Willoughby, C. E, & Parekh, M. Di Iorio E. Development of an allele-specific real-time PCR assay for discrimination and quantification of R279H mutation in EEC syndrome. J Mol Diagn. (2012)., 63.

[16] Barbaro, V, Nardiello, P, Castaldo, G, Willoughby, C. E, Ferrari, S, Ponzin, D, Amato, F, Bonifazi, E, Parekh, M, Calistri, A, & Parolin, C. Di Iorio E. A novel de novo missense mutation in TP63 underlying germline mosaicism in AEC syndrome: Implications for recurrence risk and prenatal diagnosis. Am J Med Genet A. (2012). A(8) 1957-61.

Chapter 6

Confocal Laser Scanning Microscopy as a Tool for the Investigation of Skin Drug Delivery Systems and Diagnosis of Skin Disorders

Fábia Cristina Rossetti, Lívia Vieira Depieri and
Maria Vitória Lopes Badra Bentley

Additional information is available at the end of the chapter

http://dx.doi.org/10.5772/55995

1. Introduction

Confocal laser scanning microscopy (CLSM) is a useful image tool to study the fate of delivery systems and biomolecules applied into the skin. Through the use of fluorescence probes, it is possible to evaluate their behavior, like: i) interaction with the biological system [1]; ii) cellular uptake [2-3]; iii) depth of penetration [4]; iv) penetration routes into the skin [5]; v) quantification of skin penetration of drugs and biomolecules [6]; and vi) effect of topical therapies for several skin diseases by morphological analysis of the tissue [7].

In fluorescence microscopic terms, the skin tissue is difficult to investigate because it reflects and scatters incoming light, and melanin and other chromophores, significantly attenuate visible wavelengths [8]. CLSM has the potential to overcome these limitations, since it generates high-resolution imaging, non-invasive optical sectioning and three-dimensional reconstructions, in combination with sensitivity, selectivity and versatility of fluorescence measurements [1]. In this way, the expertise of the CLSM technique provides excellent opportunities for probing and better understanding the behavior and fate of pharmaceuticals in the skin, including *in vivo* monitoring of skin drug penetration, allowing a rational development of skin delivery systems.

This chapter intends to discuss important topics for pharmaceutical researchers related to the proper use of this technique to design and optimize skin delivery systems, as well as to diagnose skin diseases. The first topic will shortly comment about the skin structure for an easier understanding of the next ones. With the advent of nanotechnology, CLSM technique can be an imaging tool for better understanding of the fate of nanocarriers in delivering drugs into the skin layers and the efficacy of them in the therapies of different skin disorders. Furthermore, our previous experience in this technique [9-10] will allow a critical discussion on the main topics in this specific subject and to update the advances in this field application.

2. The skin structure

The human skin is the largest organ in the body that has many important physiological functions, such as, protection (physical, chemical, immune, pathogen, UV radiation and free radical defenses), major participant in thermoregulation, sensory organ and performs endocrine functions (vitamin D synthesis, peripheral conversion of prohormones) [11]. Its thickness varies from 0.05 mm to 2 mm and is composed of four main layers: the stratum corneum (SC) and the viable epidermis, the outermost layers; the dermis, and the subcutaneous tissue (Figure 1)[12].

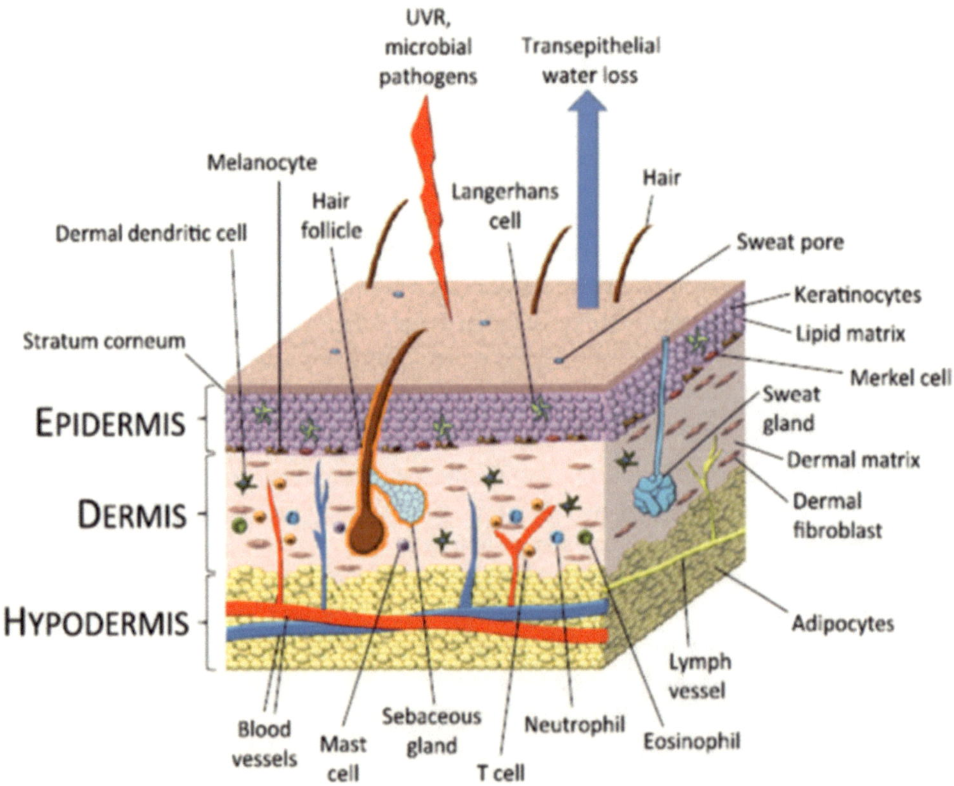

Figure 1. Schematic representation of skin structure and cell population. The skin comprises three main layers – the epidermis, dermis and hypodermis. The resident cell populations and various structures present throughout the skin allow for maintenance of an efficient barrier against water loss, and protection against threats such as ultraviolet radiation (UVR) and microbial pathogens. The blood and lymph vessels allow for the migration of immune cells in and out of the skin, so that the cell population of the skin is constantly in a state of flux, in response to the demands of the cutaneous inflammatory and immune systems. Reproduced from [13] with permission.

The SC is 10-20 μm thick, highly hydrophobic and contains 10-15 layers of interdigitated dead cells called corneocytes. It is the major barrier to penetration of drugs because its structure is highly organized [14]. Its "brick and mortar" structure is analogous to a wall. The corneocytes of hydrated keratin comprise the "bricks", embedded in a "mortar", composed of multiple lipid bilayers of ceramides, fatty acids, cholesterol and cholesterol esters. These bilayers form regions of semicrystalline, gel and liquid crystals domains. Most molecules penetrate through skin via this intercellular microroute and therefore many enhancing techniques aim to disrupt or bypass its highly organized structure [15].

The viable epidermis (~100-150 μm thick) is composed by multiple layers of keratinocytes at various stages of differentiation. Besides the keratinocytes, the epidermis also contain several other cells (melanocytes, Langerhans cells, dendritic T cells, epidermotropic lymphocytes and Merkel cells) and active catabolic enzymes (e.g. esterases, phosphatases, proteases, nucleotidases and lipases) [12].

The dermis is rich in blood vessels, nerves, hair follicles, and sebaceous and sweat glands. The elasticity of the dermis is due to the presence of collagen, elastin, glycosaminoglycans, collectively termed the extracellular matrix (ECM), as well as fibroblasts that elaborate the ECM. Dermal adipose cells, mast cells, and infiltrating leucocytes are also present in this skin layer [11-12].

3. Operational parameters for digital image capture for CLSM to assess the drug penetration into skin layers

3.1. Equipment characteristics

This technique is fluorescence-based image and offers greater resolution than fluorescence microscopy due to its point illumination and detection properties [16]. The illumination in a confocal microscopy is achieved by a collimated laser beam across the specimen [16-17]. This laser beam is reflected by a dichroic mirror and passes through the objective lens of the microscope in a focused manner on the specimen, which, and then, excites fluorescence probe in the sample. So, light is emitted at a longer wavelength which can come through the dichroic mirror and is again focused at the upper pinhole aperture (Figure 2) [16,18]. With CLSM, out-of-focus light (coming from places of the specimen above or below the focus) is cut off before the beam hits the electronic detector due to the addition of a spatial filter containing an aperture, – the pinhole or slit – the point detection. Just the light in-focus can pass through the pinhole (now termed confocal apertures), come to detector, and then form the image with more details because the blurring from out-of-focus has vanished. By using CLSM, it is possible to obtain high-resolution images (lateral, ~140 nm; axial, ~1 μm) from the samples, which increase to accuracy of the microscopic images [5,16-17].

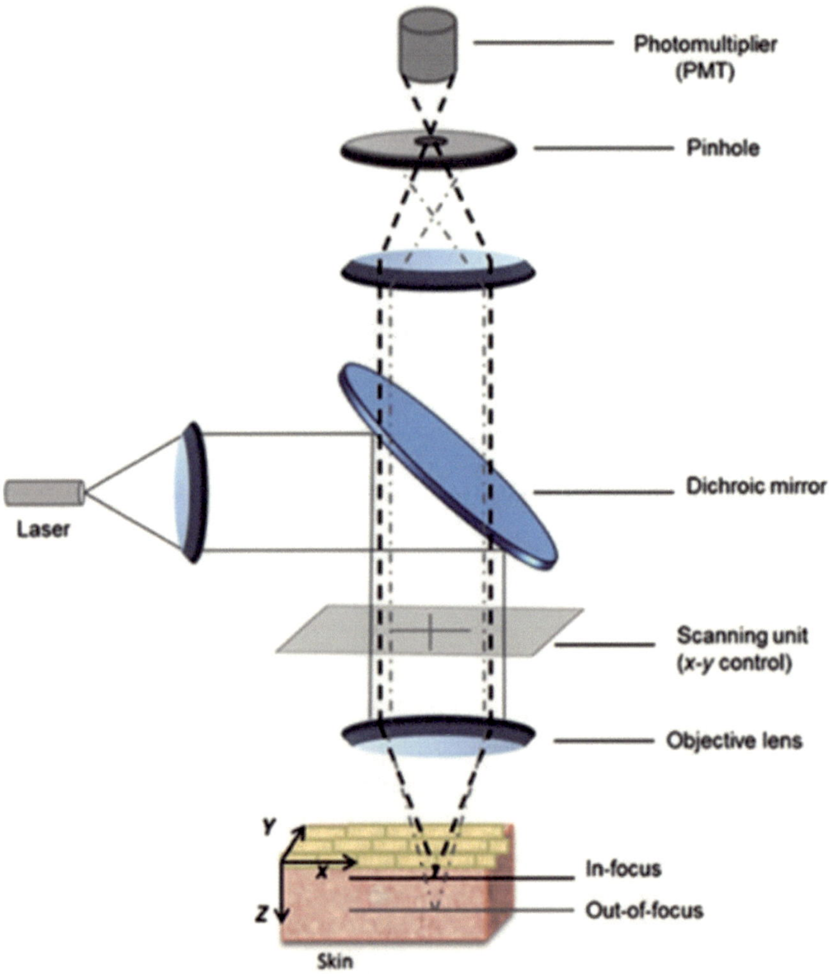

Figure 2. Schematic diagram of the principle of confocal laser scanning microscopy.

http://dx.doi.org/10.5772/55995

3.1.1. Measurement of different optical sections

Acquisition of several optical sections (*x-y* plane) taken at successive focal planes along the *z* axis is a usual practice to obtain a three-dimensional information from the skin that can be viewed as a simple image. Figure 3 shows the principle of *z* series acquisition.

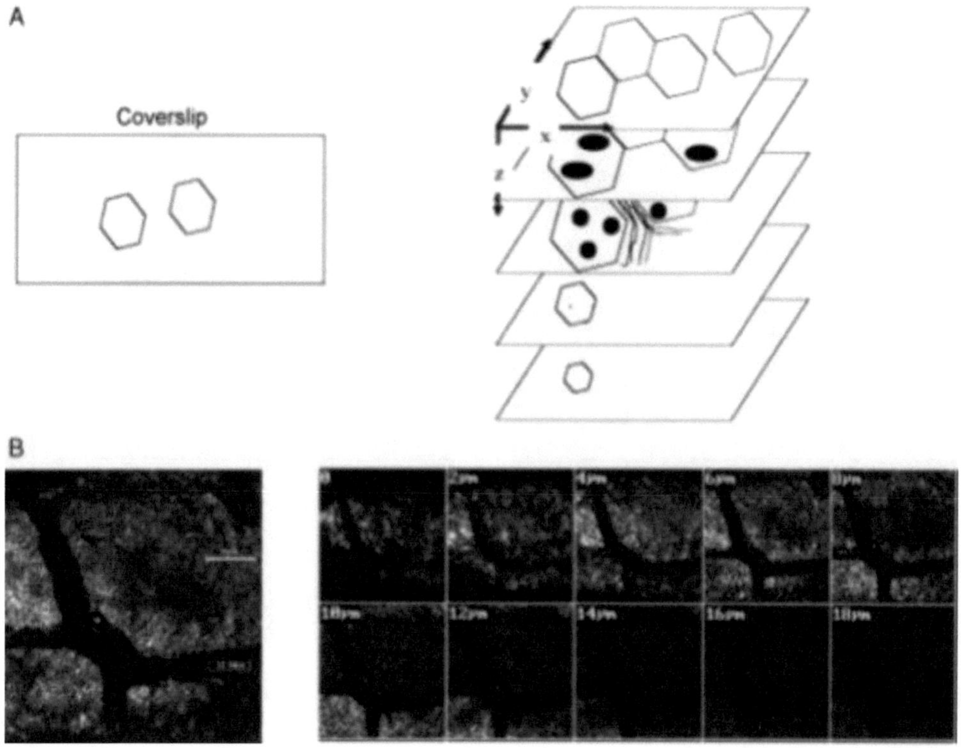

Figure 3. Confocal optical sectioning: (A) a schematic of a sample (hexagonal features) mounted on a microscope slide and covered with coverslip for a *z* series processing (sequential *x-y* sections as a function of depth (*z*); (B) confocal images of a *z* series through the porcine skin. Reproduced from [5] with permission.

Acquisition of *x-z* section is useful to obtain depth information from a specific surface. First, it is necessary to get an image in the *x-y* plane. Then, after choosing a region of interest, a horizontal line is drawn across this area in the z = 0 μm – *x-y* plane and is "optically sliced" through the digitalized image data of successive *x-y* sections; the results are (*x-z* planar) optical cross-sections (Figure 4).

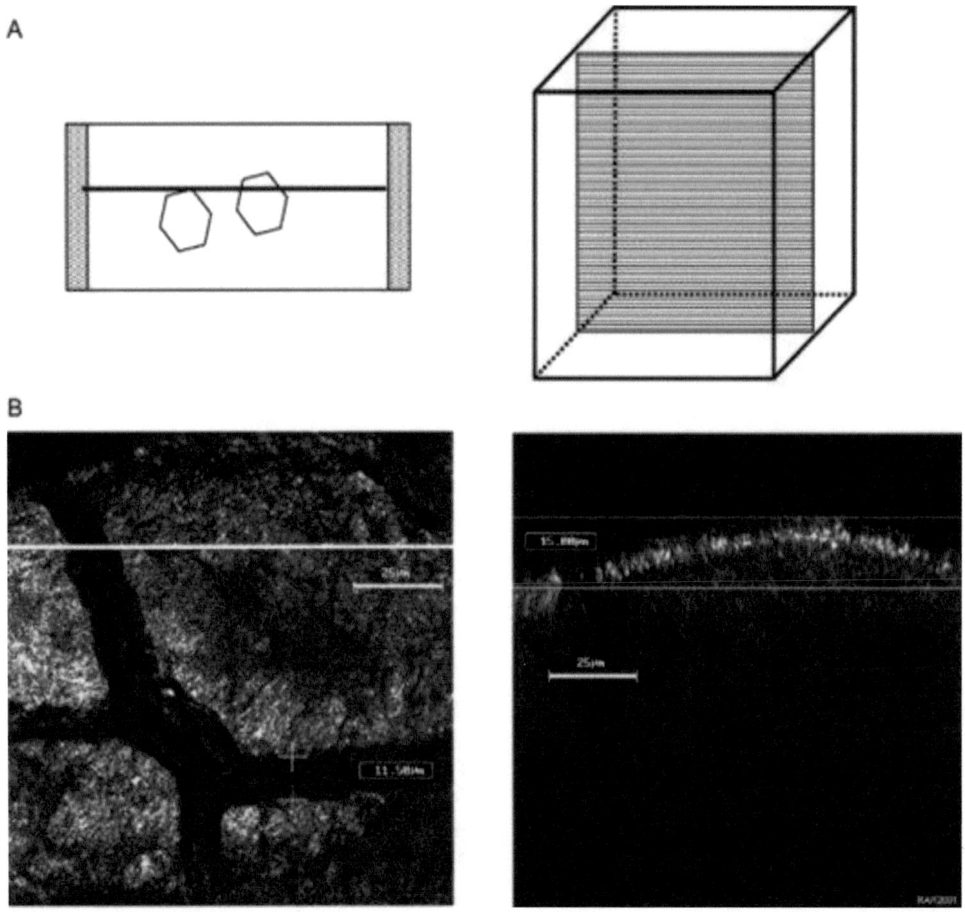

Figure 4. Confocal optical sectioning: (A) schematic of an *x-z* planar optical cross-section; (B) confocal *x-z* image of porcine skin. Reproduced from [5] with permission.

3.2. Sample processing techniques

The main goal of confocal microscopy is to explore the structure and structural relationship along the optical (z) axis as well in the *x-y* plane. For this, preservation of the tissues and cells during the preparation of the sample is necessary to obtain a reliable image. Optimum sample preparation is very dependent upon the cell or tissue type, the labeling technique and type of data to be collected [5,19].

Analyses by confocal microscopy can be done with living or preserved (fixed) samples. Working with living samples, is possible to analyze dynamic events and avoid some of the artifacts that can be introduced with preservation techniques and processing of the sample. It is easier to work with preserved samples since it does not have to concern about keeping the cells or tissue alive but; however, the impossibility to observe dynamic events and the presence of artifacts make this routine less attractive [19].

For skin samples, little or no sample preparation is necessary to visualize skin structures and the localization of fluorescent probes used within the tissue. Hence, the technique is rapid, with minimal physical perturbation or damage to the tissue. CLSM offers nearly accurate images with few artifacts contributing significantly in topical/transdermal permeation studies to elucidate and understand the mechanisms and pathways of drug penetration using different delivery technologies [5,19].

3.3. Main fluorescence probes used to assess skin penetration

Fluorescence is an important optical readout mode in biological confocal microscopy due to its sensitivity and specificity. These features can be influenced by the capability to tag onto biological systems to show your localization and by the local environment [20].

In some situations in which the drug carried by the delivery system studied has fluorescence properties, the use of a specific fluorophore is not necessary, once the drug can be excited and it emits fluorescence signals that are captured and transformed into image [9-10, 21].

Several fluorophores are used in skin visualization. The Table 1 shows the main used.

Objective of study	Fluorescence probe	$\lambda_{exc}/\lambda_{em}$ (nm)	Reference
Visualization and suggested mechanism of interaction between liposome-skin	β-carotene and 5(6)carboxyfluorescein	514/660-1250; 488/515-535	[22]
Evaluation of composition and preparation methods of vesicles in skin penetration	β-carotene and 5(6)carboxyfluorescein	514/660-1250; 488/515-535	[23]
Penetration and distribution of lipophilic probes in the hair follicle	Bodipy FL C_5, Bodipy 564/570 C_5 and Oregon Green	488 /514; 564/574; 488/514	[24]
Visualization of influence the use low-frequency ultrasound in skin penetration	Calcein	495/515	[25]
Evaluation the composition of liposomes in skin penetration behavior	Calcein , NBC-PC[1] and Rhodamine B	488/510; 488/530; 568/590	[26]
Visualization of penetration and distribution of nanoparticles in the skin treated with microneedles	Coumarin-6	488/	[27]
Visualization the ability of cell penetrating peptides to improve skin penetration	DID-oil	644/665	[28]
Evaluation the effect of polymeric nanoparticles with surface modified with oleic acid in skin permeation	DiO	484/501	[29]
Evaluation of skin penetration of nano lipid carries with surface modified by polyarginine	DiO	484/501	[30]

Objective of study	Fluorescence probe	$\lambda_{exc}/\lambda_{em}$ (nm)	Reference
Investigate the effect of pore number and depth on rate and extension of drug delivery through the skin using a novel laser microporation technology	FITC[2]	488/520	[31]
Influence and elucidation of the transport pathway of solute-water skin penetration with use low-frequency sonophoresis	FITC-dextran and Rhodamine B	488/520; 568/590	[32]
Evaluation of microneedle shape by visualization of skin penetration of fluorescent dye	Fluorescein	488/515	[33]
Visualization of skin penetration pathways of a novel micelle formulation	Fluorescein	505/530	[34]
Evaluation the ability of dendrimers to facilitate transdermal drug delivery *in vivo*	Fluorescein-PAMAM	488/<560	[35]
Visualization of nanoparticles deposition in a skin	Fluospheres®	405/420-480; 488/505-530; 633/647-754	[36]
Evaluation the influence of a liposome surface charge in skin penetration	NBD-PC and Rhodamine B	488/530; 568/590	[37]
Visualization the effect of heat on skin permeability	Nile red	543/630	[38]
Visualization the behavior of microemulsion formulation in the skin stratum corneum	Nile red	543/630	[39]
Visualization the behavior of polymeric particles for drug delivery to the inflamed skin	Nile red	543/595	[40]
Visualization of skin transport properties of model as a carrier for oligodeoxynucleotide (ODN) during iontophoresis	Oregon Green and TAMRA	488/505; 543/560	[6]
Evaluation the interaction of quantum dots nanoparticles in the skin	Quantum dots	351, 364, 488/610-632	[41]
Visualization the skin penetration behavior of ethosomes	Quantum dots (CdTe)	488/<560	[42]
Investigate skin penetration of quantum dots in human skin	Quantum dots (CdTe)	800/	[43]
Evaluation of ultrasound and sodium lauryl sulfate for increase transdermal delivery of nanoparticles	Cationic, neutral, and anionic quantum dots (CdSe/ZnS)		[44]
Visualization of ethosome penetration in skin and mechanism of action study	Rhodamine 6 GO	543/<560	[45]
Exploration of the three-dimensional structure, organization and barrier function of the stratum corneum *ex vivo* and *in vivo* after transfersome permeation	Rhodamine-DHPE[3], Texas Red-DHPE[4] and fluorescein-DHPE[5]	543/590; 583/601; 496/519	[46]

[1] NBD-PC=1-palmitoyl-2-{12-[(7-nitro-2-1,3-benzoxadiazol-4-yl)amino]dodecanoyl}-snglycero-3-phosphocholine.
[2] FITC=fluorescein isothiocyanate.
[3] Rhodamine-DHPE=1,2-dihexa-adecanoyl-sn-glycero-3-phosphoethanolamine-N-Lis-samine.
[4] Texas Red-DHPE=1,2-dihexa-adecanoyl-sn-glycero-3-phosphoethanolamine-N-Lis-samine.
[5] FI-DHPE=N-(5-fluoresceinthiocarbamoyl)-1,2-dihexadecanoyl-sn-glycero-3-phospho-ethanolamine.

Table 1. Main fluorescence probes used to assess skin penetration

Unfortunately, there is not a comprehensive list of available fluorophores that is selective for a particular parameter of the cell and that also has a suitable wavelength. The choice of an ideal fluorophore is not a simple task; however, it is possible to realize the choice of the more

http://dx.doi.org/10.5772/55995

appropriate fluorophore within a range of wavelengths. This selection is based on the objectives of the study taking into account the physicochemical properties of the fluorophore as well as the characteristics of the sample.

In the case of skin, various endogenous substances (Table 2) can interfere with analyzes because they are excited at a wavelength in the excitation band of the chosen fluorophores.

Fluorochrome	**$\lambda_{excitation}$ (nm)**	**$\lambda_{emission}$ (nm)**
Tryptophan	295	345
Tyrosine	275	300
Phenylalanine	260	280
Melanin	330-380	400-700
Keratin	375	430
Collagen	335	390-405
Elastin	360	460
FAD	390	520
NADH	290, 364	440, 475
NADPH	336	464
Flavoprotein	450-490	500-560
Porphyrins	476	625

Table 2. Endogenos substances responsible for the skin's autofluorescence [5]

To minimize autofluorescence of the skin, configuration settings of fluorescence detection must be made so that it can separate the fluorescence signal of the skin from that of the used fluorophore, avoiding any interference of the specimen [43]. The imaging using dual-channel and subsequent overlapping of images obtained at each channel is a resource used for reducing potential problems with autofluorescence from the sample [5]. Another resource is the choice of fluorophores with distinct emission spectra from those presented by endogenous fluorophores of the skin [47].

Some fluorophores can be covalently bound to biomolecular targets as well as the components of formulations without changing their spectral properties. The wide variety of derivatives of rhodamine and fluorescein, both well-known dyes, has been used with this subject, which result in greater functionality of these dyes [20].

Rhodamine is a lipophilic dye and because of its physicochemical property it is widely used to evaluate the dynamic properties of living cells such as membrane potential and ion concentration. It is also used to study the behavior and mechanisms of permeation of lipid delivery systems like liposomes. It can be incorporated into the lipid bilayers promoting a marking and also mimicking lipophilic drugs such as betamethasone, since they both exhibit a similar partition coefficient (log P). In many cases, it is associated with a hydrophilic dye as, for example, calcein, which is expected to be encapsulated in the aqueous compartment of the liposome, promoting marking and functioning as a model for hydrophilic drugs that can be carried by this type of delivery system [26,37].

On the other hand, fluorescein has a wide use due to its ability to react with many substrates. One example of this, it is its conjugation with dendrimers, which enables the study of the behavior and properties of dendrimer on skin permeation [35].

Moreover, for a successful CLMS analysis of the skin samples, the fluorophores probe must present: (i) good quantum efficiency and persistence of the signal sufficient for the instrument to achieve image data; (ii) selectivity for the target molecule; (iii) high resistance to bleaching; (iv) minimal perturbation to the sample; (v) minimizing the cross-talk when multiple fluorophores are being used together. Another important consideration is that the fluorescence intensity of depth images depends on both the sample transparence or opacity and the interference of the fluorescence of upper layers with the fluorescence signal of the adjacent layers [1]. Instrumentally, fluorescence emission collects can be improved by careful selection of objectives, detector aperture dimensions, dichromatic and barrier filters, as well as keeping the optical train in precise alignment [48].

3.4. Non–invasive assessment

Noninvasive optical techniques, such as CLSM, have become widely used in recent years because they are an efficient tool for skin characterization, as additional technique or even as a replacement for invasive biopsies of skin [49].

In this technique, images are obtained by scanning the specimen with one or more focus of beams of light. The images produced by this way are called optical section. This term refers to the method of collecting images noninvasively, which uses light to section the sample out instead of performing mechanical sections [17]. Due to this important characteristic, CLSM has been used in dermatology diagnosis by assessing the skin *in vivo*, where a comparison is made between the healthy skin aspect with the pathological state of living human skin [50-53].

Two modes are established for the use of *in vivo* CLSM: reflectance and fluorescence modes. The reflectance mode is based on detection of own endogenous contrast by refractive indices of several cellular structures, like melanin or keratin, which have both a high refractive index. Generally, a laser with near-infrared wavelengths is used for this type of measurements.

Backscattered in-focus signals are captured and transformed for the image visualization [50-51]. The fluorescence mode is based on detection of the distribution of an exogenous dye administrated before the measures. A laser beam with visible-light wavelengths is used to excite selectively the dye to produce contrast. Backscattered fluorescence signals are captured and transformed for the image visualization [50]. They are both optical non-invasive techniques that permit *in vivo* and *ex vivo* (biopsy) images without the fixing, sectioning and staining, which are common procedures for histology analysis [54].

CLSM facilitates the image of living specimens, provides data from the three-dimensional structure of the sample and improves the image resolution due to high sensitivity, selectivity and versatility in fluorescent measures resulting in a valuable opportunity to study the behavior of pharmaceutical systems.

4. Advantages and disadvantages of using CLSM as a tool in skin delivery studies

Skin delivery research involves determination of physicochemical parameters of the formulation and *in vitro* and *in vivo* release studies. The fate of the drug/carrier into the living skin, either in healthy or pathological condition, is very important information; and CLSM can be a useful tool for development of cosmetic and dermatological products. The main advantages for studies with these aims include: (i) the ability to obtain images in a noninvasive manner *in vitro* and *in vivo* conditions [5]; (ii) suitable for *in vivo* diagnosis through imaging of superficial skin layers [49]; (iii) same skin site can be imaged serially over time that is an important advantage in studies like hair follicle neogenesis following wounding where the analysis are processed directly on the skin and information about number, length and width of follicles neogenics are obtained concurrently and from one animal [55]; (iv) ability to produce serial thin (0.5 to 1.5 μm) optical sections without mechanical sections and preserving the skin structure [5,16,47]; (v) eliminates artifacts that occur during physical sectioning and staining of sample; (vi) visualization of images at multiple depths through the sample without mechanical sections which allows the visualization of the skin layers where the drugs appropriately marked or dyes were able to achieve after their topical application [5,48]; (vii) ability in monitoring the skin penetration of drugs and delivery systems appropriately marked or dyes in real time, enabling, in this ways, bioavailability studies. This is done by performing analysis immediately after topical application of the dyes, for example, and after certain periods of time. Immediately after application, the dye was observed only in the stratum corneum, distributed in the intercellular spaces of the first layer of skin. In all other times analyzed; it is possible to assess whether the dye could penetrate the skin and if succeeded, the depth achieved [49]; (viii) high-resolution images; (ix) three-dimensional reconstruction from a series of optical sections at different depths; (x) sensitivity; (xi) versatility and selectivity florescence measures; (xii) reduced blurring of the image from light scattering and improved signal-to-noise that resulting in improvement of contrast and definition; (xiii) more precise quantification of images using image analysis software; (xiv) allows to analyze of fixed and live samples under a variety of conditions and with greater clarity [1,5,16,18,47-48].

However, the limitations of fluorophores and equipments as well as inherent interference of the tissue, bring some disadvantages of CLSM technique, such as: (i) limited number of excitation wavelengths available with common lasers; (ii) possibility to cause damage to living cells and tissues due to high intensity laser irradiation; (iii) autofluorescence of skin samples which may interfere in the analysis; (iv) problems in adjusting the focus on the skin surface due to cellular components and hydration state of the skin; (iv) limited special resolution that interfere significantly *in vivo* analyzes in dermatological practice where the maximum depth of analysis is around 200 μm, allowing diagnosis of only superficial skin disorders; (v) slow scanning laser action for high quality images; (vi) quantification in concentration terms since it is possible just when the relationship between probe and the fluorescence emission is linear and when this signal is not attenuated differently for each depth localization in the sample; (vii) high cost of acquisition and operation of systems for confocal microscopy [1,5,16,18,48,53].

5. CLSM technique for assessment of drug and delivery systems penetration into the skin and diagnosis of skin disorders

5.1. CLSM technique for assessment of drug and delivery systems penetration into the skin

CLSM is a valuable technique that helps to elucidate the mechanism, depth and distribution of skin penetration for delivery systems loaded or not with drugs. Nowadays, there is an increasing attempting to determine the possible penetration pathways because, with this information in hand, researchers are able to propose some structural modifications in the delivery systems, aiming to improve the skin penetration into the target tissue and, consequently, the pharmacological action of the drug [56]. Furthermore, penetration pathway elucidation is useful to address toxicological issue, since the fate of some formulations, mainly non-biodegradable nanoparticles, can be harmful to the body [44].

The emission of fluorescence by the delivery system and/or the drug is a condition when using CLSM for assessment of drug and delivery systems penetration into the skin. For that, the delivery systems are made fluorescent using probes covalently linked to polymers or homogeneously distributed in the system. In the case of the drugs, some studies link them to probes or, in many cases; they use a model probe with similar physicochemical properties of the drug. The mechanism of skin penetration from flexible polymerosomes, vesicles composed by polymers, was determined through the *in vitro* cutaneous penetration of vesicles containing Nile Red as the model probe. The skin samples were stained with fluorescein prior to fixation. The images (Figure 5) showed a time-dependence penetration into the skin, as initially, the particles tended to penetrate between the corneocytes isles (green stained) and, further application time, the vesicles tended to penetrate via intercellular lipids and follicular regions (red stained) until a maximum depth of 60 μm [57]. CSLM showed that a fluorescein loaded micelle formulation penetrate via follicular pathway and accumulate in epidermis up to a depth of 40 μm [34].

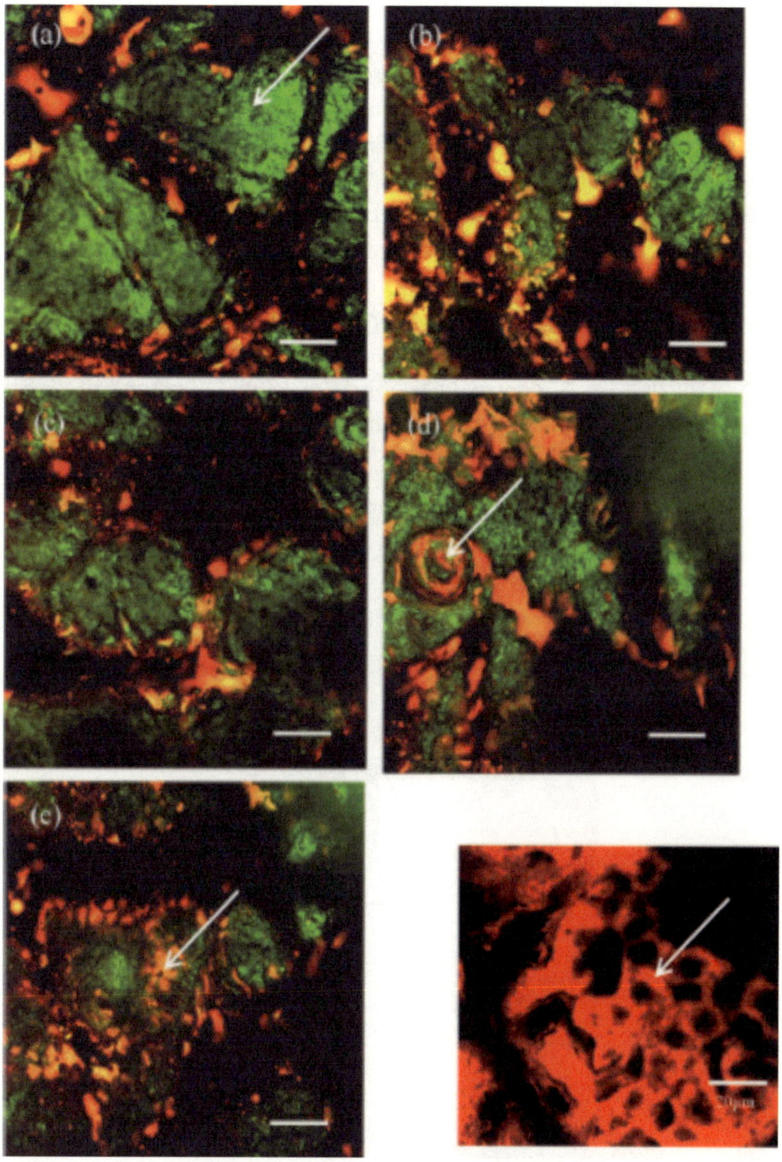

Figure 5. CLSM images showing distribution of polymer vesicles in cadaver epidermis with increasing durations: (a) 2 h, (b) 4 h, (c) 6 h, (d) 8 h and (e) 24 h. Inset depicts the accumulation of free Nile Red in the inter-corneocyte spaces. Being lipophilic in nature, the dye is seen to accumulate excessively in intercellular spaces (red stained). Vesicles initially (a) show localization in the 'furrows' between corneocyte groups (green stained) followed by the distribution in follicular (d) and intercellular spaces (e) (scale bar = 200 μm). Reproduced from [57] with permission.

It is possible to obtain CLSM images by sequential excitation using dual-labeled delivery systems, i.e., systems formulated with two different types of probes aiming to elucidate their mode of action. After *ex vivo* skin delivery, by sequential excitation of dual-labeled liposomes and vesicles carrying diclofenac, it was possible to show that the vesicles penetrated intact down to the epidermis and the fluorescence intensity was higher and predominantly accumulated in the inter-corneocytes spaces [22]. Teixeira et al. (2010) [58], taking the advantage that the vitamin A (retinyl palmitate) fluoresces due to the presence of cromophore in its structure, obtained dual-labeled images after *in vitro* studies using elastic polymeric nanocapsules, marked with Nile blue, carrying the vitamin. The images (Figure 6) showed that the nanocapsule did not penetrate the skin carrying the vitamin, but a deep permeation (around 30 μm) of both was observed, which suggests that the drug present in deep skin layers was released from nanocapsules in the superficial skin layers. The uniform permeation of both labels suggests an intercellular permeation as the main mechanism for this type of nanocapsules. The visualization of dually labeled nanoparticles by the combination of multiphoton and CLSM in human skin biopsies showed that the polymeric nanoparticles did not penetrate the skin, whereas the dye Texas red, used to mimic a drug loaded in the nanoparticles, slowly penetrated the skin up to the stratum granulosum [59].

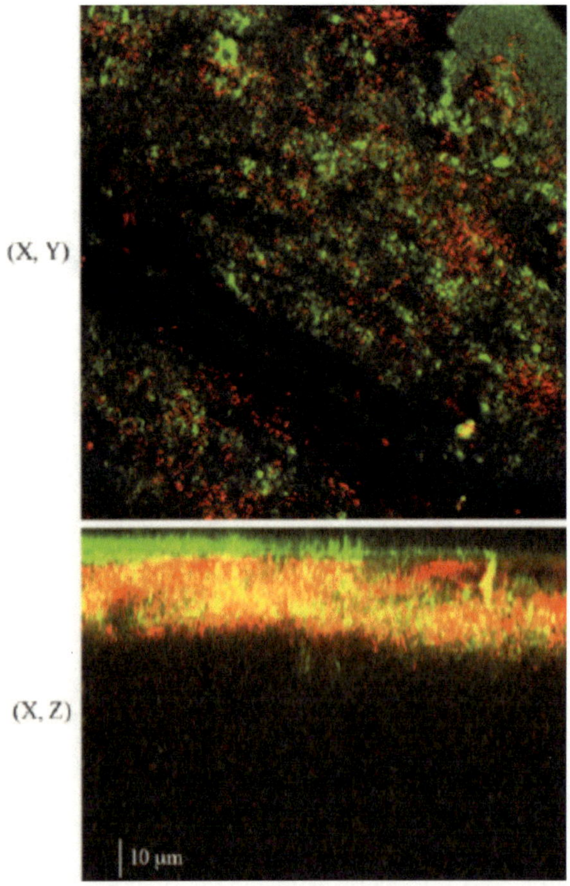

Figure 6. Confocal images of Nile blue-PLA nanocapsules. (a) xy image of the skin surface (initial fluorescence) and (b) deeper layers into skin, cross-sectional (xz mode) image. Red fluorescence, polymeric shell; green fluorescence, retinyl palmitate. Reproduced from [58] with permission.

Besides to determine the penetration in different skin layers, it is also possible to determine in which cellular site the drug preferentially penetrates. In this situation, Borowska et al. (2012) [35] stained skin sections with special dyes of high affinity to the nucleus before the *in vitro* skin penetration study using 8-MOP in formulations containing polyamidoamine dendrimers, conjugated with the probe fluorescein. CLSM images (Figure 7) revealed that 8-MOP encapsulated in dendrimers was able to penetrate into deep skin layers, epidermis and dermis, compared to standard formulation. In addition, the probe fluorescence (fluorescein) presented in the nucleus (blue stained) revealed that 8-MOP accumulated mostly in this cellular site important for its phototherapeutic activity.

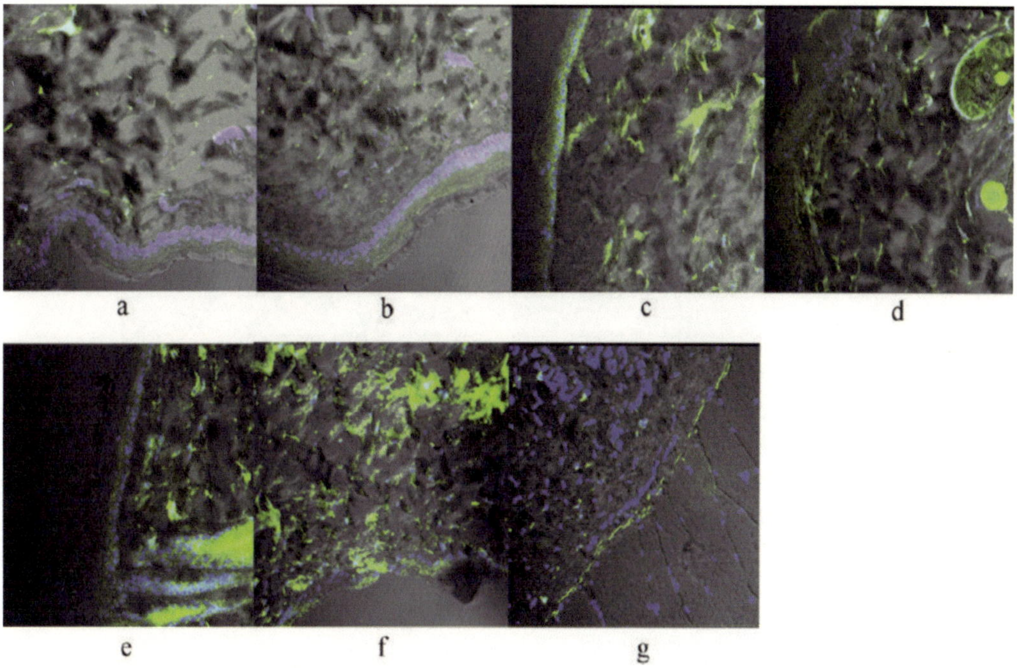

Figure 7. Distribution of 8-MOP (green) in rat's skin samples obtained by confocal microscopy following skin application of tested 8-MOP formulation. Cellular nuclei were counterstained with 7 AAD (blue): (a) 8-MOP after 1 h; (b) 8-MOP after 2 h; (c) 8-MOP-G3 PAMAM after 1 h; (d) 8-MOP-G3 PAMAM after 2 h; (e) 8-MOP-G4 PAMAM after 1 h; (f) 8-MOP-G4 PAMAM after 2 h; (g) 8-MOP-G4 PAMAM after 2 h – all skin layers and subcutaneous fatty tissue. Reproduced from [35] with permission.

In order to compare the skin penetration enhancement ability of different carriers and formulations, probes are visualized in the skin samples by CLSM after permeation studies. The fluorescence signal of a probe in liposome and transfersome containing valsartan was scanned at different skin depths after *in vitro* skin permeation studies. The results showed an increased skin penetration up to the dermis when valsartan was loaded in transfersomes, compared to rigid liposomes. The fluorescence liposomes was visualized up to 50 μm, while transfersomes were assessed up to 150 μm with a high fluorescence intensity and homogeneous skin distribution, evidencing the transdermal potential of transfersomes compared to liposomes [56]. CLSM showed an improved and homogeneous skin penetration of the probe rhodamine B in the role epidermis using β- cyclodextrin composite ethosomal gel carrying the drug clotrimazole compared to gel and ethosomal gel formulations [60]. The *in vitro* percutaneous permeation of ethosomes, containing the drug 5- fluorouracil and labeled with rhodamine 6GO, in human hypertrophic scar and normal skin was assessed by CLSM. The images (Figure 8) showed a higher fluorescence intensity throughout the hypertrophic skin than normal skin compared to hydroethanolic solution, evidencing that the skin penetration of ethosomes was superior than the control in this type of skin [45]. Labeled (Nile red) solid lipid nanoparticles loaded with tacrolimus also showed improved skin penetration in various layers of skin (in the order of 5–6 times) compared to control, an ointment formulation [61].

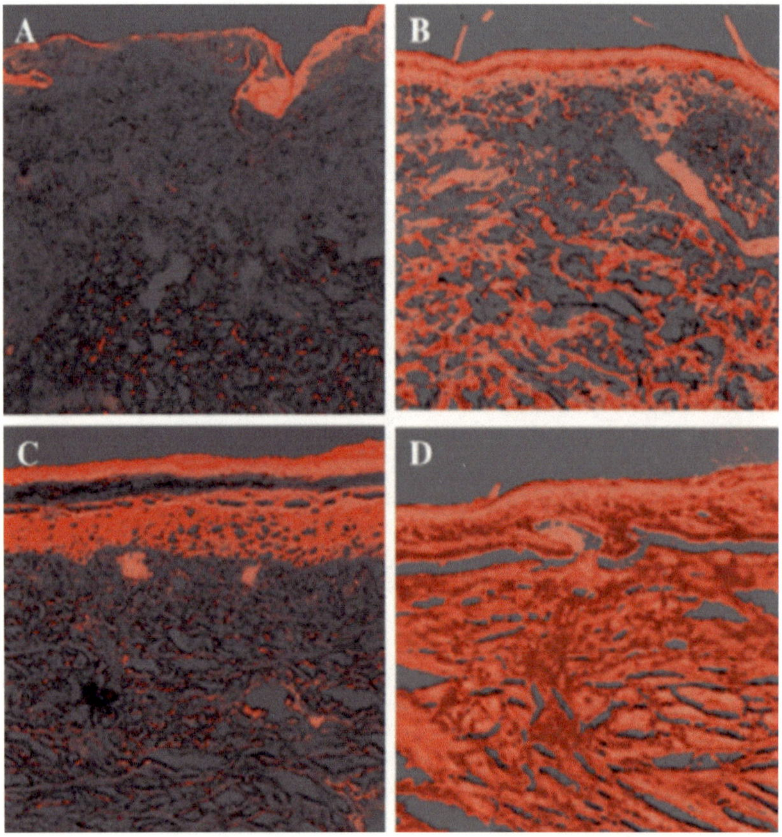

Figure 8. CLSM images of cross-sections of (A) human skin after application of rhodamine 6GO–hydroethanolic solution; (B) human skin after application of rhodamine 6GO–ethosomes; (C) human hypertrophic scar after application of rhodamine 6GO–hydroethanolic solution; and (D) human hypertrophic scar after application of rhodamine 6GO–ethosomes for 24 hours. Each image represents a 500 μm × 500 μm area. Reproduced from [45] with permission.

CLSM is extensively used to elucidate whether or not non-biodegradable nanoparticles penetrate the skin. Aiming to investigate the distribution of this type of nanoparticles (20 nm and 200 nm), covalently linked with FITC, across excised porcine skin, and to elucidate the pathways of penetration into/through the cutaneous barrier, Alvarez-Roman et al. (2004) [62] obtained CSLM images showing that FITC-nanoparticles preferentially accumulated in the hair follicles and furrows that demarcate clusters of corneocytes (Figure 9). Furthermore, images from the xz plane showed that, independently of size, both nanoparticles did not penetrate the skin being localized at the furrows. Similar results were obtained by Campbell et al. (2012) [36] who verified polystyrene nanoparticles in the top layers of the stratum corneum up to a depth of 2-3 μm.

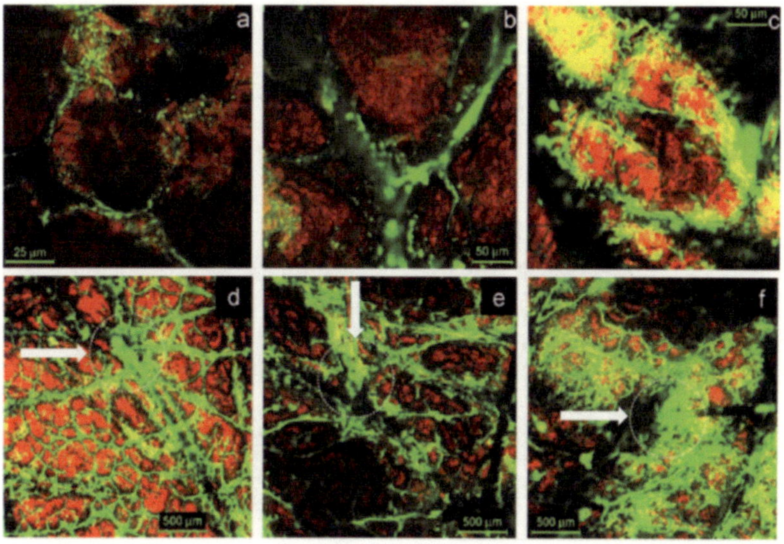

Figure 9. Dual label *x–y* images of the skin surface subsequent to treatment with FITC-nanoparticles (20 nm) for (a) 0 min, (b) 1 h, and (c) 2 h. The green fluorescence in figures a and c corresponds to furrows where nanoparticles accumulated. Follicular localization of FITC-nanoparticles subsequent to application of nanoparticles (200 nm) for (d) 30 min, (e) 1 h, and (f) 2 h. The white circles correspond to hair follicles. The red color corresponds from the fluorescence emanating from the porcine skin and the yellow-green from the FITC-nanoparticles. Reproduced from [62] with permission.

Photodynamic therapy studies take the advantage that the drugs (photosensitizers) used for cancer treatment self-fluoresce and CLSM technique is extensively used to assess the skin localization of the photosensitizer after *in vitro* topical application of a delivery system, once its skin penetration, besides other factors, is responsible for the success of the treatment. De Rosa et al. (2000, 2004) [9,63] verified by CLSM images (Figure 10) that the skin penetration and accumulation of protoporphyrin IX, an endogenously photosensitizer obtained from 5-aminolevulinic acid (5-ALA) by the biosynthetic pathway of heme, depended on the presence of penetration enhancer substance in the vehicle and 5-ALA derivative used. The presence of 5-ALA in ethosomes influenced the penetration depth and fluorescence intensity obtained from protoporphyrin IX after *in vivo* skin penetration [64]. CLSM also confirmed the improved biodistribution of a photosensitizer, zinc pthalocyanine tetrasulfonate, when using a microemulsion [10].

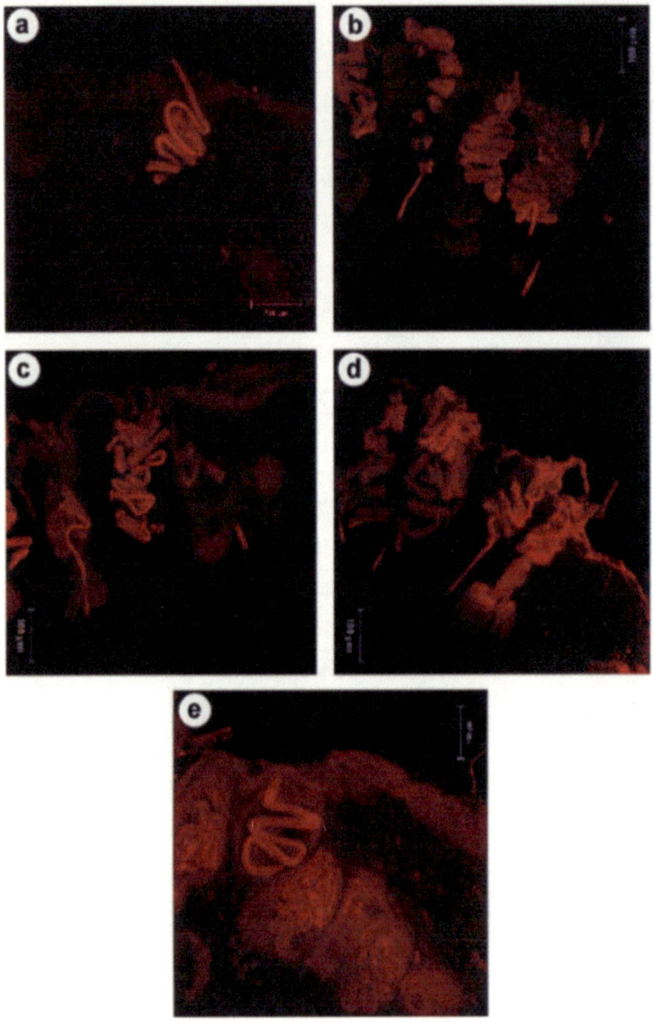

Figure 10. CLSM images of mechanical cross sections of hairless mouse skin (perpendicular series), optically sectioned 10 μm below the cutting surface: (a) control (untreated skin); skin treated with: (b) 10% 5-aminolevulinic acid (w/w); (c) 20% Dimethylsulphoxide (DMSO) (w/w); (d) 10% 5-aminolevulinic acid + 10% DMSO; and (e) 10% 5-aminolevulinic acid (w/w) + 20% Dimethylsulphoxide (w/w). All applied formulations contained 3% EDTA (w/w). Compared to controls (Figure 10 (a)), increases of red fluorescence in skins treated with 5-aminolevulinic acid (Figure 10 (b)), Dimethylsulphoxide (Figure 10 (c)), or with associations of these substances (Figures 10 (d) and 10 (e)), can be clearly seen as points with intense red fluorescence, indicating a PpIX accumulation. The association of 10% 5-aminolevulinic acid + 20% DMSO provided higher accumulation of PpIX, being considered more adequate for topical photodynamic therapy. Reproduced from [9] with permission.

CLSM is also used to assess the skin penetration of drugs after pre-treatment of the skin by physical methods, such as, sonophoresis, iontophoresis, microneedles and laser. Zhang et al. (2010) [27] showed that the pre-treatment of skin with microneedles permits the skin penetration of PLGA nanoparticles in the epidermis and dermis, and this would benefit a sustained drug release in the skin, supplying, in this way, the skin with drug over a prolonged period.

The depth of skin penetration from the drug heparin (FITC-labeled) was assessed *in vitro* in skins that were previously subjected to pretreatment using enhancement strategies by physical methods, such as sonophoresis, iontophoresis and microneedles. The study showed (Figure 11) that microneedles pretreatment was the only enhancement strategy that permitted heparin to reach the epidermis and deeper dermal layers, since FITC-labeled heparin was observed to follow microchannels formed by the microneedle device [65]. The influence of the low fluence fractional laser on the penetration of high molecular weight model drug, a polypeptide, FITC and FITC-labeled dextran (MWs of 4 and 150 kDa), was assayed by CSLM, whose image of skin showed a fluorescence increase in the upper and middle dermis, as well as in hair shafts and hair sheaths, evidencing a transfollicular route for high molecular weight substances [66].

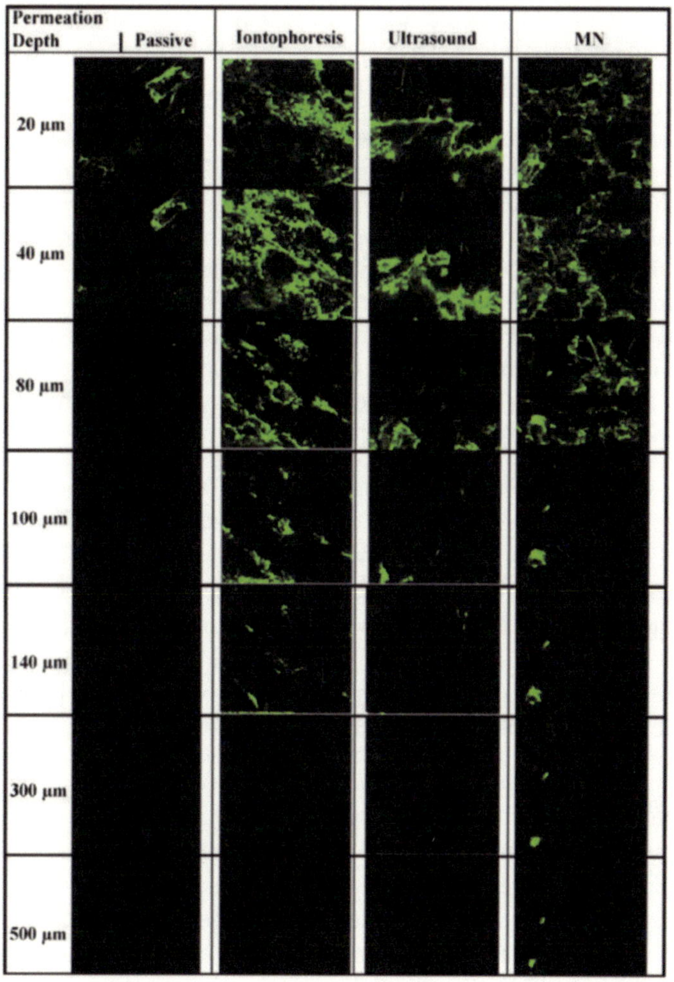

Figure 11. CLSM (X10 objective) of the permeation of FITC-labeled heparin across hairless rat skin at various depths from the surface of the stratum corneum. Heparin transported across microchannel can be seen as a florescent green color up to a depth of 500 µm. Reproduced from [65] with permission.

5.2. Non–invasive method for diagnosis of skin disorders and diseases

As mentioned before, CLSM can be used in fluorescence and reflectance modes for studies of skin disorders and diseases. It allows optical en face sectioning with quasihistological resolution and good contrast within living intact human tissue. The resolution allows for imaging of nuclear, cellular and tissue architecture of epidermis and the underlying structures, including connective tissue, inflammatory infiltrates, tumour cells, capillaries, and even circulating blood cells, without a biopsy [67].

Technically, confocal microscopes used in dermatology are not very different from their counterparts used in basic sciences. Typically, confocal microscopy employs lasers as sources of illumination due to their capability to generate monochromatic, coherent beams [54]. Confocal reflectance microscopy (CRM) in dermatology uses a near-infrared laser at 830 nm operating at a power of less than 20 mW, which is harmless for the tissue [67], whereas fluorescent confocal microscopes use He/Ne, Kr or Ar lasers to illuminate fluorescent samples in a wavelength range of 400-700 nm [68].

The company Lucid Inc. (Rochester, NY, USA) introduced in 1997 the first generation confocal microscope called VivaScope 1000. This microscope had a bulky configuration impeding convenient attachment to certain anatomical areas and imaging was very time consuming. The most widely used confocal microscope for imaging of human skin, the VivaScope 1500, was commercialized in the year 2000. This device is considerably smaller, more flexible and portable in contrast to its stationary predecessor. Further developments are the VivaScope 3000, a handheld confocal microscope for imaging of difficult to access areas, and the *ex vivo* VivaScope 2500 designed for imaging of excised tissue, especially in Mohs surgery as well as Vivascope 1500 Multilaser, which combines fluorescence laser scanning microscopy and confocal reflectance microscopy (CRM) modes [7].

In dermatology, CRM has several uses, such as: i) to diagnose diseases non-invasively *in vivo* [69-72]; ii) to non-invasive monitoring of treatment response *in vivo* and permits early detection of subclinical disease [73]; iii) to improve the accuracy of clinical diagnosis [74]; iv) to improve clinical discrimination between benign and malignant lesions [75]; v) to evaluate the same skin site over time because it produces no tissue damage [76]; vi) to assess the boundaries of the lesion pre- or post-surgery [54]; vii) to evaluate the dynamics of structural and cellular changes that take place during the occurrence of the disease [77]; and viii) to study physiopathologic processes non-invasively over time [78-79].

Despite the benefits, CRM has some limitations, such as: i) not all lesions image well; ii) subject motion occurs frequently; iii) the depth of imaging is limited at present to only 200 to 500 μm (epidermis and superficial dermis); iv) the presence of refractive structures may also decrease contrast and make melanocyte visualization difficult; v) lesions with a thick epidermis or certain anatomic sites, such as the palm or sole, will image very superficially; vi) images are viewed in horizontal sections rather than vertical sections (as in conventional histopathology), which makes direct comparisons difficult and requires experience to properly interpret images [67,74]; and vii) the relative high cost of CRM (approximately $50,000). Although the upfront cost of the device is high, the supplies to image individual lesions cost only about $1 per lesion, allowing imaging of multiple lesions with minimal increased cost to the patient per lesion [80].

Recently, there was a formation of an international CRM group (www.skinconfocalmicroscopy.org) constituted by clinicians who use this technique for diagnosis and monitoring, researchers and experts in many aspects of CRM to dermatology. By providing a forum for free communication of results, the establishment of meaningful collaborations and formation of trained personnel, this group aims to spread the use of CRM in dermatology [54].

5.2.1. CRM to diagnose melanocytic lesions

Langley et al., in 2001 [74], were the first to publish a study of benign and malignant melanocytic lesions by *in vivo* confocal microscopy. In the study, they discovered apparent differences in the CRM characteristics of nevi, dysplastic nevi, and melanomas (Figure 12), indicating that this tool could be helpful in the clinical discrimination of benign and malignant melanocytic lesions. The diagnostic applicability of CRM in melanocytic skin tumours, determined by evaluating sensitivity, specificity, positive and negative predictive value was first described by Gerger et al. in 2005 [81]. In 2006, Langley et al [75] determined that CRM has sensitivity and specificity compared to dermoscopy for the diagnosis of melanoma.

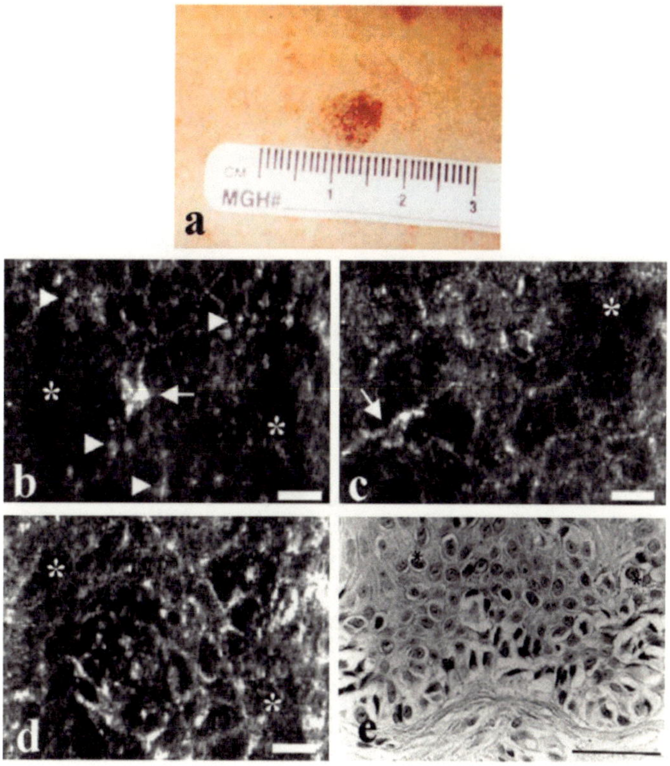

Figure 12. Clinical photograph of superficial spreading melanoma, invasive to anatomic level II, with a measured depth of 0.44 mm on the right posterior shoulder of 47-year-old man. This lesion was clinically diagnosed as dysplastic nevus, but by CLSM showed features consistent with melanoma. b-d, Confocal images of melanomas. Key features noted in melanomas as viewed by CLSM were loss of keratinocyte cell border (asterisks in B-D); bright, granular, highly

refractile particles (arrowheads in b); and atypical stellate cells (arrows in b and c). Scale bar = 25 μm. e, Histological section of intraepidermal component of superficial spreading malignant melanoma showing confluence of tumor cells at dermoepidermal junction as well as individual cells of varying levels of the epidermis, so-called "pagetoid" spread (asterisks). Scale bar = 50 μm. Reproduced from [74] with permission.

Benign nevi can be differentiated from melanomas by CRM and criteria proved to be valuable [81-84]. CRM features can distinguish lentigo maligna (Figure 13) from benign macules of the face such as solar lentigo, ephelis, actinic keratosis, and flat seborrheic keratosis [70,75,85].

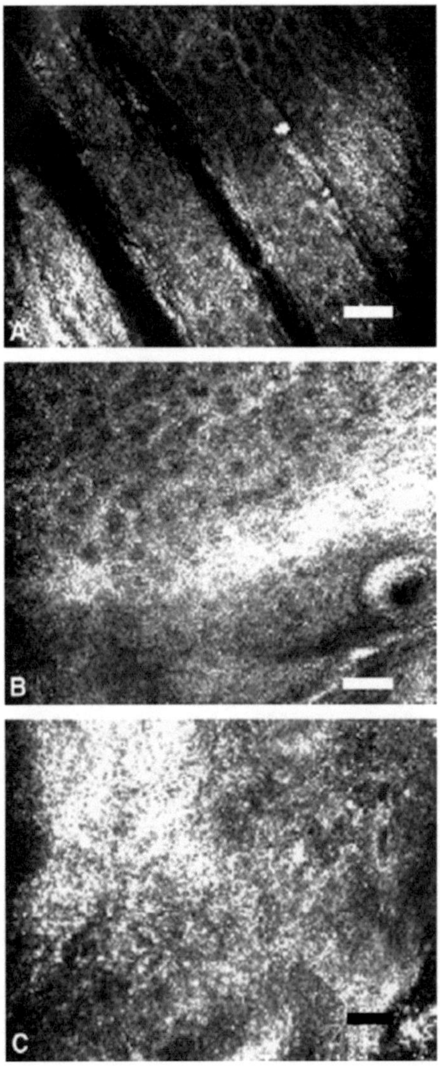

Figure 13. A, CRM image of control skin at the stratum spinosum displaying well-organized keratinocyte cell borders. B, CRM image of lentigo at the level of the stratum spinosum. The well-organized honey combed pattern of keratinocyte cell borders is preserved. C, CRM image of lentigo maligna at the stratum spinosum level. There is loss of keratinocyte cell borders, and a grainy image is noted. (Scale bar, 50 μm.). Reproduced from [75] with permission.

http://dx.doi.org/10.5772/55995

Even non-melanotic lesions can be recognized by CRM because of the presence of melanosomes and melanin granules in their cytoplasm [70,86-87].

5.2.2. CRM and non–melanocitic lesions

The clinical diagnosis of actinic keratosis by CRM (Figure 14) was first determined by Aghassi et al. (2000) [78]. Later, others confirmed the usefulness of this tool [73,88-89], although it has not been able to unequivocally differentiate actinic keratosis from squamous cell carcinomas [90].

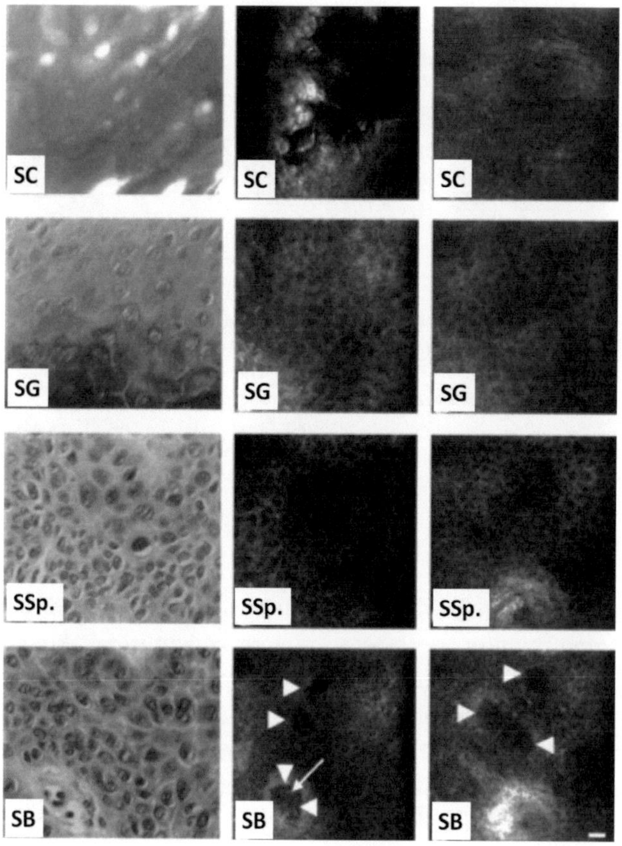

Figure 14. Left column is conventional histopathology of actinic keratosis (AK), center is CRM of AK, and right column is CRM of adjacent normal skin. Sections were obtained by means of conventional histopathology of AK (hematoxylin-eosin stain; original magnification ×20; 0.4 numerical aperture, dry objective lens), CM of AK (×30, 0.9 numerical aperture, water immersion objective lens, scale bar = 25 μm),and CM of adjacent normal skin (original magnification ×30, 0.9 numerical aperture, water immersion objective lens, scale bar = 25 μm). SC, Stratum corneum. Irregular hyperkeratosis of AK is evident on conventional histopathology and CRM, contrasting with smooth surface of normal skin. SG, Stratum granulosum. Conventional histopathology and CRM demonstrate uniform, evenly spaced, broad keratinocytes both in AK and normal skin. In CRM images, nuclei appear dark in contrast to bright, refractile cytoplasm. SSp., Stratum spinosum. Conventional histopathology and CRM of AK show enlarged, pleomorphicnuclei with haphazard orientation, contrasting with small, uniform, evenly spaced nuclei from normal skin. SB, Stratum basale. Conventional histopathology and CRM of AK show enlarged, pleomorphic nuclei with haphazard orientation, contrasting with small, uniform, evenly spaced nuclei from normal skin. In CRM images, dermal papillae appear as well-demarcated, dark holes in epidermis (arrowheads), containing blood vessels (arrow). Reproduced from [78] with permission.

CRM offers a sensitive and specific tool for the noninvasive diagnosis of basal cell carcinoma *in vivo* (Figure 15) [69,91-92] showing good correlation with histology.

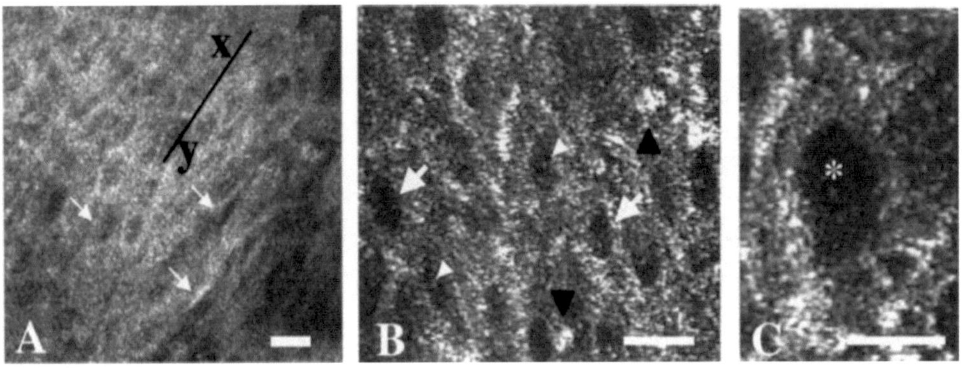

Figure 15. Real-time *in vivo* confocal images of a basal cell carcinoma (BCC) showing a uniform population of elongated cells (arrows, A and B) oriented along the same principal axis (x-y). Mononuclear cells (black arrowheads, B) are seen intermixed with BCC cells (arrows, B) with some of the BCC cells showing nucleoli (white arrowheads, B). C, High-power magnification of one BCC cell demonstrating a large elongated nucleus with a low refractility (asterisk), surrounded by a bright cytoplasm. (A-C, Hematoxylin-eosin stain; original magnifications: A, x30, 0.9 numerical aperture (NA) water-immersion objective lens; B and C, x100, 1.2 NA water-immersion objective lens. Scale bar, 25 μm). Reproduced from [91] with permission.

Non-melanoma skin cancers (NMSC) were evaluated by fluorescence confocal microscopy aiming to diagnose and monitor the lesions in reference to normal skin and correlation with routine histology. The results suggest that fluorescence confocal microscopy may allow a systematic, noninvasive histomorphometric evaluation of actinic keratosis and basal cell carcinoma, potentially aiding in the detection of subclinical actinic keratosis and early therapeutic management [93]. CRM was also evaluated to diagnose NMSC early and the authors concluded that *in vivo* CRM is a promising and innovative technology for the early diagnosis of skin cancer, which may ultimately play an important role in skin cancer screening and prevention as well as in the early detection of progression or recurrence after therapy [94].

5.2.3. CRM and other skin diseases

In vivo CRM resolves changes at the cellular and subcellular level comparable to that obtained with standard histology. In this case, this tool was successfully used to evaluate the dynamics of structural and cellular changes that take place during the occurrence of allergic contact dermatitis [77]. Furthermore, it is a promising tool for dynamic, noninvasive assessment and may help to differentiate acute and induced contact dermatitis (Figure 16) [76,79].

http://dx.doi.org/10.5772/55995

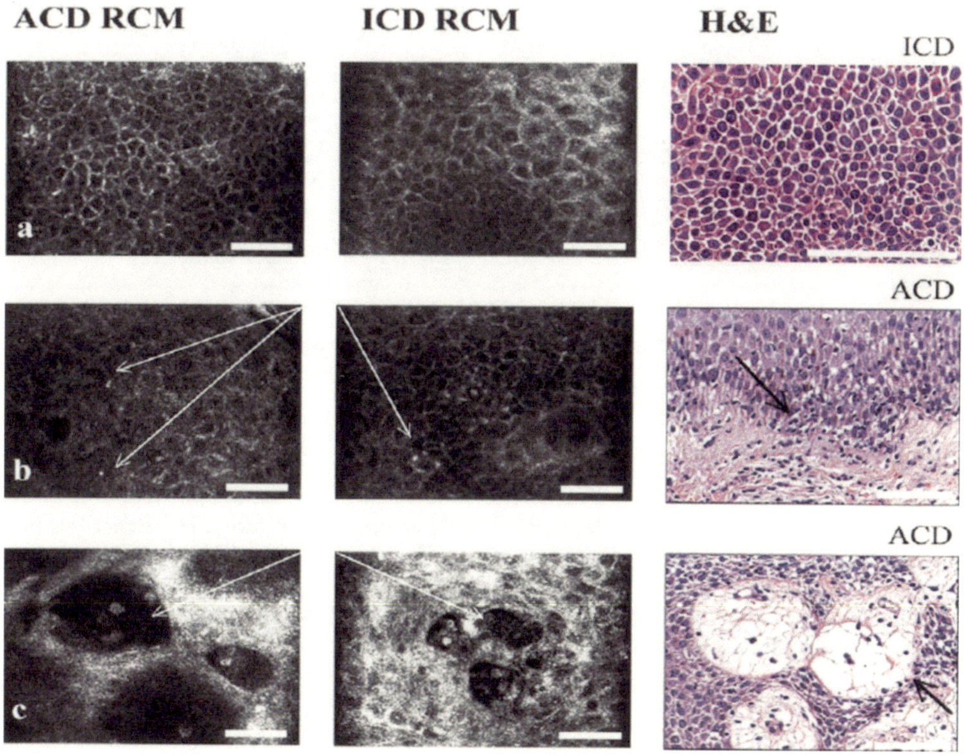

Figure 16. Features common to allergic contact dermatitis and irritant contact dermatitis observed with reflectance confocal microscopy CRM and correlated by routine histology. a, Spongiosis: increased intercellular brightness apparent on CRM. b, Inflammatory cell infiltrate: bright structures 12- to 15-μm size interspersed between keratinocytes. Arrows denote inflammatory cells. c, Intraepidermal vesicle (arrow) formation: dark spaces in epidermis containing inflammatory cells and necrotic keratinocytes. Scale bars = 50 μm. H&E, hematoxylin-eosin stain. Reproduced from [76] with permission.

CRM was also used to investigate lesions of progressive macular hypomelanosis (Figure 17) showing that the ''pigmented ring'' is intact, but its melanin content is decreased compared with the surrounding normal skin [95].

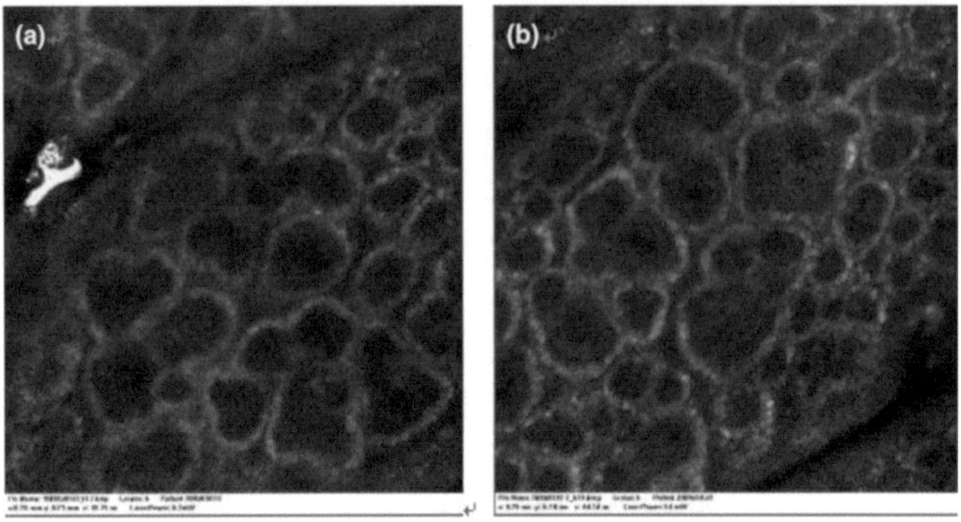

Figure 17. a) Lesion areas and (b) perilesional normal skin were observed under CRM. The ''pigmented ring'' around dermal papilla show completed, but compared with the surrounding normal skin its melanin content was decreased (the same horizon). Reproduced from [95] with permission.

Chiavèrini et al., 2001 [96] have shown crystalline deposits of cystine within the papillary dermis (Figure 18) by *in vivo* CRM, in patients with infantile cystinosis. The study concluded that CRM scoring of dermal cystine deposition could be used as a marker of treatment response complementary to the leukocyte cystine level.

CRM is a valuable tool to evaluate histological features of psoriasis (Figure 18), and the findings correlate with histology [97-99].

CRM imaging elucidates and monitors the dynamic pathophysiologic response, such as those that occur subsequent to laser treatment of cutaneous lesions, like cherry angiomas [78] and sebaceous gland hyperplasia [100-101]. Moreover, it facilitates the discrimination of sebaceous hyperplasia from basal cell carcinoma *in vivo* [102].

Noninvasive diagnosis of pemphigus foliaceus by CRM was consistent with the routine histology of the preexisting lesions [52] and criteria for diagnosing pemphigus (Figure 20) was established [72].

Diagnostic CRM criterion for mycosis fungoids was first determined by Agero et al. in 2007 [103] and later on confirmed by Lange-Asschenfeldt et al. (2012) [104].

Others diseases diagnosed by CRM *in vivo* were: discoid lupus erythematosus [105]; dermatophyte infections [106-108]; lichen sclerosus [109] and folliculitis [110].

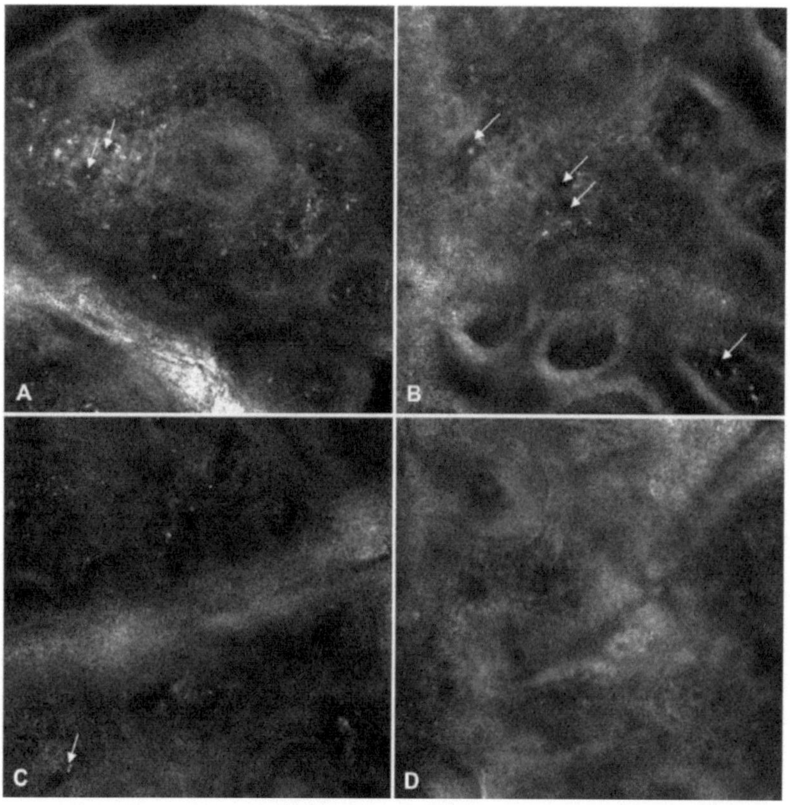

Figure 18. *In vivo* CRM of skin in infantile cystinosis. Confocal images of skin: block image of 4 x 4 mm. Numerous (+++) (A), few (++) (B), and some (+) (C) bright particles were seen around or in vessels (arrows) at level of dermis, in patients with cystinosis compared with control subjects (D). Reproduced from [96] with permission.

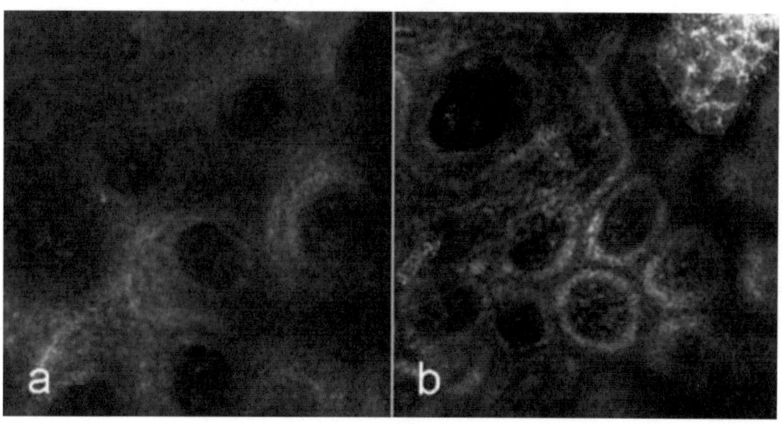

Figure 19. CRM image (0.5 × 0.5 mm) of the basal layer: absence of papillary rings in plaque psoriasis skin (a), compared to normal skin taken on symmetrical anatomical sites (b). Reproduced from [98] with permission.

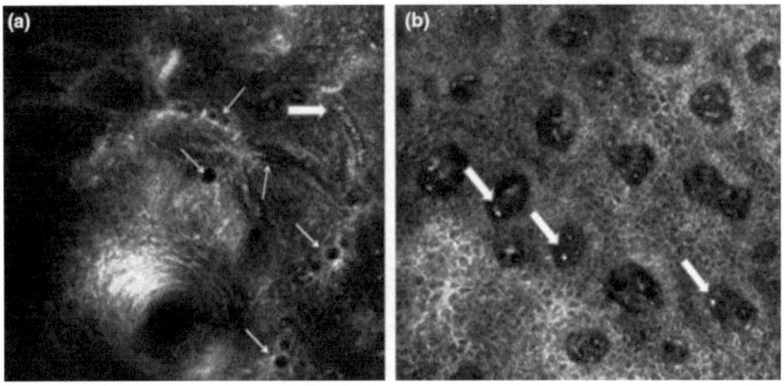

Figure 20. Dilated blood vessels (thin arrows) containing highly refractive peripheral blood cells (thick arrows) in pemphigus vulgaris (a) and pemphigus foliaceous (b) (0.5 x 0.5 mm). Reproduced from [72] with permission.

6. Pathways involved in penetration of drugs and biomolecules in the skin

The SC is the outmost layer of the skin and provides an efficient barrier for the ingress of extraneous substances and egress of endogenous substances [111]. The packing of the lipids (ceramides, cholesterol and free fatty acids) present in the SC are responsible for this barrier as they are arranged surrounding the corneocytes forming a "brick and mortar" model with the corneocytes as the bricks and the lipids as the mortar [112].

Drugs and biomolecules can penetrate the skin to the viable tissue by three main routes: i) the intercellular (between the corneocytes, through the lipid matrix); ii) the transcellular (through the corneocytes); and iii) the shunt (through the skin appendages, like hair follicles, sweat ducts and sebaceous glands). Generally, the intercellular pathway is predominantly for most molecules, and the shunt pathway is important for polymers and colloidal nanoparticles and large polar molecules [15].

CLSM helps to elucidate not only the extension and depth of skin penetration of drugs and biomolecules [10,45] but also the main pathways involved in their penetration. In this sense, the use of CLSM with other physicochemical techniques such as Raman [113], infrared spectroscopy [39], differential scanning calorimetry and small- and wide-angle X-ray scattering [114], occlusive studies [115], dermatopharmacokinetic analysis [39] and transmission electron microscopy [116], help to further elucidate the exact drug/skin interaction, i.e., the mechanism of skin penetration. In addition, these techniques also permit to determine the skin interaction of delivery systems [34,117-120] and the skin modification promoted by physical methods [32,121-123], aiding to clarify the exact mechanism by which these techniques increase the skin penetration of drugs and biomolecules.

Interesting study was carried out with microemulsion containing Nile red dye, showing that this delivery systems lead to a dye accumulation between the coneocytes of the role stratum corneum, evidencing an intercellular penetration route (Figure 21). Moreover, to address whether the microemulsion penetrated the stratum corneum or not, a dermatopharmacokinetic analysis of the microemulsion components across this skin layer helped to clarify that every component diffuses separately [39].

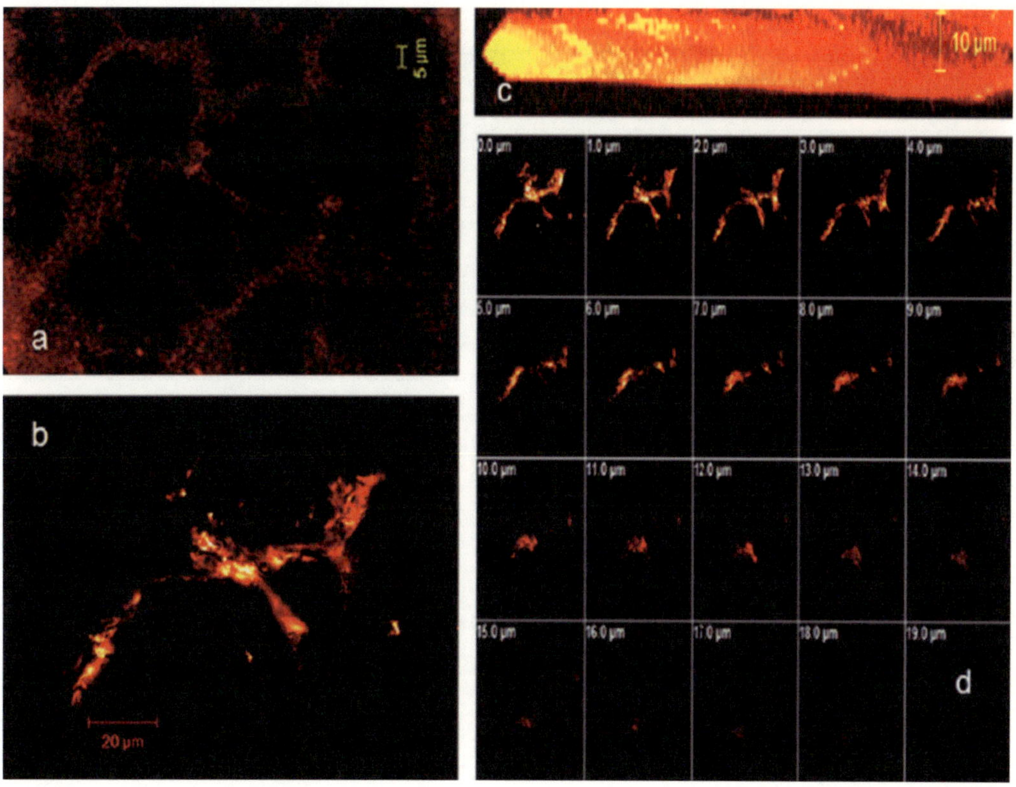

Figure 21. CLSM images of the microemulsion fluorolabelled with Nile red (a) and (b) Normal xy images, (c) Reconstructed xz image after sectioning through z-plane using one projection at 0º angle (sectioned from 0 to 20 µm with 1 µm increments) and (d) z-stack image. Reproduced from [39] with permission.

Table 3 shows the main skin permeation pathways uncovered by CLSM studies using different delivery systems, probes and methods of application.

Main Permeation Pathways	Delivery system/probe	Mechanism of skin application	Reference
Follicular	Ethosomes/β-carotene	Passive delivery	[117]
	Micelle/Fluorescein	Passive delivery	[34]
	Polystyrene nanoparticles/FITC	Passive delivery	[62]
	Nanoparticles/5-fluoresceinamine	Passive delivery	[118]
	Polypeptide, FITC and FITC-labeled dextran	Skin pre-treatment with fractional laser	[66]
Intercellular (lipid pathway)	Liposomes/fluorescein–DHPE	Passive delivery	[119]
	Ethosomes/Bacitracin-FITC	Passive delivery	[120]
	Posintro™ nanoparticles/Acridine	Passive delivery	[116]
	Elastic polymeric nanocapsules/ retinyl palmitate	Passive delivery	[58]
		Passive delivery	[114]
	Biphasic vesicles/Alexa 594-labeled Interferon α	Passive delivery	[22]
	Vesicles/β-carotene and carboxyfluorescein	Passive delivery	[39]
	Microemulsion/Nile red		
Transcellular (corneocytes pathway)	FITC label oligonucleotide	Electroporation	[121]
Mixed Inter and Transcellular	Calcein dissolved in phosphate buffer	Iontophoresis and high voltage pulsing	[122]
	Fluorescein isothiocyanate dextran dissolved in phosphate buffer	Low-frequency sonophoresis	[32]
Channels in the stratum corneum	Elastic liposomes/Rhodamine red	Passive delivery	[124]
Lacunar regions of the intercellular lipids and follicular	Quantum dots (cationic, neutral and anionic)	Ultrasound and sodium lauryl sulphate	[44]
Intercellular and follicular	Flexible polymerosomes/Nile red	Passive delivery	[57]

Table 3. Skin permeation pathways uncovered by CLSM studies

7. Future perspectives

The use of confocal microscopy in recent years has revolutionized many aspects in the pharmaceutical and medical sciences. It is an useful tool that aids to develop and improve delivery systems aimed to reach certain targets in the skin; it permits to elucidate the mechanisms of penetration of these systems in relevant skin models and to uncover the cellular uptake of drugs; and help to diagnose cutaneous pathologies. The ability to obtain non-invasively images with high resolution and sensitivity, at different depths of the sample and three-dimensional reconstruction are important advantages of this technique. The assessment of various delivery systems *in vitro* by CLSM provides a better characterization of their physicochemical properties and their behavior and the ability to view them *ex vivo* and *in vivo* allows us to elucidate their mechanisms *in situ*. With this, the rational development of topical delivery systems becomes more effective.

The use of CLSM in the reflectance mode to diagnose skin disorders is relatively recent and the challenge in this area is related to the presence of substantial clinical studies that compares morphological confocal features with histology and establishes their correlations, the formation of trained and experienced personnel and the cost of the equipment. Although its usefulness related to the collection of images fast and relatively ease, some limitations of the technique related to the equipment capability still limits its wide use in dermatology.

In this context, improvements in equipment, fluorescent probes and confocal microscopy techniques will enhance and promote new opportunities for studies involving skin, cosmetic and pharmaceutical products, since it is clear the important contribution of CLSM in this application area.

Author details

Fábia Cristina Rossetti, Lívia Vieira Depieri and Maria Vitória Lopes Badra Bentley*

*Address all correspondence to: vbentley@usp.br

Faculdade de Ciências Farmacêuticas de Ribeirão Preto, Universidade de São Paulo, Avenida do Café, s/n, Ribeirão Preto, SP, Brazil

References

[1] Pygall, S. R, Whetstone, J, Timmins, P, & Melia, C. D. Pharmaceutical applications of confocal laser scanning microscopy: The physical characterisation of pharmaceutical systems. Advanced Drug Delivery Reviews (2007).

[2] White, P. J, Fogarty, R. D, Liepe, I. J, Delaney, P. M, Werther, G. A, & Wraight, C. J. Live confocal microscopy of oligonucleotide uptake by keratinocytes in human skin grafts on nude mice. Journal of Investigative Dermatology (1999).

[3] Ghartey-tagoe, E. B, Morgan, J. S, Ahmed, K, Neish, A. S, & Prausnitz, M. R. Electroporation-mediated delivery of molecules to model intestinal epithelia. International Journal of Pharmaceutics (2004).

[4] Verma, D. D, Verma, S, Blume, G, & Fahr, A. Liposomes increase skin penetration of entrapped and non-entrapped hydrophilic substances into human skin: a skin penetration and confocal laser scanning microscopy study. European Journal of Pharmaceutics and Biopharmaceutics (2003).

[5] Alvarez-román, R, Naik, A, Kalia, Y. N, Fessia, H, & Guy, R. H. Visualization of skin penetration using confocal laser scanning microscopy. European Journal of Pharmaceutics and Biopharmaceutics (2004).

[6] Brus, C, Santi, P, Colombo, P, & Kissel, T. Distribution and quantification of polyethylenimine oligodeoxynucleotide complexes in human skin after iontophoretic delivery using confocal scanning laser microscopy. Journal of Controlled Release (2002).

[7] Hofmann-wellenhof, R, Pellacani, G, Malvehy, J, & Soyer, H. P. editors. Reflectance Confocal Microscopy for Skin Diseases. Springer-Verlag Berlin Heidelberg; (2012).

[8] Cullander, C. The laser-scanning confocal microscope. In: Wilhelm K-P., Elsner P., Berardesca E., Maibach HI. (eds.) Bioengineering of the Skin: Skin Surface and Analysis. CRC Press, Boca Raton; (1997). , 23-37.

[9] De Rosa, F. S, Marchetti, J. M, Thomazini, J. A, & Tedesco, A. C. Bentley MVLB. A vehicle for photodynamic therapy of skin cancer: influence of dimethylsulphoxide on aminolevulinic acid *in vitro* cutaneous permeation and *in vivo* protoporphyrin IX accumulation determined by confocal microscopy. Journal of Controlled Release (2000). , 5.

[10] Rossetti, F. C, Lopes, L. B, Carollo, A. R, Thomazini, J. A, & Tedesco, A. C. Bentley MVLB. A delivery system to avoid self-aggregation and to improve *in vitro* and *in vivo* skin delivery of a phthalocyanine derivative used in the photodynamic therapy. Journal of Controlled Release (2011).

[11] Menon, G. New insights into skin structure: scratching the surface. Advanced Drug Delivery Review (2002). SS17., 3.

[12] Foldvari, M. Non-invasive administration of drugs through the skin: challenges in delivery system design. Pharmaceutical Science & Technology Today (2000).

[13] Kendall, A. C, & Nicolaou, A. Bioactive lipid mediators in skin inflammation and immunity. Progress in Lipid Research (2013).

[14] Elias, P. M, & Friend, D. S. The permeability of the skin. Physiological Reviews (1971).

[15] Barry, B. W. Novel mechanisms and devices to enable successful transdermal drug delivery. European Journal of Pharmaceutical Sciences (2001).

[16] Verma, D. D, & Fahr, A. Confocal Laser Scanning Microscopy: An Excellent Tool for Tracking Compounds in the Skin. In: Smith WE., Maibach HI. (ed.) Percutaneous Penetration Enhancers. Taylor & Francis Group; (2006). , 335-357.

[17] Paddock, S. W. Principles and Practices of Laser Scanning Confocal Microscopy. Molecular Biotechnology (2000).

[18] Meng, F, Liao, B, Liang, S, Yang, F, Zhang, H, & Songe, L. Morphological visualization, componential characterization and microbiological identification of membrane fouling in membrane bioreactors (MBRs). Journal of Membrane Science (2010).

[19] Jerome, W. G, Fuseler, J, & Price, R. L. Specimen Preparation. In: Price RL., Jerome WG., (ed.) Basic Confocal Microscopy. Springer; (2011). , 61-77.

[20] Tsienv, R. Y, Ernst, L, & Waggoner, A. Fluorophores for Confocal Microscopy: Photophysics and Photochemistry. In: Pawley JB. (ed.) Handbook of Biological Confocal Microscopy. Springer Science Business Media; (2006). , 338-352.

[21] Souza, J. G, Gelfuso, G. M, Simão, P. S, & Borges, A. C. Lopez RFV. Iontophoretic transport of zinc phthalocyanine tetrasulfonic acid as a tool to improve drug topical delivery. Anti-Cancer Drugs (2011).

[22] Manconi, M, Caddeo, C, Sinico, C, Valenti, D, Mostallino, M. C, Biggio, G, & Fadda, A. M. Ex vivo skin delivery of diclofenac by transcutol containing liposomes and suggested mechanism of vesicle-skin interaction. European Journal of Pharmaceutics and Biopharmaceutics (2011).

[23] Manconi, M, Caddeo, C, Sinico, C, Valenti, D, Mostallino, M. C, Lampis, S, Monduzzi, M, & Fadda, A. M. Penetration enhancer-containing vesicles: Composition dependence of structural features and skin penetration ability. European Journal of Pharmaceutics and Biopharmaceutics (2012).

[24] Grams, Y. Y, Alaruikka, S, Lashley, L, Caussin, J, Whitehead, L, & Bouwstr, J. A. Permeant lipophilicity and vehicle composition influence accumulation of dyes in hair follicles of human skin. European Journal of Pharmaceutical Sciences (2003).

[25] Herwadkar, A, Sachdeva, V, Taylor, L. F, Silver, H, & Banga, A. K. Low frequency sonophoresis mediated transdermal and intradermal delivery of ketoprofen. International Journal of Pharmaceutics (2012).

[26] Gillet, A, Lecomte, F, Hubert, P, Ducat, E, Evrard, B, & Piel, G. Skin penetration behaviour of liposomes as a function of their composition. European Journal of Pharmaceutics and Biopharmaceutics (2011).

[27] Zhang, W, Gao, J, Zhu, Q, Zhang, M, Ding, X, Wang, X, Hou, X, Fan, W, Ding, B, Wu, X, Wang, X, & Gao, S. Penetration and distribution of PLGA nanoparticles in the human skin treated with microneedles. International Journal of Pharmaceutics (2010).

[28] Patlolla, R. R, Desai, P. R, Belay, K, & Singh, M. S. Translocation of cell penetrating peptide engrafted nanoparticles across skin layers. Biomaterials (2010).

[29] Shah, P. P, Desai, P. R, & Singh, M. Effect of oleic acid modified polymeric bilayered nanoparticles on percutaneous delivery of spantide II and ketoprofen. Journal of Controlled Release (2012).

[30] Shah, P. P, Desai, P. R, Channer, D, & Singh, M. Enhanced skin permeation using polyarginine modified nanostructured lipid carriers. Journal of Controlled Release (2012).

[31] Bachhav, Y. G, Summer, S, Heinrich, A, Bragagna, T, Böhler, C, & Kalia, Y. N. Effect of controlled laser microporation on drug transport kinetics into and across the skin. Journal of Controlled Release (2010).

[32] Morimoto, Y, Mutoh, M, Ueda, H, Fang, L, Hirayama, K, Atobe, M, & Kobayashi, D. Elucidation of the transport pathway in hairless rat skin enhanced by low-frequency sonophoresis based on the solute-water transport relationship and confocal microscopy. Journal of Controlled Release (2005).

[33] Bal, S. M, Kruithof, A. C, Zwier, R, Dietz, E, Bouwstra, J. A, Lademann, J, & Meinke, M. C. Influence of microneedle shape on the transport of a fluorescent dye into human skin *in vivo*. Journal of Controlled Release (2010).

[34] Bachhav, Y. G, Mondon, K, Kalia, Y. N, Gurny, R, & Möller, M. Novel micelle formulations to increase cutaneous bioavailability of azole antifungals. Journal of Controlled Release (2011).

[35] Borowska, K, Wolowiec, S, Rubaj, A, Glowniak, K, Sieniawska, E, & Radej, S. Effect of polyamidoamine dendrimer G3 and G4 on skin permeation of 8-methoxypsoralene-*In vivo* study. International Journal of Pharmaceutics (2012).

[36] Campbell CSJContreras-Rojas LR., Delgado-Charro MB., Guy RH. Objective assessment of nanoparticle disposition in mammalian skin after topical exposure. Journal of Controlled Release (2012).

[37] Gillet, A, Compère, P, Lecomte, F, Hubert, P, Ducat, E, Evrard, B, & Piel, G. Liposome surface charge influence on skin penetration behavior. International Journal of Pharmaceutics (2011).

[38] Park, J, Lee, J, Kim, Y, & Prausnitz, M. R. The effect of heat on skin permeability. International Journal of Pharmaceutics (2008).

[39] Hathout, R. M, Mansour, S, Geneidi, A. S, & Mortada, N. D. Visualization, dermatopharmacokinetic analysis and monitoring the conformational effects of a microemulsion formulation in the skin stratum corneum. Journal of Colloid and Interface Science (2011).

[40] Abdel-Mottaleb MMAMoulari B., Beduneau A., Pellequer Y., Lamprecht A. Nanoparticles enhance therapeutic outcome in inflamed skin therapy. European Journal of Pharmaceutics and Biopharmaceutics (2012).

[41] Zhang, L. W, Yu, W. W, Colvin, V. L, & Monteiro-riviere, N. A. Biological interactions of quantum dot nanoparticles in skin and in human epidermal keratinocytes. Toxicology and Applied Pharmacology (2008).

[42] He, R, Cui, D, & Gao, F. Preparation of fluorescence ethosomes based on quantum dots and their skin scar penetration properties. Materials Letters (2009).

[43] Gratieri, T, Schaefer, U. F, Jing, L, Gao, M, & Kostka, K. H. Lopez RFV., Schneider M. Penetration of Quantum Dot Particles Through Human Skin. Journal of Biomedical Nanotechnology (2010).

[44] Lopez RFVSeto JE., Blankschtein D., Langer R. Enhancing the transdermal delivery of rigid nanoparticles using the simultaneous application of ultrasound and sodium lauryl sulfate. Biomaterials (2011).

[45] Zhang, Z, Wo, Y, Zhang, Y, Wang, D, He, R, Chen, H, & Cui, D. *In vitro* study of ethosome penetration in human skin and hypertrophic scar tissue. Nanomedicine: Nanotechnology, Biology, and Medicine (2012).

[46] Cevc, G, Schätzlein, A, & Richardsen, H. Ultradeformable lipid vesicles can penetrate the skin and other semi-permeable barriers unfragmented. Evidence from double label CLSM experiments and direct size measurements. Biochimica et Biophysica Acta (2002).

[47] Smith, C. L. Basic Confocal Microscopy. Current Protocols in Neuroscience (2011). , 1-2.

[48] Claxton, N. S, Fellers, T. J, & Davidson, M. W. Microscopy, Confocal. Encyclopedia of Medical Devices and Instrumentation (2006). , 1-37.

[49] Lademann, J, Otberg, N, Richter, H, Meyer, L, Audring, H, Teichmann, A, Thomas, S, Knüttel, A, & Sterry, W. Application of optical non-invasive methods in skin physiology: a comparison of laser scanning microscopy and optical coherent tomography with histological analysis. Skin Research and Technology (2007).

[50] Meyer, L. E, Otberg, N, Sterry, W, & Lademann, J. *In vivo* confocal scanning laser microscopy: comparison of the reflectance and fluorescence mode by imaging human skin. Journal of Biomedical Optics (2006). , 11(4), 0440121-0440127.

[51] González, S, & Gilaberte-calzada, Y. *In vivo* reflectance-mode confocal microscopy in clinical dermatology and cosmetology. International Journal of Cosmetic Science (2008).

[52] Angelova-fischer, I, Pfeuti, T, Zillikens, D, & Rose, C. *In vivo* confocal laser scanning microscopy for non-invasive diagnosis of pemphigus foliaceus. Skin Research and Technology (2009).

[53] Koehler, M. J, Speicher, M, Lange-asschenfeldt, S, Stockfleth, E, Metz, S, Elsner, P, Kaatz, M, & König, K. Clinical application of multiphoton tomography in combination with confocal laser scanning microscopy for *in vivo* evaluation of skin diseases. Experimental Dermatology (2011).

[54] González, S. Confocal Reflectance Microscopy in Dermatology: Promise and Reality of Non-Invasive Diagnosis and Monitoring. Actas Dermosifiliogr (2009).

[55] Fan, C, Luedtke, M. A, Prouty, S. M, Burrows, M, Kollias, N, & Cotsarelis, G. Characterization and quantification of wound-induced hair follicle neogenesis using *in vivo* confocal scanning laser microscopy. Skin Research and Technology (2011).

[56] Ahad, A, Aqil, M, Kohli, K, Sultana, Y, Mujeeb, M, & Ali, A. Formulation and optimization of nanotransfersomes using experimental design technique for accentuated transdermal delivery of valsartan. Nanomedicine: Nanotechnology, Biology, and Medicine (2012).

[57] Rastogi, R, Anand, S, & Koul, V. Flexible polymerosomes-An alternative vehicle for topical delivery. Colloids and Surfaces B: Biointerfaces (2009).

[58] Teixeira, Z, Zanchetta, B, Melo, B. A. G, Oliveira, L. L, Santana, M. H. A, Paredes-gamero, E. J, Justo, G. Z, Nader, H. B, Guterres, S. S, & Durán, N. Retinyl palmitate flexible polymeric nanocapsules: Characterization and permeation studies.Colloids and Surfaces B: Biointerfaces (2010).

[59] Stracke, F, Weiss, B, Lehr, C. M, Konig, K, Schaefer, U. F, & Schneider, M. Multiphoton Microscopy for the Investigation of Dermal Penetration of Nanoparticle-Borne Drugs. Journal of Investigative Dermatology doi:sj.jid.5700374. http://www.nature.com/jid/journal/vaop/ncurrent/abs/5700374a.htmlacessed 01 october (2012).

[60] Akhtar, N, Pathak, K, & Cavamax, W. Composite Ethosomal Gel of Clotrimazole for Improved Topical Delivery: Development and Comparison with Ethosomal Gel. AAPS PharmSciTech (2012). , 13(1), 344-355.

[61] Pople, P. V, & Singh, K. K. Targeting tacrolimus to deeper layers of skin with improved safety for treatment of atopic dermatitis. International Journal of Pharmaceutics (2010).

[62] Alvarez-román, R, Naik, A, Kalia, Y. N, Guy, R. H, & Fessi, H. Skin penetration and distribution of polymeric nanoparticles. Journal of Controlled Release (2004).

[63] De Rosa, F. S, Lopez, R. F. V, Thomazini, J. A, Tedesco, A. C, Lange, N, & Bentley, M. V. L. B. *In vitro* Metabolism of 5-ALA Esters Derivatives in Hairless Mice Skin Homo-

genate and *in vivo* PpIX Accumulation Studies. Pharmaceutical Research (2004). , 21(12), 2247-2252.

[64] Fang, Y, Tsai, Y, Wu, P, & Huang, Y. Comparison of aminolevulinic acid-encapsulated liposome versus ethosome for skin delivery for photodynamic therapy. International Journal of Pharmaceutics (2008). , 5.

[65] Lanke, S. S. S, Kolli, C. S, Strom, J. G, & Banga, A. K. Enhanced transdermal delivery of low molecular weight heparin by barrier perturbation. International Journal of Pharmaceutics (2009).

[66] Lee, W, Shen, S, Al-suwayeh, S. A, Yang, H, Yuan, C, & Fang, J. Laser-assisted topical drug delivery by using a low-fluence fractional laser: Imiquimod and macromolecules. Journal of Controlled Release (2011).

[67] Gerger, A, Wellenhof, R. H, Samonigg, H, & Smolle, J. *In vivo* confocal laser scanning microscopy in the diagnosis of melanocytic skin tumours. British Journal of Dermatology (2009).

[68] Rajadhyaksha, M, Grossman, M, Esterowitz, D, Webb, R. H, & Anderson, R. R. *In vivo* confocal scanning laser microscopy of human skin: melanin provides strong contrast. Journal of Investigative Dermatology (1995).

[69] Nori, S, Rius-díaz, F, Cuevas, J, Goldgeier, M, Jaen, P, Torres, A, & González, S. Sensitivity and specificity of reflectance-mode confocal microscopy for *in vivo* diagnosis of basal cell carcinoma: A multicenter study. Journal of the American Acaddemy of Dermatology (2004).

[70] Guitera, P, Pellacani, G, Crotty, K. A, Scolyer, R. A, Li, L, Bassoli, S, Vinceti, M, Rabinovitz, H, Longo, C, & Menzies, S. W. The Impact of *In vivo* Reflectance Confocal Microscopy on the Diagnostic Accuracy of Lentigo Maligna and Equivocal Pigmented and Nonpigmented Macules of the Face. Journal of Investigative Dermatology (2010).

[71] Koller, S, Gerger, A, Ahlgrimm-siess, V, Weger, W, Smolle, J, & Hofmann-wellenhof, R. *In vivo* reflectance confocal microscopy of erythematosquamous skin diseases. Experimental Dermatology (2009).

[72] Kurzeja, M, Rakowska, A, Rudnicka, L, & Olszewska, M. Criteria for diagnosing pemphigus vulgaris and pemphigus foliaceus by reflectance confocal microscopy. Skin Research and Technology (2012).

[73] Ulrich, M, Krueger-corcoran, D, Roewert-huber, J, Sterry, W, Stockfleth, E, & Astner, S. Reflectance confocal microscopy for noninvasive monitoring of therapy and detection of subclinical actinic keratoses. Dermatology (2010). , 220(1), 15-24.

[74] Langley, R. G. B, Rajadhyaksha, M, Dwyer, P. J, Sober, A. J, Flotte, T. J, & Anderson, R. R. Confocal scanning laser microscopy of benign and malignant melanocytic skin lesions *in vivo*. Journal of the American Academy of Dermatology (2001).

[75] Langley, R. G. B, Burton, E, Walsh, N, Propperova, I, & Murray, S. J. *In vivo* confocal scanning laser microscopy of benign lentigines: comparison to conventional histology and *in vivo* characteristics of lentigo maligna. Journal of the American Academy of Dermatology (2006).

[76] Swindells, K, Burnett, N, Rius-diaz, F, González, E, Mihm, M. C, & González, S. Reflectance confocal microscopy may differentiate acute allergic and irritant contact dermatitis *in vivo*. Journal of the American Academy of Dermatology (2004).

[77] González, S, González, E, White, M, Rajadhyaksha, M, & Anderson, R. R. Allergic contact dermatitis: Correlation of *in vivo* confocal imaging to routine histology. Journal of the American Academy of Dermatology (1999a).

[78] Aghassi, D, Anderson, R. R, & González, S. Confocal laser microscopic imaging of actinic keratoses *in vivo*: A preliminary report.Journal of the American Academy of Dermatology (2000).

[79] Astner, S, González, E, Cheung, A, Díaz, F. R, Doukas, A. G, William, F, & González, S. Non-invasive evaluation of the kinetics of allergic and irritant contact dermatitis. Journal of Investigative Dermatology (2005).

[80] Ferris, L. K, & Harris, R. J. New Diagnostic Aids for Melanoma. Dermatologic Clinics (2012).

[81] Gerger, A, Koller, S, Kern, T, Massone, C, Steiger, K, Richtig, E, Kerl, H, & Smolle, J. Diagnostic applicability of *in vivo* confocal laser scanning microscopy in melanocytic skin tumors. Journal of Investigative Dermatology (2005).

[82] Pellacani, G, Cesinaro, A. M, & Seidenari, S. Reflectance-mode confocal microscopy of pigmented skin lesions-improvement in melanoma diagnostic specificity. Journal of the American Academy of Dermatology (2005a).

[83] Pellacani, G, Cesinaro, A. M, & Seidenari, S. *In vivo* assessment of melanocytic nests in nevi and melanomas by reflectance confocal microscopy. Modern Pathology (2005b).

[84] Pellacani, G, Guitera, P, Longo, C, Avramidis, M, Seidenari, S, & Menzies, S. The impact of *in vivo* reflectance confocal microscopy for the diagnostic accuracy of melanoma and equivocal melanocytic lesions. Journal of Investigative Dermatology (2007).

[85] Ahlgrimm-siess, V, Massone, C, Scope, A, Fink-punches, A, Richtig, E, Wolf, I. H, Koller, S, Gerger, A, Smolle, J, & Hoffmann-wellenhor, R. Reflectance confocal microscopy of facial lentigo maligna and lentigo maligna melanoma: a preliminary study. British Journal of Dermatology (2009).

[86] Busam, K. J, Hester, K, Charles, C, Sachs, D. L, Antonescu, C. R, Gonzalez, S, & Halpern, A. C. Detection of clinically amelanotic malignant melanoma and assessment of its margins by *in vivo* confocal scanning laser microscopy. Archives of Dermatology (2001).

[87] Curiel-lewandrowski, C, Williams, C. M, Swindells, K. J, Tahan, S. R, Astner, S, Frankenthaler, R. A, & González, S. Use of *in vivo* confocal microscopy in malignant melanoma: an aid in diagnosis and assessment of surgical and nonsurgical therapeutic approaches. Archives of Dermatology (2004).

[88] Ulrich, M, Maltusch, A, Röwert-huber, J, González, S, Sterry, W, Stockfleth, E, & Astner, S. Actinic keratoses: non-invasive diagnosis for field cancerisation. British Journal of Dermatology (2007). , 156(3), 13-7.

[89] Ulrich, M, Maltusch, A, Rius-diaz, F, Röwert-huber, J, González, S, Sterry, W, Stockfleth, E, & Astner, S. Clinical applicability of *in vivo* reflectance confocal microscopy for the diagnosis of actinic keratoses. Dermatologic Surgery (2008).

[90] Rishpon, A, Kim, N, Scope, A, Porges, L, Oliviero, M. C, Braun, R. P, Marghoob, A. A, Fox, C. A, & Rabinovitz, H. S. Reflectance Confocal Microscopy Criteria for Squamous Cell Carcinomas and Actinic Keratoses. Archives of Dermatology (2009). , 145(7), 766-772.

[91] González, S, & Tannous, Z. Real-time, *in vivo* confocal reflectance microscopy of basal cell carcinoma. Journal of American Academy of Dermatology (2002).

[92] Sauermann, K, Gambichler, T, Wilmert, M, Rotterdam, S, Stücker, M, Altmeyer, P, & Hoffmann, K. Investigation of basal cell carcinoma by confocal laser scanning microscopy *in vivo*. Skin Research and Technolgy (2002).

[93] Astner, S, Dietterle, S, Otberg, N, Röwert-huber, H. J, Stockfleth, E, & Lademann, J. Clinical applicability of *in vivo* fluorescence confocal microscopy for noninvasive diagnosis and therapeutic monitoring of nonmelanoma skin cancer. Journal of Biomedical Optics (2008).

[94] Ulrich, M, Lange-asschenfeldt, S, & González, S. *In vivo* reflectance confocal microscopy for early diagnosis of nonmelanoma skin cancer. Actas Dermosifiliográficas (2012). http://dx.doi.org/10.1016/j.ad.2011.10.017.

[95] Wu, X, Xu, A, Song, X, Zheng, J, Wang, P, & Shen, H. Clinical, pathologic, and ultrastructural studies of progressive macular hypomelanosis. International Journal of Dermatology (2010).

[96] Chiaverini, C, Kang, H, Sillard, L, Berard, E, Niaudet, P, Guest, G, Cailliez, M, Bahadoran, P, Lacour, J. P, Ballotti, R, & Ortonne, J. P. *In vivo* reflectance confocal microscopy of the skin: A noninvasive means of assessing body cystine accumulation in infantile cystinosis. Journal of the American Academy of Dermatology 10.1016/j.jaad. (2011).

[97] González, S, Rajadhyasksha, M, Rubinstein, G, & Anderson, R. R. Characterization of psoriasis *in vivo* by confocal reflectance microscopy. Journal of Medicine (1999).

[98] Ardigo, M, Cota, C, Berardesca, E, & González, S. Concordance between *in vivo* reflectance confocal microscopy and histology in the evaluation of plaque psoriasis. Journal of the European Academy of Dermatology and Venereology (2009).

[99] Wolberink, E. A. W, Van Erp, P. E. J, & Teussink, M. M. van de Kerkhof P.C.M., Gerritsen M.J.P. Cellular featuresof psoriatic skin: imaging and quantification using *in vivo* reflectance confocal microscopy. Cytometry Part B (2011). B) , 141-149.

[100] González, S, White, W. M, Rajadhyaksha, M, Anderson, R. R, & González, E. Confocal Imaging of Sebaceous Gland Hyperplasia *In vivo* to Assess Efficacy and Mechanism of Pulsed Dye Laser Treatment. Lasers in Surgery and Medicine (1999b).

[101] Aghassi, D, González, E, Anderson, R. R, Rajadhyaksha, M, & González, S. Elucidating the pulsed-dye laser treatment of sebaceous hyperplasia *in vivo* with real-time confocal scanning laser microscopy. Journal of the American Academy of Dermatology (2000b).

[102] Propperova, I, & Langley, R. G. B. Reflectance-Mode Confocal Microscopy for the Diagnosis of Sebaceous Hyperplasia *In vivo*. Archives of Dermatology (2007).

[103] Agero, A. L, Gill, M, Ardigo, M, Myskowski, P, Halpern, A. C, & González, S. *In vivo* reflectance confocal microscopy of mycosis fungoides: a preliminary study. Journal of the American Academy of Dermatology (2007).

[104] Lange-asschenfeldt, S, Babilli, J, Beyer, M, Ríus-diaz, F, González, S, Stockfleth, E, & Ulrich, M. Consistency and distribution of reflectance confocal microscopy features for diagnosis of cutaneous T cell lymphoma. Journal of Biomedical Optics (2012).

[105] Ardigo, M, Zieff, J, Scope, A, Gill, M, Spencer, P, Deng, L, & Marghoob, A. A. Dermoscopic and reflectance confocal microscope findings of trichoepithelioma. Dermatology (2007).

[106] Hongcharu, W, Dwyer, P, González, S, & Anderson, R. R. Confirmation of onychomychosis by confocal microscopy. Journal of the American Academy of Dermatology (2000).

[107] Turan, E, Erdemir, A. T, Gurel, M. S, & Yurt, N. A new diagnostic technique for tinea incognito: *In vivo* reflectance confocal microscopy. Report of five cases. Skin Research and Technology (2012). Jun 7. doi:j.x., 1600-0846.

[108] Hui, D, Xue-cheng, S, & Ai-e, X. Evaluation of reflectance confocal microscopy in dermatophytosis. Mycoses (2012). Sep 10. doi:j.x, 1439-0507.

[109] Kreuter, A, Gambichler, T, Sauermann, K, Jansen, T, Altmeyer, P, & Hoffmann, K. Extragenital lichen sclerosus successfully treated with topical calcipotriol: evaluation by *in vivo* confocal scanning laser microscopy. British Journal of Dermatology (2002).

[110] González, S, Rajadhyaksha, M, González-serva, A, White, W. M, & Anderson, R. R. Confocal reflectance imaging of folliculitis *in vivo*: correlation with routine histology. Journal of Cutaneous Pathology (1999c).

[111] Scheuplein, R. J, & Blank, I. H. Permeability of the skin. Physiological Reviews (1971).

[112] Elias, P. M. Epidermal lipids, barrier function, and desquamation. Journal of Investigative Dermatology (1983). s-49s.

[113] Desai, P, Patlolla, R. R, & Singh, M. Interaction of nanoparticles and cell-penetrating peptides with skin for transdermal drug delivery. Molecular Membrane Biology (2010)., 27(7), 247-259.

[114] Foldvari, M, Badea, I, Wettig, S, Baboolal, D, Kumar, P, Creagh, A. L, & Haynes, C. A. Topical Delivery of Interferon Alpha by Biphasic Vesicles: Evidence for a Novel Nanopathway across the Stratum Corneum. Molecular Pharmaceutics (2010)., 7(3), 751-762.

[115] Teeranachaideekul, V, Boonme, P, Souto, E. B, Müller, R. H, & Junyaprasert, V. B. Influence of oil content on physicochemical properties and skin distribution of Nile red-loaded NLC. Journal of Controlled Release (2008)., 128(2), 134-41.

[116] Madsen, H. B, Ifversen, P, Madsen, F, Brodin, B, Hausser, I, & Nielsen, H. M. *In vitro* Cutaneous Application of ISCOMs on Human Skin Enhances Delivery of Hydrophobic Model Compounds Through the Stratum Corneum. The AAPS Journal (2009)., 11(4), 728-739.

[117] Lopez-pinto, J. M, Gonzalez-rodriguez, M. L, & Rabasco, A. M. Effect of cholesterol and ethanol on dermal delivery from DPPC liposomes. International Journal of Pharmaceutics (2005).

[118] Lademann, J, Richter, H, Teichmann, A, Otberg, N, Blume-peytavi, U, Luengo, J, Wei, B, Schaefer, U. F, Lehr, C. M, Wepf, R, & Sterry, W. Nanoparticles- An efficient carrier for drug delivery into the hair follicles. European Journal of Pharmaceutics and Biopharmaceutics (2007).

[119] Van Kuijk-meuwissen, M, Mougin, L, Junginger, H. E, & Bouwstra, J. A. Application of vesicles to rat skin *in vivo*: a confocal laser scanning microscopy study. Journal of Controlled Release (1998).

[120] Godin, B, & Touitou, E. Mechanism of bacitracin permeation enhancement through the skin and cellular membranes from an ethosomal carrier. Journal of Controlled Release (2004).

[121] Regnier, V, & Préat, V. Localization of a FITC-labeled phosphorothioate oligodeoxynucleotide in the skin after topical delivery by iontophoresis and electroporation. Pharmaceutical Research (1998)., 15(10), 1596-602.

[122] Prausnitz, M. R, Gimm, J. A, Guy, R. H, Langer, R, Weaver, J. C, & Cullander, C. Imaging Regions of Transport across Human Stratum Corneum during High-Voltage

and Low-Voltage Exposures. Journal of Pharmaceutical Sciences (1996). , 85(12), 1363-1370.

[123] Seto, J. E, Polat, B. E, Lopez, R. F. V, Blankschtein, D, & Langer, R. Effects of ultrasound and sodium lauryl sulfate on the transdermal delivery of hydrophilic permeants: Comparative *in vitro* studies with full-thickness and split-thickness pig and human skin. Journal of Controlled Release (2010). , 145(1), 26-32.

[124] Dubey, V, Mishra, D, Asthana, A, & Jain, N. K. Transdermal delivery of a pineal hormone: Melatonin via elastic liposomes. Biomaterials (2006).

Chapter 7

Allergic Contact Dermatitis to Dental Alloys: Evaluation, Diagnosis and Treatment in Japan — Reflectance Confocal Laser Microscopy, an Emerging Method to Evaluate Allergic Contact Dermatitis

Emi Nishijima Sakanashi, Katsuko Kikuchi,
Mitsuaki Matsumura, Miura Hiroyuki and
Kazuhisa Bessho

Additional information is available at the end of the chapter

http://dx.doi.org/10.5772/55720

1. Introduction

Traditionally, different types of metal alloys have been used in restorative dentistry. The common criterion for all these materials is their permanent existence in the oral cavity for a prolonged period of time and this exposure may sensitize patients. The clinical manifestations of contact allergy to dental alloy are not uniform. Diseases such as pustulosis palmaris et plantaris, lichen planus, systemic or palmoplantar eczema, symptoms like glossodynia, cheilitis related with ions released from these metals are well documented [1-6]. Furthermore, the Japanese Ministry of Health and Welfare reported in 1997, allergy affects approximately 30% of the population in Japan and recently, the frequency of dental metal allergy has risen significantly [7-9].

Between metals, Nickel is the most allergic element in Japan, however; in recent years Palladium showed high positivity to path test. This might be to the well-known high toxicity of Nickel and the fluent use of Gold-Silver-Palladium alloys since the 80's that is covered by the national health insurance in Japan for dental restorations [10-17].

Allergic reactions induced by the metals are described according to the classification presented by Coombs and Gell [18]. The sensitizing metals are haptens which are not themselves able to act as antigens. There is evidence that combination of the metals with circulating or tissue protein gives rise to new antigens. Type IV hypersensitivity reaction of the skin takes place following exposure to the metals, and the diagnosis of metal-induced allergic diseases is usually made on the basis of allergological tests with metal antigens including radioallergosorbent test for specific antibody, skin patch test, and blood test such as lymphocyte transformation test.

Patch testing is the primary tool to diagnose allergens causing allergic contact dermatitis. On the other hand, this method is strongly dependent on the experience of the observer, and distinguishing irritant and doubtful positive from positive patch test reaction for different metal reagents remains difficult[19-23]. For that, in clinical diagnostics, as well as in routine dermatology, the need for more accurate non-invasive diagnosis is increased. Reflectance confocal laser microscopy (RCLM) has been used to provide a virtual window into tissues *in vivo* without staining process or destruction of the skin and it is a useful device to observe and measure living skin on time. More recently, diagnostic criteria with RCLM were investigated to characterize features of skin reactions and may be a promising new technology for longitudinal noninvasive studies of contact dermatitis. Moreover, RCLM can reliably visualize cutaneous changes at subclinical degrees of contact dermatitis, which suggests a possible role for RCLM as an adjunctive tool in allergic skin diagnosis.

2. Allergic reaction to metals

Gell and Coombs developed their widely accepted classification of hypersensitivity reactions into four types (Table 1).

Classification	Effector Mechanism	Typical Clinical Manifestations
Type I (Immediate)	IgE	anaphylaxis, angioedema, urticaria
Type II (Cytotoxic)	IgM, IgG, complement, phagocytosis	Cytopenia, nephritis
Type III (Immune complex)	IgM, IgG, complenent, precipitins	Serum sickness, vasculitis
Type IV (Delayed)	T lymphocytes	Contact dermatitis
Other (Idiopatic)	varies	Nonspecific rash

Table 1. Gell and Coombs classification schema of hypersensitivity reactions.

The reactions can be viewed as describing broad strategies that the body uses in order to combat classes of allergen agents.

Metal allergy is classified as Type IV hypersensitivity reaction; that compromise T cells (CD4 and CD8), macrophages, natural killer cells and other leucocytes and destruction of host cells ensues, by a combination of apoptotic death and cytotoxicity.

Allergic reaction to metals is presented as dermatitis by external skin exposure or by intestinal absorption in ingestion of food containing high mount of the allergen. Dental metal alloys frequently induce local symptoms such as oral lichen planus, gingivitis, cheilitis on mucosa in direct contact or systemic symptoms like palmoplantar pustulosis, and eczema (Figure 1).

http://dx.doi.org/10.5772/55720

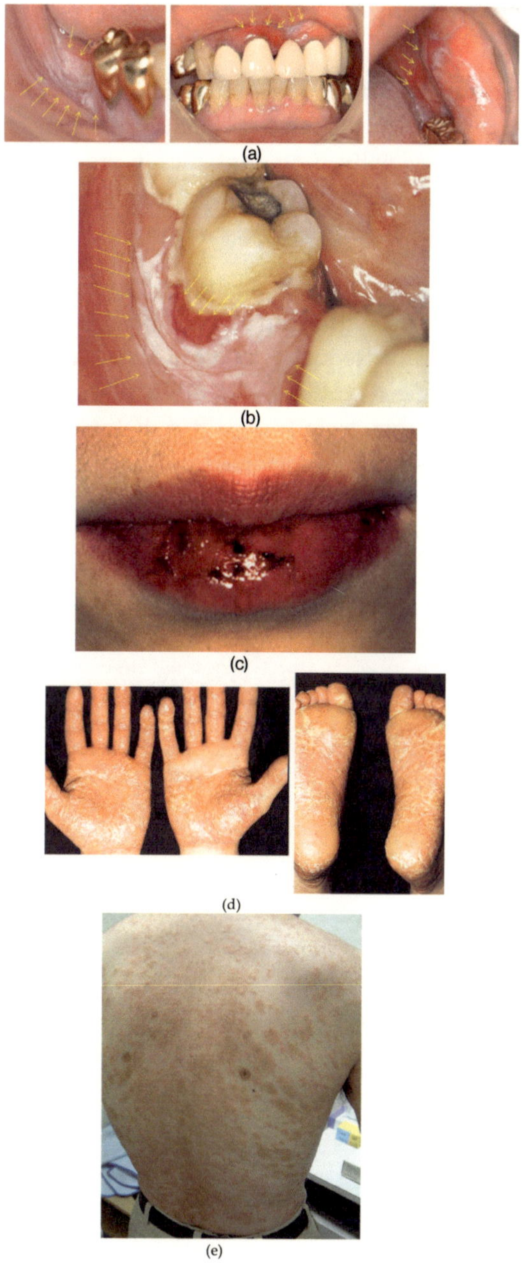

Figure 1. Symptoms of Dental metal allergy. a). Oral lichen planus on the buccal gingiva and mucosa caused by gold alloy restorations: erythematous, erosive lesions are showed between white lacy streaks on the oral mucosa; b). Oral lichen planus cuaused by mercury from amalgam filling: white lacy streaks are observed on the cervical mucosa of the lower left second molar; c). Cheilitis caused by nickel from orthodontic device: lip inflammation; d). Pustulosis palmaris et plantaris caused by palladium from Gold-Silver-Palladium alloy restoration: chronic recurrent pustular dermatosis with a background of erythema, scaling and fissuring of the skin; e). Eczema caused by mercury from amalgam filling: dryness and recurring skin rashes are observed.

3. Dental metal allergy in japan

Dental metal allergy produced by mercury in dental amalgam was first described by Fleischmann in 1928, symptoms of which included stomatitis and anal eczema. In Japan, Nakai reported in 1960 gingivitis related with chrome and nickel, and Nakayama in 1972, oral lichen planus by mercury from detal amalgam filling[1, 2]. Although, various symptoms associated with different metals have been reported in many countries.

The increase of high percentage of allergy in Japan, lead to the Japanese Ministry of Health and Welfare jointly with the Tokyo Medical and Dental University (TMDU) with cooperation of other 12 universities, analyze the frequency of metal allergy related with dental alloys, the demographic and epidemiologic distribution of the dental allergic population, and, the mechanism of allergic reaction from years 1989 to 1991. Between the results, over the 20% of the patients with skin or mucosa diseases showed positive reaction by patch test to dental metal present in the oral cavity [24]. Since them, the number of patients who visited Dental Allergy Clinic of TMDU has risen significantly (Figure 2). Figures 3 and 4 showed the positive rates to metals by patch test between gender, from 1998-2002(n: 881) and 2003-2007(n: 1112). This study indicates that the metal to which most patients reacted was Nickel(1998 to 2002:24,7%; 2003 to 2007:36,8%), Cobalt(1998 to 2002:17,6%; 2003 to 2007:17,7%), Mercury(1998 to 2002:14,4%; 2003 to 2007:19,4%), Chrome (1998 to 2002:12,5%; 2003 to 2007:8,1%)and Palladium(1998 to 2002:9,6%;2003 to 2007:15,9%).

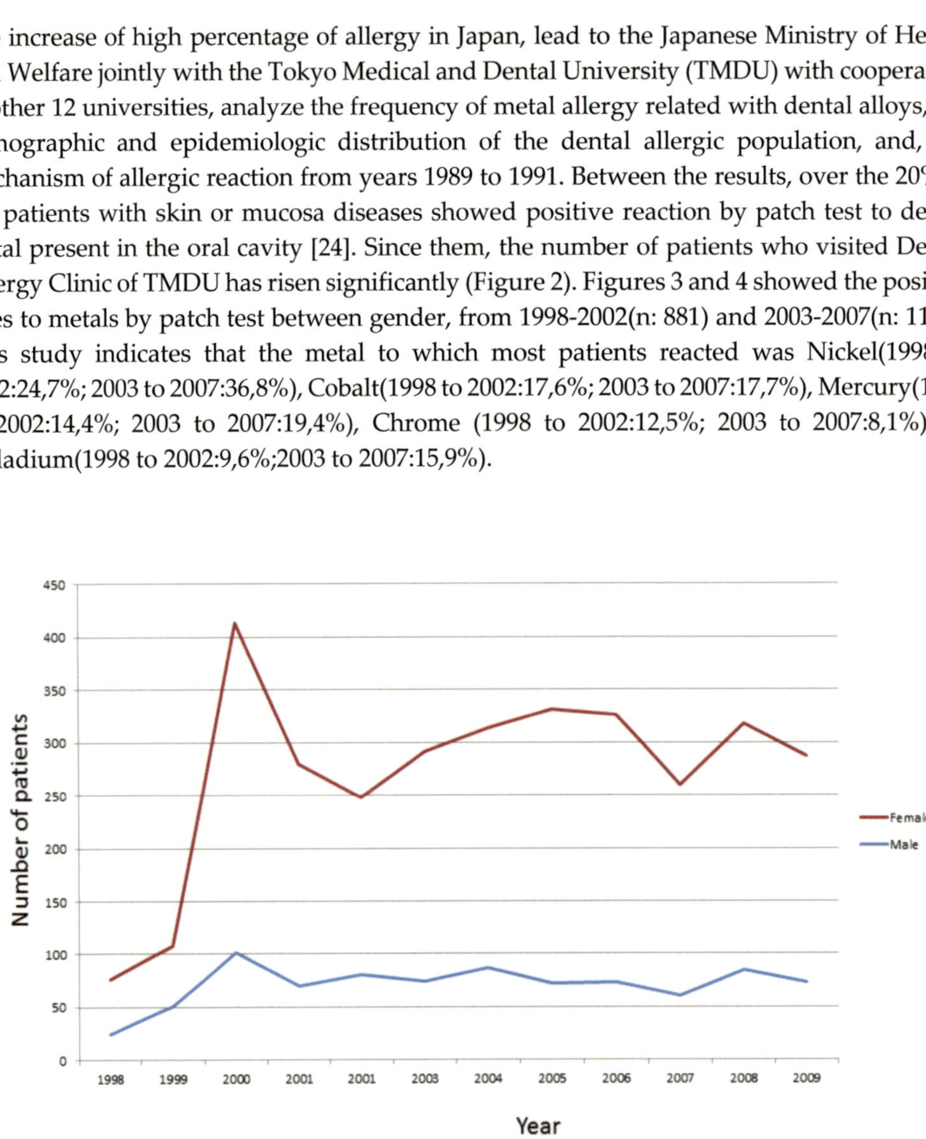

Figure 2. The number of outpatients who visited Dental Allergy Clinic of Tokyo Medical and Dental University's hospital from years 1998 to 2009.

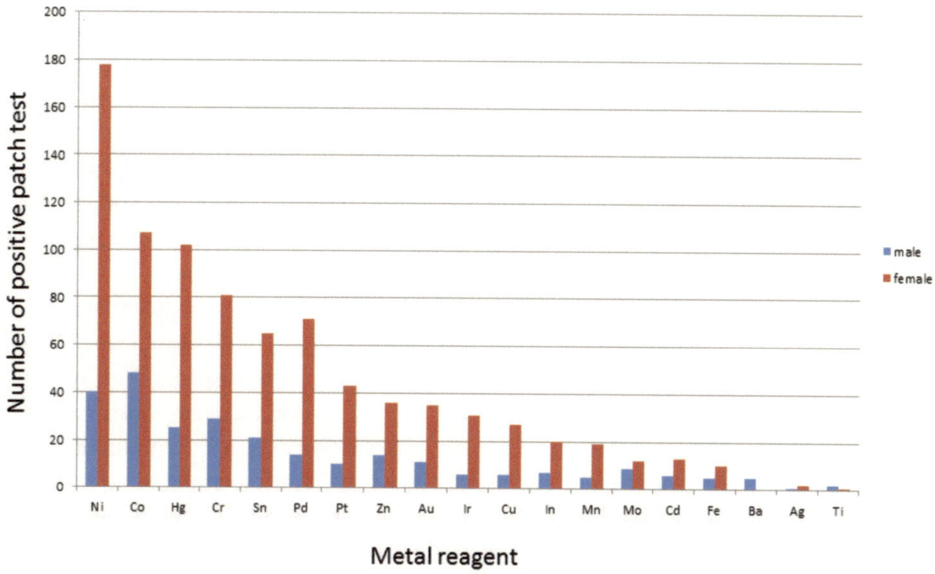

Figure 3. The positive rates to metals by patch test between genders, from years 1998 to 2002. Total of 881 outpatients; 697 females, 184 males.

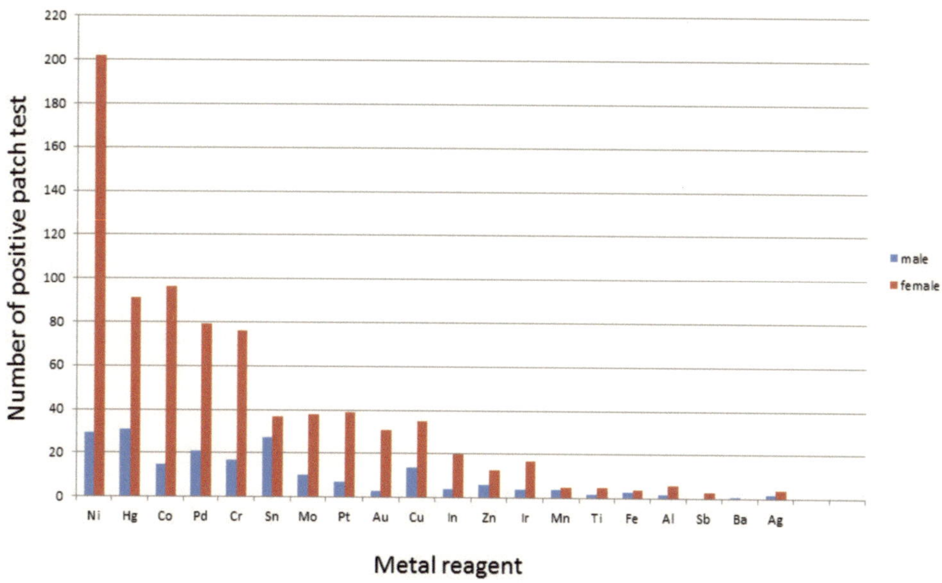

Figure 4. The positive rates to metals by patch test between genders, from years 2003 to 2007. Total of 1112 outpatients; 893 females, 219 males

An important finding was the high positive rate to Palladium than that from 1998 to 2002. This increase is speculated because of the opportunity for Palladium sensitization. Although Gold-Silver-Palladium alloy is the primary choice for metallic restorations and fixed prostheses for the Japanese health insurance system, so many citizens could be exposed to this metal. Past studies showed that Palladium is unstable in the oral cavity, releasing metal content into the saliva that could cause a serious allergic reaction [17, 25, 26]. The cross-reaction between nickel and palladium is also reported [27]. As well as dentistry, Palladium is increasingly used in industry and in the manufacture of fine jewelry, so sensitivities to this metal must be carefully considered. On the other hand, the positive rate to Mercury showed lower positivity in recent years. This is speculated for the decreasing use of dental filling amalgams, thermometers and mercurochrome. Between genders, there were significantly more women reacting to dental metals than men in all ages. This might be caused by the difference of life style, in particular wearing jewelries and accessories such as ear piercing since young ages. However, another possibility should be considered; the patch test reagent for metals in its different approaches and materials could lead to variations in the results and the selection criteria of the tests should be considered and to need to structured to obtain more accurate result of allergic contact dermatitis and differentiate from irritant skin reactions.

4. Diagnosis of metal hypersensitivity

A diagnosis of hypersensitivity to metal is usually done by epicutaneous patch testing; because of it is delayed type of reaction. The patch testers with the respectively metal reagents are applied on the skin and removed 2 days later on day 2 (48hs). Skin readings are performed on day 2, 3 (72hs), and, 7 (1 week) according to the clinical scoring criteria recommended by the International Contact Dermatitis Research Group (Figure 5).

The instrument used for clinical assessment is a combination of vision and feel in the form of palpation with the clinician's finger, and this measurement is a totally subjective method based on the examiner's knowledge and experience; and interestingly, past reports of patch test readings have shown the disagreement on scoring among examiners under the same conditions [28].

5. Reflectance confocal laser microscopy

The Reflectance Confocal Laser Microscopy based on an optical fiber system, allows the non-invasive *in vivo* determination and analysis of different levels in the skin with a high, quasi-histological resolution and in real time up to a depth of 300μm, namely upper dermis in normal human skin except palms and soles; and it's represents a useful measurement for determining an individual's skin hypersensitivity [29-31].

-	negative
?+ 	doubtful
+ 	faint macular, erythema only weak(non-vesicular)positive
++ 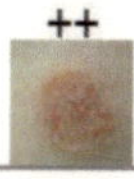	Strong(vesicular) positive erythema, infiltration, possibly papules
+++ 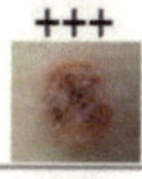	extreme positive bullous reaction erythema, infiltration, papules, vesicles
R+ 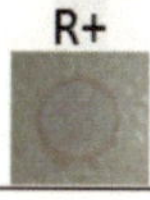	Irritant Reaction of different types Ring reaction

Figure 5. Severity of skin readings for patch test according to the clinical scoring criteria recommended by the International Contact Dermatitis Research Group.

5.1. The human skin

Under physiologic conditions, human skin maintains a number of structural, sensory, mechanical and metabolic functions. The uppermost layer is the stratum corneum which is formed by flat dead cells-the corneocytes and intercellular lipids which are located between those corneocytes. Among others properties, the stratum corneum represents an important barrier to the environment and protects the body from water loss and the penetration of harmnful elements and microorganisms. The average of thickness of stratum corneum except palms and soles is between 15 to 30 μm, however there are significant topographical variations depending on body site and exogenous factors such as mechanical stressors. The stratum granulosum and spinosum lie directly below the stratum corneum and consist of living cells with central nucleus, which differ significantly from the corneocytes in structure size and morphology.

The number of inflammatory, irritative or allergic processes is associated with significant disruption of skin barrier function and skin conditions with established associations to a dysfunctional skin barrier include wound healing, and contact dermatitis. The routine histological sections obtained from skin biopsies but the preparation process leads to significant tissue shrinkage, delipidation and artifacts due to tissue processing, fixation and staining. An exact determination of the actual thickness of the skin has therefore not been possible. Additionally, the process of obtaining tissue biopsies remains highly invasive, i.e., it is painful and leaves a scar.

5.2. In vivo reflectance confocal laser microscopy

Non-invasive imaging modalities have received increased attention in recent years. In contrast to vertical histological sections obtained from biopsy specimen, RCLM evaluates skin in horizontal sections at a near-cellular resolution comparable to routine histology without preparing tissue processing and staining. In a RCLM, near-infrared light from a diode laser is focused on cellular structures having different refraction indexes, and this reflected light is captured and recomposed into a two-dimensional gray scale image by computer software (Figure 6).

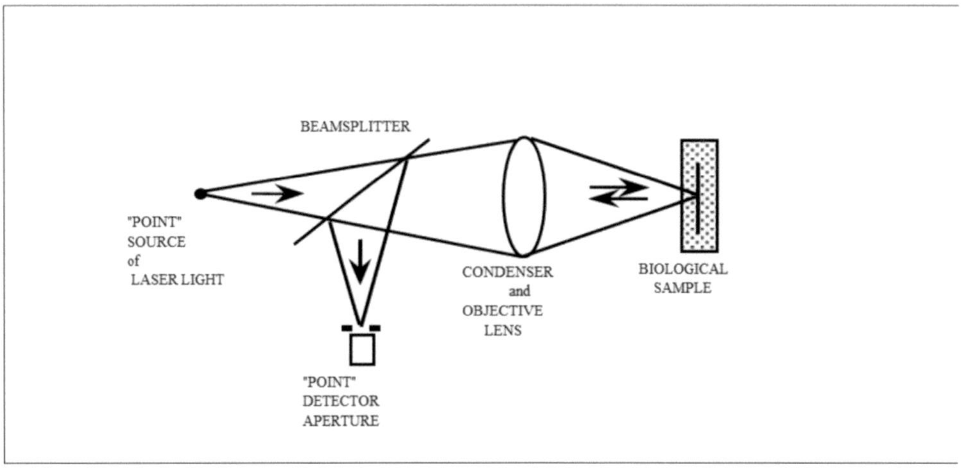

Figure 6. Reflectance Confocal Laser Microscopy

In dermatology, *in vivo* RCLM has been utilized in investigating benign and malignant tumors of melanocytes and keratinocytes, inflammatory skin lesions including allergic contact dermatitis (ACD) or irritant contact dermatitis (ICD), pigmentary disorders, vascular lesions and other skin conditions including normal skin [32- 34].

Moreover, RCLM allows a descriptive and qualitative cellular and morphologic analysis of skin barrier function, by visualizing individual cornecytes, cell-to-cell cohesiveness at the level of the stratum corneum and features of disruption.

The efficiency of the RCLM in grading the severity of allergic skin reactions and the correlation with routine visual patch testing results was also evaluated. A commercially available RCLM (Vivascope 1500 Plus, Lucid Inc, Henrietta, NY) was used to produce horizontal (surface) images of skin sites with X,Y, and vertical (in depth) images with Z plane micrometer screws. This device use a diode laser at 830nm with a power less than 16mW at tissue level. The X30 water-immersion lens of numerical aperture 0.9 was applied to the skin. It was immobilized with a tissue ring and template fixture device to provide standardized mechanical contact with the RCLM. In each of the skin sites analyzed, a systematic $4mm^2$ X-Y mapping was performed and 5 images were captured in Z plane per 1μm in depth, beginning at the stratum corneum and going through the epidermis and into the upper reticular dermis. The suprabasal epidermis (from the surface of the stratum corneum with presence of corneocytes to the bottom of the cells in the uppermost portion of the stratum basale) were used to calculate thickness before and after patch testers were removed on day 2, 3, and, 7. In positive patch test reaction group, an overall increase in suprabasal epidermal thickness, intercellular edema, acanthosis and vesicle formation was observed, especially in metal elements with strong proliferative response of keratinocytes and T-cell involvement. On the other hand, in doubtful positive patch test reaction group, the RCLM images showed different aspects, such as irritant reaction with superficial disruption in the stratum corneum, or, an increase of suprabasal epidermis thickness at level which could recognized as positive reaction (Figure 7a to 7c, 8a to 8e).

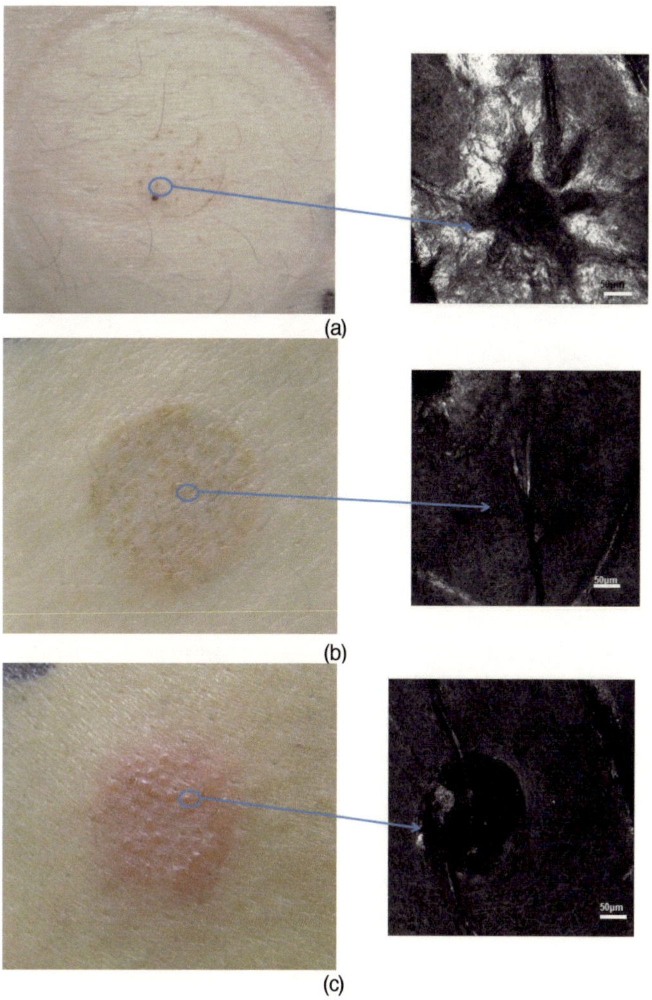

Figure 7. Clinical images and corresponding RCLM images of patch test on day 3. a). Irritant reaction with punctate erythema and slightly hemorrhagic around hair follicle openings; b). Doubtful positive reaction shows spongiogis I stratum supinosum by RCLM; c). Positive reaction, with vesicle formation in stratum supinosum.

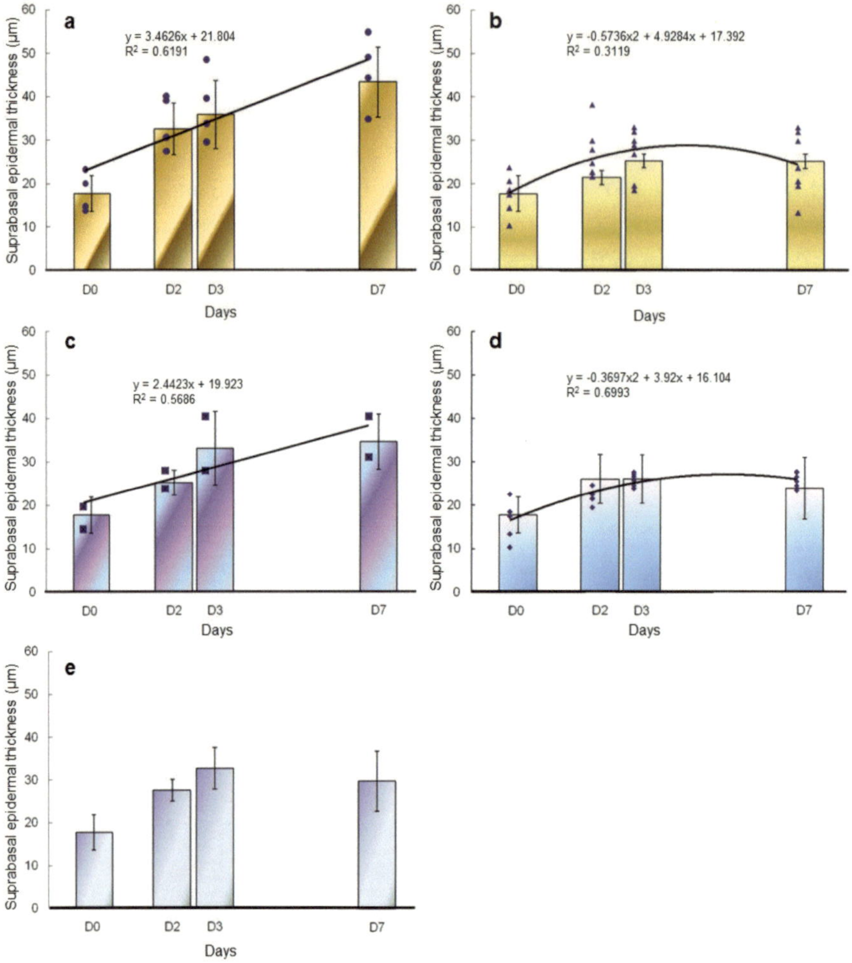

Figure 8. Regression analysis of suprabasal epidermal thickness and elapsed period of time by days on patch test. a). Nickel positive reaction; b). Nickel negative reaction; c). Cobalt positive reaction; d). Cobalt negative reaction; e). Cobalt doubtful positive reaction.

This study demonstrated the potential of the RCLM to assist in more accurate interpretation of patch test result between allergic, doubtful and irritant reaction, differences and the degree of the skin reaction between allergen, rather than by using visual assessment alone [35].

5.3. Future possibilities of RCLM

In vivo RCLM offers quasi-histological resolution tissue images in real-time and does no damage to the tissue. Thus, it is suitable for monitoring longitudinal follow-up study for overtime. Furthermore, *in vivo* RCLM has obvious advantages as compared conventional skin biopsy in investigating esthetic dermatology because it is acceptable to patients who does not want scar.

Skin biopsy remains the golden rule for microscopic diagnostic up to now in dermatology. Although stacked horizontal gray scale images give us different information form that vertical colored sections of traditional microscopy, reading gray scale horizontal images is challenging for dermatologists or pathologists that are not experts in RCLM. Therefore, a prolonged and relevant training is needed before the dermatologist becomes confident with the obtained RCLM images. With the RCLM, each horizontal scan line is generated by each facet of the quickly rotating multi-faceted polygonal mirror, and the vertical shift occurs during the oscillations of the galvanometer. Scanning is performed at a rate of 9 frames per second, and the resulting image is equivalent to an en face 0,5mm x 0,5mm horizontal microscopic section. Moreover, this device can capture a series of focal planes at different depths by changing the focal length of the beam, and it has the capability of imaging 300μm below the skin surface. A series of images is typically captured from the top of the stratum corneum, through the epidermis and down into the dermis, forming a vertical image stack (Figure 9a, 9b, 9c, 9d, 9e, 9f).

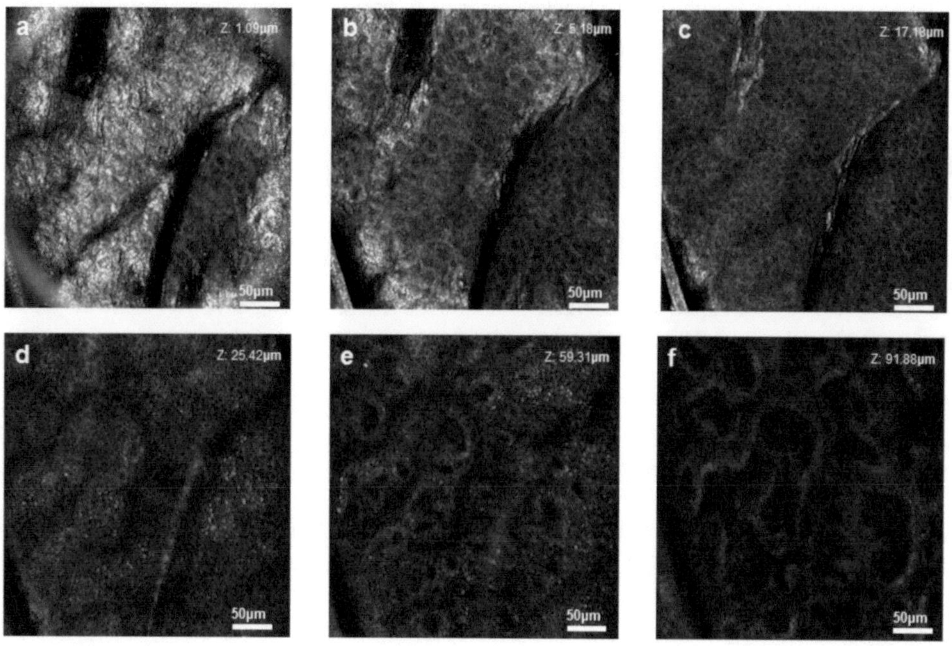

Figure 9. Horizontal section in XY planes of the RCLM images. a). stratum corneum; b). stratum granulosum; c). stratum supinosum above papillary dermis; d). stratum supinosum between papillary dermis; e). stratum basale; f). interface between epidermis and dermis.

From individual images, cellular size can be measured, and, cell organelles and microstructures, such as melanin, keratin, and collagen provided contrast in images by differences in refractive index, appearing brighter than other cell structures. Other cellular details such as inflammatory infiltrates, dendritic cells, and capillaries can also be imaged by RCLM. Total light is reflected back when structures appear with, while no reflection is represented by black becoming an important guide for the clinician.

Although there are limitations as described below, it might become a real-time diagnostic or adjunctive tool to determine the suspicious lesion or to delineate tumor margins [36].

The dominanting antigen-presenting cells in the epidermis are the Langerhans' cells, which constitutively express the MHC class II molecules and the invariant chain. To differential epidermal expression of the invariant chain in different contact dermatitis

5.4. Limitations of RCLM

One of the main limitations of this technique is the fundamental inability to images deep objects in the dermis in normal skin. In the skin of palms and soles, even living layers of epidermis is hardly observed due to its thick stratum corneum. In addition, acanthosis due to intercellular edema, vesicle formation and keratinocyte proliferation can also restrict the visualization. Moreover, not late-stage, grown-up tumor but early-stage tumor is suitable for examining due to its depth limitations.

Furthermore, higher and better contrast is desired in order to distinguish different cells and determine pathological characteristics.

Author details

Emi Nishijima Sakanashi[1], Katsuko Kikuchi[2], Mitsuaki Matsumura[3], Miura Hiroyuki[3] and Kazuhisa Bessho[1]

1 Oral and Maxillofacial Surgery Department, School of Medicine, Kyoto University, Kyoto, Japan

2 Department of Dermatology, Tohoku University Graduate School of Medicine, Sendai, Japan

3 Fixed Prosthodontics, Department of Restorative Sciences, Division of Oral Health Sciences, Graduate School, Tokyo Medical and Dental University, Tokyo, Japan

References

[1] Fleischmann, P. Zur Frage der Gefärlichkeit Kleinster Quecksilbermenger. DtMedWschr (1928). , 54, 304-7.

[2] Nakayama, H. clinical cases of oral lichen planus probably related with detal metal allergy. Otolaryngology (1972). , 44, 239-47.

[3] Schmalz, G, & Garhammer, O. Biological interactions of dental cast alloys with oral tissues. Dent Mater (2002). , 18, 396-406.

http://dx.doi.org/10.5772/55720

[4] Kosugi, M, Ishihara, K, & Okuda, K. Implication of responses to bacterial heat shock proteins, chronic microbial infections, and dental metal allergy in patients with pustulosis palmaris et plantaris. Bull Tokyo Dent Coll (2003)., 44, 149-58.

[5] Nakayama, H. New aspects of metal allergy. Acta Dermatovenerol Croat (2002)., 10, 207-19.

[6] Lomaga, M. A, Polak, S, Grushka, M, & Walsh, S. Results of patch testing in patients diagnosed with oral lichen planus. J Cutan Med Surg (2009)., 13, 88-95.

[7] Evaluation of Year 2004 research operation Available at: http://wwwmhlw.go.jp/shingi/2004/10/sc06.hyml#11 Accessed September 3, (2012)., 100-6.

[8] Fisher, A. A. The role of the patch testing. In: Contact Dermatitis. Philadelphia: Lea and Febiger, (1986)., 1986, 9-29.

[9] SuzukiMetal Allergy in Dentistry: Detection of allergen metals with X-ray fluorescence spectroscope and its application toward allergen elimination. Int J Prosthodont (1995)., 8, 351-69.

[10] Sinigaglia, F, Scheidegger, D, Garotta, G, Scherper, R, & Lanzavecchia, S. Isolation and characterization of Ni-specific clones from patients with Ni-contact dermatitis. J Immunol (1985)., 135, 3929-32.

[11] Kapsenberg, M. L, Res, P, Bos, J. D, Schootemijer, A, Teunissen, M. B, & Van Schooten, W. Nickel-specific T lymphocytes clones from patients with nickel-contact dermatitis lesion in man: heterogeneity based on requirement of dendritic antigen-presenting cell subsets. Eur J Immuol (1987)., 7, 861-5.

[12] Saglam, A. M, Baysal, V, & Ceylan, A. M. Nickel and cobalt hypersensitivity reaction before and after orthodontic therapy in children. J Contemp Dent Pract (2004)., 15, 79-90.

[13] Mei-eng, T, & Yu-hung, W. Multiple allergies to metal alloys. Dermatol Sinica (2001)., 29, 41-43.

[14] Adachi, A. Metal contact allergy ad systemic metal allergy. J Environ Dermatl Cutan Allergol (2009)., 3, 413-22.

[15] Garwkrodger, D. J, Lewis, F. M, & Shah, M. Contact sensitivity tonickel and other metal in jewelry reactors. J Am Acad Dermatol (2000)., 43, 31-6.

[16] Kanerva, L, Rantanen, T, Aalto-korte, K, et al. A multicenter study to patch test reaction with dental sdreening series. Am J Contact Dermatitis. (2001)., 12, 83-7.

[17] Durosao, O, & Azhary, R. A. A. year restrospective study on palladium sensitivity. Dermatitis (2009)., 20, 208-13.

[18] Gell, P. G. H, & Coombs, R. R. A. (1963). The classification of allergicreactions underlying disease. In Clinical Aspects of Immunology (Coombs, R.R.A. and Gell, P.G.H., eds) Blackwell Science

[19] Nosbaum, A, Vocanson, M, Rozieres, A, Hennino, A, & Nicolas, J. F. Allergic and irritant contact dermatitis. Pathophysiology and inmunological diagnosis. Eur J Dermatol (2009). , 19, 325-32.

[20] Nethercott, J. R. Practical problems in the use of patch testing in the evaluation of patients with contact dermatitis. In: Weston WL, Mackie RM, Provost TT, eds. Current Problems in Dermatology. St. Louis, MO:Mosby, (1990). , 1990, 101-3.

[21] Nethercott, J. R. Sensitivity and specificity of patch tests. Am J Contact Dermatitis (1994). , 5, 136-42.

[22] Brasch, J, Henseler, T, Aberer, W, Bäuerle, G, Frosch, P. J, Fuchs, T, Fünfstück, V, Kaiser, G, Lischka, G. G, Pilz, B, et al. Reproducibility of patch tests. A multicenter study of synchronous left-versus right-sided patch tests by the German Contact Dermatitis Research Group. J Am Acad Dermatol. (1994). , 31(4), 584-91.

[23] Sarma, N. Late reaction, presistent releaction and doubtful allergic reaction: The problems of interpretation. Indian J Dermatol (2009). , 54, 56-8.

[24] Inoue, M. The Status Quo of Metal Allergy and Measures Against it in Dentistry, J.Jpn. Prosthodont.Soc (1993). , 37, 1127-1138.

[25] Filon, F. L, Uderzo, D, & Bagnato, E. Sensitization to pallafium: a 10-year evaluation. Am J Contact Dermat (2003). , 14, 78-81.

[26] Vicenzi, C, Tosti, A, Guerra, L, et al. Contact dermatitis to palladium; a study of 2,300 patients. Am J Contact Dermat (1995). , 6, 110-2.

[27] Hindsén, M, Spirén, A, & Bruze, M. Cross-reactivity between nickel and palladium demonstrated by systemic administration of nickel. Contact Dermat (2005). , 53(1), 2-8.

[28] Rietschel, R. L, Fowler, J. F, & Fisher, A. A. Fisher`s contact dermatitis. Lippincott Williams and Wilkins, (2001).

[29] Rajadhyaksha, M, Grossman, M, Esterowitz, D, Webb, R. H, & Anderson, R. R. In vivo confocal scanning laser microscopy of human skin: melanin provides strong contrast. J Invest Dermatol (1995). , 104, 946-52.

[30] Rajadhyaksha, M, Gonzalez, S, Zavislan, J. M, Anderson, R. R, & Webb, R. H. In vivo confocal scanning laser microscopy of human skin II: advances in instrumentation and comparison with histology. J Invest Dermatol (1999). , 113, 293-303.

[31] Huzaira, M, Rius, F, Rajadhyaksha, M, Anderson, R. R, & Gonzalez, S. Topographic variations in normal skin, as viewed by in vivo reflectance confocal microscopy. J Invest Dermatol (2001). , 116, 846-52.

http://dx.doi.org/10.5772/55720

[32] Gerger, A, Koller, S, Kern, T, Massone, C, Steiger, K, Richtig, E, Kerl, H, & Smolle, J. Diagnostic applicability of in vivo confocal laser scanning microscopy in melanocytic skin tumors.J Invest Dermatol. (2005). , 124(3), 493-8.

[33] Astner, S, González, S, & Gonzalez, E. Noninvasive evaluation of allergic and irritant contact dermatitis by in vivo reflectance confocal microscopy.Dermatitis. (2006). , 17(4), 182-91.

[34] Archid, R, Patzelt, A, Lange-asschenfeldt, B, Ahmad, S. S, Ulrich, M, Stockfleth, E, Philipp, S, Sterry, W, & Lademann, J. Confocal laser-scanning microscopy of capillaries in normal and psoriatic skin. J Biomed Opt. (2012).

[35] Nishijima Sakanashi E Matsumura M, Kikuchi K, Ikeda M, Miura H. A comparative study of allergic contact dermatitis by patch test versus reflectance confocal laser microscopy, with nickel and cobalt. Eur J Dermatol (2010). , 20(6), 705-11.

[36] Piergiacomo Calzavara-Pinton et alReflectance Confocal Microscopy for In Vivo Skin Imaging. Photochemistry and Photobiology, (2008).

Chapter 8

Confocal Endomicroscopy

Anjali Basil and Wahid Wassef

Additional information is available at the end of the chapter

http://dx.doi.org/10.5772/55959

1. Introduction

Confocal endomicroscopy is a recently developed endoscopic technology that allows for histological analysis of tissue in vivo. Conventional endoscopy involves identifying lesions grossly followed by biopsy for histological analysis. Confocal endomicroscopy allows for the performance of real time biopsy during endoscopy by observation of the mucosal layer of the gastrointestinal tract at the cellular level. Images are displayed in real-time during the examinations.

Confocal laser endomicroscopy was developed in 2004 (Pentax Corporation, Montvale, New Jersey) and was cleared by the Food and Drug Administration in 2004 for use. Two systems are currently available and have been approved by the FDA, tip-integrated confocal laser endoscope and a flexible fiber-based confocal miniprobe [1].

The technique of confocal endomicroscopy has been used for diagnosis of upper gastrointestinal disorders such as Barrett's esophagus, gastric carcinoma, Celiac disease. It also has application in diagnosis of lower gastrointestinal and biliary tract disorders such colon polyps, Ulcerative colitis and pancreaticobiliary strictures.

2. Technical overview

2.1. Basic principle of confocal microscopy

CLE is based on low-power blue laser light based tissue exposure and fluorescence. The laser light is focused on an area of interest and back scattered light is then refocused onto the detection system by the lens. The back scattered light passes through a pinhole aperture which increases the resolution of the image. A dye is used to provide the contrast for adequate visualization. Serial optical sections are obtained oriented parallel to the tissue surface. The confocal setup is shown in Figure 1 [2].

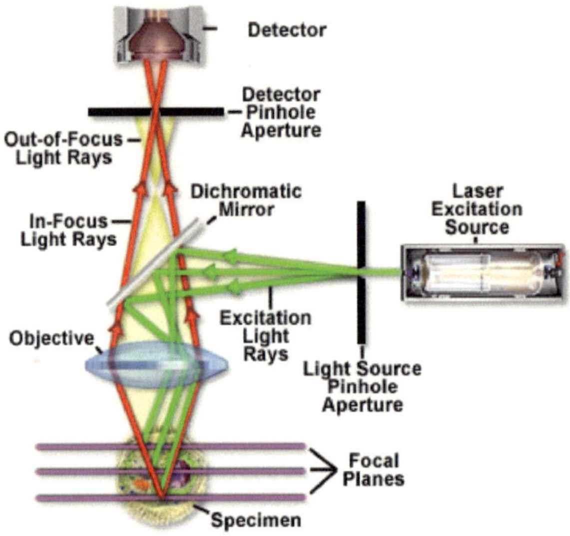

Figure 1. [2] Confocal microscope setup

2.2. Types of endoscopes

There are two clinically available CLE systems, the eCLE and the pCLE. The eCLE system is integrated into the Pentax endoscope with the confocal imaging window at the distal tip of the endoscope as shown in Figure 2 [3]. The pCLE system-the Cellvizio system (pCLE; Mauna Kea Technologies, Paris, France) is probe based and is inserted through the accessory channel of traditional endoscope. These fiber-bundle probes provide faster image acquisition. The pCLE system has a plane of visualization of up to 250 microns while for the probe based system it is 70 to 130 microns.

2.3. Staining techniques

Florescent dyes used for staining in confocal endomicroscopy include flourescein, acriflavine hydrochloride and cresyl violet. Flourescein is the most widely used dye. Acriflavine is not FDA approved as it is potentially mutagenic due to binding to nucleic acids. Acriflavine also has limited penetration and gives uneven staining. Cresyl violet has limitations due to the fact that it does not demonstrate well the vascular structure of the tissue being studied. Flourescein is used intravenously and is nontoxic. It quickly distributes throughout the body allowing for endoscopy to be performed immediately after injection. It also provides good contrast for visualization of the vasculature and subcellular structures. Flourescein's nontoxic nature, rapid visualization post injection and its ability to provide good contrast for visualization of vasculature make it the most widely used dye.

2.4. The confocal endoscopy system

There are two FDA approved confocal endomicroscopy systems. The eCLE system uses a fiber optic cable integrated at the tip of the endoscope to convey laser light. The confocal microscope in the eCLE system is integrated in the tip of the endoscope. The probe-bases CLE (pCLE) uses a fiber optic probe bundle inserted into the port of a standard endoscope to convey light from a confocal microscope situated outside the patient. Laser light collected from the tissue in focus by the confocal microscope is transferred to the photodetector. The intensity of the photosignal is measured to create a two-dimensional image. The image depth can be adjusted by moving the microscope optics.

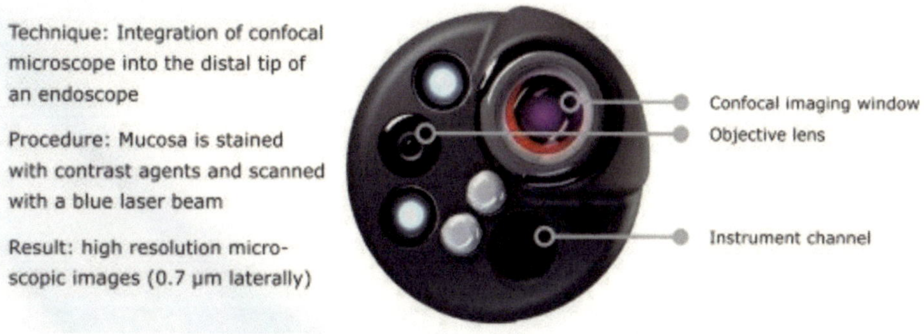

Figure 2. [3].Confocal Endoscope

2.5. Practical aspects of confocal endomicroscopy

Patient preparation for confocal endomicroscopy is the same as the preparation for conventional endoscopy. The procedure is performed under moderate sedation or general anesthesia depending of patient co-morbidities and the type of endoscopic procedure i.e. colonoscopy, upper endoscopy, endoscopic retrograde cholangiopancreatography or endoscopic ultrasound. Procedure times are similar to as those for the conventional endoscopic procedures. In addition to the contraindications for conventional endoscopies, confocal endoscopy also has the additional contraindication of allergy to contrast agents.

3. Application of confocal endomicroscopy for various clinical modalities of the gastrointestinal tracT

3.1. Barrett's esophagus

Barrett's esophagus is a premalignant condition in patients with long standing gastroesophageal reflux disease. Current guidelines for screening for Barrett's esophagus in patients with chronic gastroesophageal reflux disease recommend that random four-quadrant biopsies be performed at 1cm or 2cm intervals based of suspicion of high grade intraepithelial neoplasia. The random biopsy method has a high likelihood of sampling error as only a fraction of the epithelium is biopsied.

Confocal endomicroscopy allows for real time biopsies of target tissues identified by fluorescein staining. Kiesslich et al. [4] studied confocal imaging as a method for diagnosis of Barrett's esophagus. They found it to be a reliable modality to identify different epithelial cell types at the squamocolumnar junction. In patients with neoplasia, the CLE image showed large black cells with loss of architecture and variable size of lumina. As per the Miami Classification – described by a consensus of pCLE users to standardize image criteria, the normal squamous epithelium appears as flat cells without crypts or villi and with bright vessels. Villiform structures and dark, irregularly thickened epithelial borders and dilated irregular vessels were seen in Barrett's esophagus with high grade intraepithelial neoplasia. Figure 3 [5] demonstrates a comparison of histopathological images and confocal images.

pCLE also has a role in treatment management of Barrett's esophagus. It aids in the appropriate localization of lesions and hence prediction of pathology with targeted biopsies [6]. A possible therapeutic role for pCLE has also been shown by providing real time information to guide therapy during endoscopy.

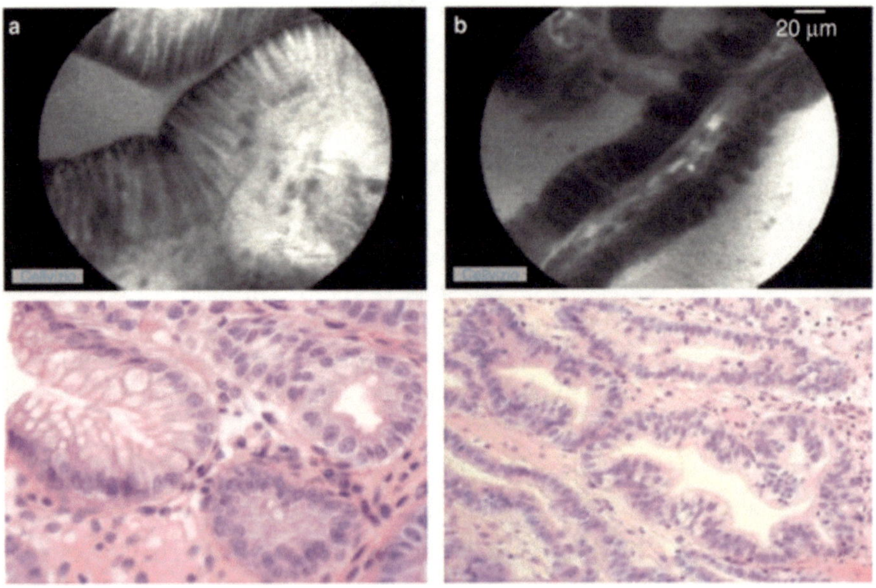

Figure 3. Comparison between pCLE images and histopathological images. (a) Non-dysplastic Barrett's mucosa seen on probe-based confocal laser endomicroscopy (pCLE): representative images are from different samples. Features include regular epithelial surface, easily identifiable goblet cells, equidistant glands, glands that are equal in size and shape, normal size cells without enlargement, and regular and equidistant cells. (b) Dysplastic Barrett's mucosa as seen on pCLE: representative images are from different samples. Features of dysplasia include saw-toothed epithelial surface, not easily identifiable goblet cells, non-equidistant glands, unequal size and shape of glands, enlarged cells, and irregular and non-equidistant cells.

3.2. Gastric cancer

In the stomach, confocal endomicroscopy can detect gastric cancer associated co-morbidities including Helicobacter pylori, intestinal metaplasia and gastritis. It can differentiate between these co-existing conditions and gastric cancer so that target biopsies can be performed. Confocal endomicroscopy can define villous and crypt architecture. Zhang et al. [7] devised a classification system based on gastric pit patterns as shown in Figure 4 [7]. They evaluated 137 patients with CLE based on gastric pit patterns. A sensitivity of 90% and specificity of 99% was found for detecting gastric cancer.

The classification of gastric pit patterns by confocal endomicroscopy

Category	The appearance of pit patterns by confocal endomicroscopy	Distribution area	Diagram
Type A	Round pits with round opening	Normal mucosa with fundic gland	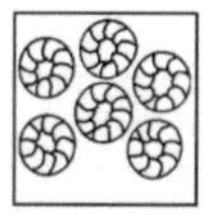
Type B	Non-continuous short rod-like pits with short thread-like opening	Corporal mucosa with chronic inflammation	
Type C	Continuous short rod-like pits with slit-like opening	Normal mucosa with pyloric gland	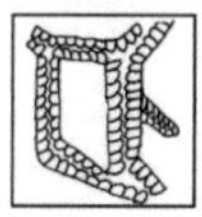
Type D	Elongated and tortuous branch-like pits	Antral mucosa with chronic inflammation	
Type E	The number of pits decreasing and pits prominently dilating	Chronic atrophy gastritis	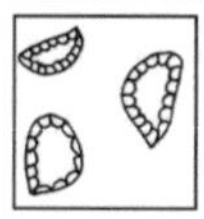
Type F	Villus-like appearance, interstitium in the centre and goblet cells appearing	Intestinal metaplastic mucosa	
Type G			
Type G_1	Normal pits disappearing, with the appearance of diffusely atypical cells	Signet-ring cell carcinoma and poorly differentiated tubular adenocarcinoma	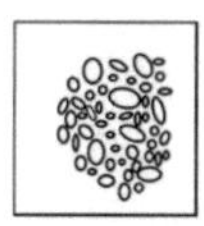
Type G_2	Normal pits disappearing, with the appearance of atypical glands	Differentiated tubular adenocarcinoma	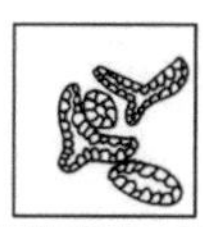

Figure 4. [7] Gastric pit patterns-G1 and G2 correspond to pathological patterns for gastric cancer.

3.3. Colon polyps

Pathological analysis is the only method to accurately distinguish benign and malignant polyps. This is performed postpolypectomy. Confocal allows us to perform a histological analysis prior to polypectomy with an in vivo diagnosis. Polyps with low malignant potential can be left behind while those with malignant potential can be removed. Confocal endomicroscopy differentiates hyperplastic and neoplastic polyps based on the appearance of dark thickened epithelium and dysplasia. Figure 5 [8] shows a comparison between the confocal laser microscopy image and hematoxylin and eosin staining of a hyperplastic polyps. A classification system, the Mainz classification, shown in Table 1 [9] was developed based on the appearance of the vessel and crypt architecture. Molecular imaging during confocal endoscopy is also being used for diagnosis of colonic malignancies. A labeled peptide is applied topically which binds to a neoplastic area. These could provide a means for surveillance for gastrointestinal neoplasms.

Grade	Vessel architecture	Crypt architecture
Normal	Hexagonal, honeycomb appearance	Regular luminal openings, homogenous layer of epithelial cells
Regeneration	Hexagonal, honeycomb appearance with no or mild increase in the number of capillaries	Star-shaped luminal crypt openings or focal aggregation of regular-shaped crypts with a regular or reduced amount of goblet cells
Neoplasia	Dilated and distorted vessels; irregular architecture with little or no orientation to adjunct tissue	

Table 1. Mainz classification

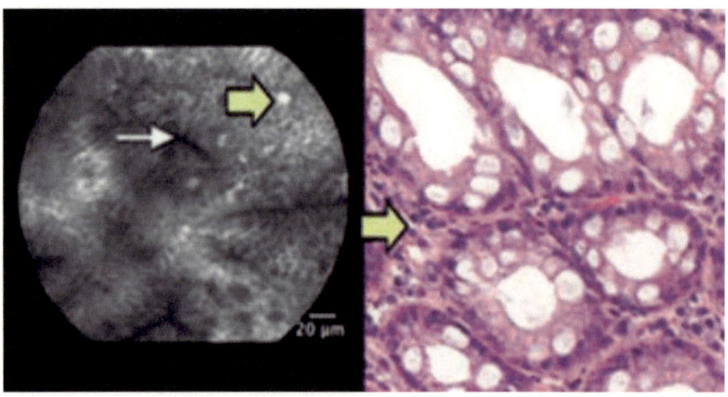

Figure 5. [8] Confocal image of hyperplastic polyp and hematoxylin and eosin image. Thin arrow shows dark goblet cells and thick arrows in both images show the small vessels.

http://dx.doi.org/10.5772/55959

3.4. Inflammatory colitis

Confocal endomicroscopy may be used for diagnosis and cancer surveillance in patients with inflammatory bowel disease. Endomicroscopy in combination with panchromoendoscopy was compared to white light endoscopy in a prospective trial for surveillance. The accuracy of endomicroscopy in predicting intraepithelial neoplasia was found to be 97.8% [4]. Confocal endomicroscopy can also diagnose inflammatory changes in the mucosa.

Confocal endomicroscopy has also been used in microscopic colitis with success. Microscopic colitis is frequently difficult to diagnose due to the patchy distribution. CLE enables targeted biopsies of abnormal mucosa. Similarly confocal endomicroscopy has also found a use in graft-vs-host disease post bone marrow biopsy in patients with diarrhea [10].

3.5. Pancreaticobiliary strictures

Pancreaticobiliary strictures have varying etiologies and maybe benign or malignant. Several methods are used to diagnose the nature of the stricture, i.e., benign or malignant. These include biopsy, brushing and needle aspiration. The diagnostic yield for malignant biliary strictures is limited as most cholangiocarcinomas arise in the bile duct wall. Confocal endomicroscopy provides real-time images. The pCLE is used for evaluation of the biliary system. Due to its small size, it is optimal for biliary imaging. The confocal miniprobe is passed through the channel of a side-viewing endoscope.

A classification for biliary and pancreatic findings on pCLE were devised in 2012-MIAMI Classification [11]. Characteristics most suggestive of malignancy were thick white or dark bands, or dark clumps. Figure 6 demonstrates thick bands as seen via confocal microscopy [12]. Meining et al. tested the validity of the Miami classification by conduction a blinded consensus review of pCLE videos from 47 patients. The sensitivity for the criteria was found to be 97%, specificity was 33%, and positive and negative predictive values of 80% each [12]. Current data on pCLE for diagnosis of biliary strictures is limited. Further studies are needed to determine the accuracy of the confocal miniprobe in diagnosing malignant biliary strictures and differentiating these from inflammatory strictures.

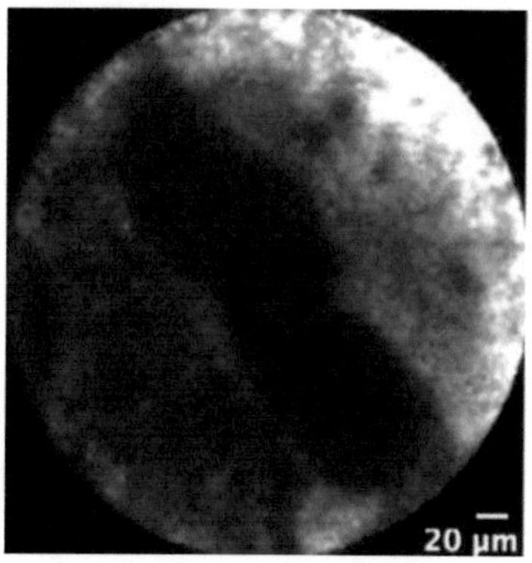

Figure 6. [12] Image from probe based confocal laser endomicroscopy of thick dark band

4. Limitations

Although confocal endomicroscopy is a novel and emerging technology, it has its technical and practical limitations. There is a learning curve for the endoscopists as evidenced by Wallace et at. [13] who evaluated the accuracy of confocal endomicroscopy of 9 international endoscopists in patients with Barrett's esophagus associated dysplasia. They found an overall accuracy for diagnosis of high grade dysplasia in Barrett's esophagus to be 90.5%, sensitivity of 88% and specificity of 94%.

Buchner et al. [14] studied the learning curve for pCLE in the diagnosis of colorectal neoplasia and found that most GI endoscopists required 2 hours of training and review of approximately 50-70 cases with high quality pCLE images to become proficient.

5. Overview of comparable modalities

Several modalities with similar applications as confocal endomicroscopy are being developed and used for detection of gastrointestinal pathologies. These include Optical coherence tomography, Virtual chromoendoscopy and magnification endoscopy.

Optical coherence tomography uses a wavelength of light to provide real-time sectional imaging of the mucosa, pit patterns and glandular architecture. It is however limited due to inability to assess large surface areas.

Chromoendoscopy uses dye application as a method to enhance the architecture of the mucosa for visualization. The dyes used include acetic acid, methylene blue and indigo carmine. Chromoendoscopy is not routinely used as it is time consuming due to the preparation and the performance of the procedure. Its use is limited to tertiary care centers due to the experience required for its use.

Magnification endoscopy uses a movable lens on the endoscope which allows the endoscopist to alter the degree of magnification to provide higher resolution. Magnification endoscopes have a pixel density of 850,000 in comparison to conventional endoscopes which have a density of 100,000 to 200,000. Magnification endoscopy allows for histologic identification by increasing the resolution of the image.

6. Summary

Confocal endomicroscopy is a developing method for diagnosis of various gastrointestinal disorders such as Barrett's esophagus, gastric cancer, inflammatory bowel disorder, colon polyps and pancreaticobiliary strictures. It provides real-time images to aid in the diagnosis and management for these conditions. Confocal microscopy also achieves a more targeted biopsy of the abnormal tissue to expedites the therapeutic planning and decisions regarding endoscopic intervention possibly eliminating the need for repeated procedures. Future trends in confocal endomicroscopy include the wide spread use of molecular imaging with labeled peptides to aid in a more accurate diagnosis of malignancies and for therapeutic planning.

Author details

Anjali Basil and Wahid Wassef

UMass Memorial Medical Center, Worcester, MA, USA

References

[1] Paull, P. E, Hyatt, B. J, & Wassef, W. et al., "Confocal Laser Microscopy-A Primer for Pathologists". Arch Pathol Lab Med. (2011). , 135, 1343-1348.

[2] Nikon MicroscopyuThe source for microscopy education. Confocal Microscopy, Basic concepts. http://www.microscopyu.comaccessed October (2012).

[3] Medical ExpoCatalog search. Confocal Endomicroscopy-Pentax. http://www.pentax.de/de/lifecare.phpaccessed October (2012).

[4] Keisslich, R, Goetz, M, Lammersdorf, K, et al. Chromoscopy-guided endomicroscopy increases the diagnostic yield of intraepithelial neoplasia in ulcerative colitis". Gastroenterology. (2007). , 132, 874-882.

[5] Gaddam, S, Mathur, S. C, Singh, M, et al. Novel Probe-Based confocal Laser Endomicroscopy Criteria and Interobserver Agreement for the Detection of Dysplasia in Barrett's Esophagus". Am J Gastro. (2011). , 106(11), 1961-9.

[6] Neumann, H, Langner, C, & Neurath, M. F. et al., "Confocal endomicroscopy for diagnosis of Barrett's esophagus: Frontiers in Oncology. (2012). , 2, 1-6.

[7] Zhang, J. N, Li, Y. Q, et al. Classification of gastric pit patterns by confocal endomicroscopy". Gastrointest Endosc. (2008).

[8] Wallace, M, Lauwers, C. Y, Chen, Y, et al. Miami classification for probe-based confocal laser endomicroscopy". Endoscopy. (2011). , 43, 882-891.

[9] Usui, V. M, & Wallace, M. B. Confocal endomicroscopy of colorectal polyps," Gastroenterology research and Practice. pages., 2012

[10] Goetz, M, & Kiesslich, R. Advances of endomicroscopy for gastrointestinal physiology and diseases" AJP-Gastrointest Liver Physiol. (2010). , 298, 797-806.

[11] Thelen, A, Scholz, A, & Benckert, C. et al., "Tumor-associated lymphangiogenesis correlates with lymph node metastases and prognosis in hilar cholangiocarcinoma," Annals of Surgical Oncology. (2008). , 15(3), 791-799.

[12] Meining, A, Shah, R. J, & Silva, A. et al., "Classification of probe-based confocal laser endomicroscopy findings in pancreaticobiliary strictures," Endoscopy. (2012). , 44, 251-257.

[13] Wallace, M. B, Sharma, P, & Lightdale, C. et al., "Preliminary accuracy and interobserver agreement for the detection of intraepithelial neoplasia in Barrett's esophagus with probe-based confocal laser endomicroscopy," Gastrointestinal Endoscopy. (2010). , 72(1), 19-24.

[14] Buchner, A. M, Gomez, V, & Heckman, M. G. et al., "The learning curve of in vivo probe-based confocal laser endomicroscopy for prediction of colorectal neoplasia", Gastrointestinal Endoscopy. (2011). , 73(3), 556-560.

Section 2

Applications in the Biological Sciences

Chapter 9

Three-Dimensional Visualization and Quantification of Structural Fibres for Biomedical Applications

Magnus B. Lilledahl, Gary Chinga-Carrasco and
Catharina de Lange Davies

Additional information is available at the end of the chapter

http://dx.doi.org/10.5772/55891

1. Introduction

Confocal laser scanning microscopy (CLSM) combined with the broad range of new fluorescent probes, has become an indispensable tool in basic and applied research within a variety of areas. Extending CLSM to multiphoton microscopy (MPM) and nonlinear optical microscopy (NLOM) opens further possibilities. MPM can be used to increase the penetration into the sample, and is thus ideal for thick samples and *in-vivo* imaging. In this chapter we describe briefly the principles of MPM and non-linear imaging techniques such as second harmonic generation (SHG) and Coherent anti-Stokes Raman scattering (CARS) microscopy. These imaging techniques are especially useful to study structural molecules such as collagen and cellulose. Image analysis is crucial to extract quantitative information about various parameters, for instance the structure of molecules in the sample. Such quantitative information can be used as input parameters in mathematical models describing mechanical properties of tissue. Here we describe 3D Fourier transformation to obtain structural information, gradient techniques, used to characterize the orientation, and distance transforms to measure the thickness of fibres. Additionally, biomedical applications of collagen and cellulose imaging will be presented.

2. Confocal laser scanning and multiphoton microscopy

CLSM is widely used to image cells *in vitro*. Relatively thick samples up to approximately 50 μm can be imaged. By using a confocal pinhole, out-of-focus scattered light is rejected and depth selective imaging is achieved. The absorption and scattering of photons through the tissue limit the corresponding penetration depth. Furthermore, the confocal pinhole rejects scattered light from the focal plane, reducing the detection efficiency. Extending CLSM to MPM makes it possible to penetrate further into the sample, and imaging several hundred μm, has been reported [1].

Two-photon excitation arises from the simultaneous absorption of two photons and requires a high photon density. The excitation is thus intrinsically confocal and no pinhole in front of the detector is required. To obtain a sufficient number of photons in the focal volume, high power, pulsed femtosecond IR lasers are used. The absorption cross section depends on the square of the excitation intensity, and absorption occurs only in the focal volume [2], [3]. Thus, no absorption or photo-bleaching occurs above or below the focal plane, as illustrated in Figure 1. The reduced absorption of photons through the tissue and the use of IR excitation light which is less scattered than visible light, is responsible for the possibility of imaging deeper into the thick sample. Furthermore, the detection of emitted photons can be more efficient than in CLSM as no pinhole is needed in front of the detector and thereby more scattered photons can reach the photomultiplier tubes. 3D imaging of thick specimens, based on 3D reconstruction of 2D optical slices, is superb in MPM compared to CLSM, as no bleaching occurs outside the focal plane. 3D imaging several hundred μm into the sample is regularly reported and imaging down to 1 mm in brain tissue has been achieved [1] - [4]. Most fluorescent organic dyes, quantum dots, and reporter proteins can be excited in a two-photon process, although the absorption spectra are very different from single photon excitation spectra [5]. The excitation is followed by emission of photons from the same excited state as after single photon excitation, thus the emitted fluorescence has the same spectrum in the two cases.

Figure 1. Schematic illustration of the difference between one- and two-photon absorption. On the left is two-photon absorption which only occurs in the small focal volume. On the right, the one-photon absorption occurs above and below the focal plane.

The pulsed IR laser can be used for SHG, another non-linear process. SHG requires that the second order optical susceptibility is non-zero, thus occurring only in non-centrosymmetric molecules [6]. Two photons combine their energy and emit a photon with exactly twice the energy of the two incoming photons, or equivalently, at half the wavelength [7]. The interaction is illustrated in the Jablonski diagram in Figure 2. Thus, placing a narrow bandpass filter at half the wavelength of the excitation light in front of the detector, the SHG signal is easily detected. It should be emphasized that in the SHG process no energy is lost and SHG does not involve an excited state, as opposed to the two-photon excitation fluorescence where some of the incoming energy is lost during relaxation of the excited state. Consequently, SHG does not involve excitation of molecules and no phototoxicity occurs. The SHG signal is polarization sensitive and its intensity depends on the angle between the fibres and the polarization of the excitation light. Therefore, polarization sensitivity studies of SHG from tissues can provide information on the fibre organization. Furthermore, the SHG signal is anisotropic and can be detected in both the forward and backscattered direction. The SHG signal in the forward direction is generally dominating, but if the size of the scattering structures along the optical axis is less than the wavelength, the intensity of the backscattered signal will be equal or even larger than the forward signal [8].

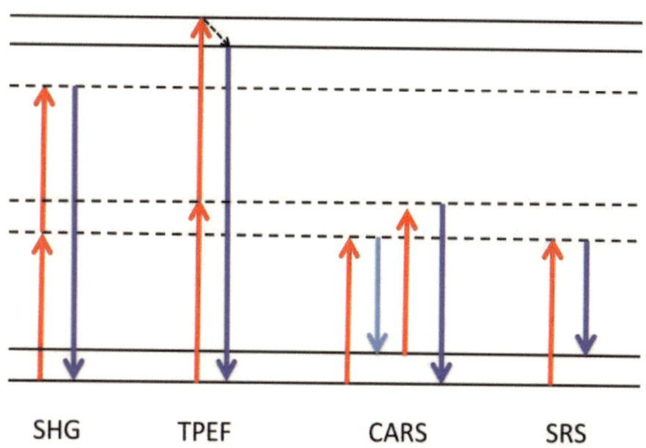

Figure 2. Jablonski diagram illustrating several nonlinear processes. SHG: Two photons are converted to a single photon with twice the energy, corresponding to half the wavelength. TPEF: Two photons are absorbed simultaneously to excite a molecular energy state which can emits fluorescene as if excited by a single molecule. CARS: Three beams interact with a vibrational mode to emit photons at slightly longer wavelengths. SRS: Two beams interact with a vibrational level to transfer energy from one beam to the other.

A major advantage of SHG is that no exogenous labeling of the sample is needed. This label-free technique can be used to image materials with a non-centrosymmetric molecular organization. Examples of such molecules are protein fibres such as collagen, microtubules and actomyosin, as well as cellulose. Combining SHG, two photon excitation fluorescence (TPEF) and conventional CLSM makes it possible to obtain information in the same sample of various molecular and structural constituents (see applications below).

A recent addition to the family of nonlinear optical microscopy is coherent anti-Stokes Raman scattering (CARS) and stimulated Raman scattering (SRS) microscopy [9], [10]. These techniques allow for imaging contrast based on molecular vibrational states, similar to information derived from spontaneous Raman spectroscopy. These methods are slightly more technically complex as two pulsed lasers, temporally and spatially overlapped, must be employed. In CARS, the detection is at a different wavelength than the excitation lasers, so filter-based detection is used. For SRS, the signal is actually a decrease or increase in the laser light intensity so detection of the weak signal in the presence of the laser noise is challenging. A technique called modulation transfer is used where one of the laser beams is modulated and a corresponding modulation in the signal is detected using a lock-in amplifier. CARS and SRS are mainly used for imaging lipids which give a very strong signal, but recent work have shown the potential for imaging other biologically relevant molecules, *e.g.* cellulose [11]. CARS and SRS, together with TPEF and SHG provide the opportunity to image a wide range of structural fibres without exogenous stains. This means that dynamic process (*e.g.* in tissue engineering or biomass processing) may be monitored over time.

3. NLOM and structural fibres

Structural fibres are abundant in nature. They provide support for the cells to develop complex organs. Not only do the fibres provide mechanical support, but they are also modifiers for cellular behaviour. They control the diffusion of signalling molecules and act as signalling molecules themselves, for example through mechanobiological transduction.

NLOM is especially useful when imaging structural fibres in biological thick samples. The thicker 3D volumes generated will facilitate the subsequent analysis of 3D structural information. Also, being able to perform optical sectioning through thicker sections than in CLSM, fewer serial sections have to be cut mechanically, which reduces workload and potential for sectioning artefacts. When using traditional fluorophores for providing imaging contrast, the main advantage of nonlinear microscopy is the reduced out-of-focus bleaching and the availability of non-descanned detection.

However, the main advantage of NLOM for imaging structural properties lies in the possibility of using endogenous signals from the molecule in question to generate contrast in the image. This is highly advantageous as there are limitations in quantitation caused by uneven labelling or fading of the fluorescence. Furthermore, since the signal is generated from the molecule itself, and not some fluorescent molecule attached to it, a range of optical signals (polarization, spectral, lifetime) can be used to extract even sub-resolution information about the molecule. For example, SHG has been used to measure the helical pitch angle of myosin and collagen with a high degree of accuracy [12]. Fortunately, nature has also provided bright nonlinear endogenous signals from several highly relevant structural molecules, e.g. collagen [7] with SHG, elastin [13] with TPEF and cellulose [11] with CARS.

3.1. Collagen

Collagen is the most abundant protein in the extracellular matrix (ECM) in animals. Collagen forms fibres with high tensile strength and thus accounts for much of the mechanical strength of tissue. Collagen exists in many different varieties but the main common feature is a polypeptide with repeating sequences of three amino acids. Typically, every third amino acid is glycine. The repeating sequence can thus be described as gly-X-Y where the amino acids in the X and Y positions vary. Three of these polypeptides then form a triple stranded helix to create a collagen molecule. Various combinations of the three strands yields different collagen types which are denoted by roman numerals (I,II, III etc.). Some of these types have a molecular structure which causes them to aggregate into even larger structures called collagen fibrils (~20-500 nm in diameter), and the fibrils form fibre collagens (e.g. type I and II) (~500-3000 nm in diameter). Due to their semi-crystalline ordering and non-centrosymmetric structure of the collagen molecule, these fibrils are efficient emitters of SHG. In addition to the SHG intensity, the full tensor nature of the second order susceptibility and the ratio of forward to backscattered light may be used to extract sub-resolution information about the molecules. Several works have illustrated how the orientation of the collagen molecules can be determined from a single pixel and how the forward/backward scattering ratio can be used to determine the size and sub-resolution structure of collagen fibrils [8].

3.2. Elastin

Elastin fibres are aggregates of monomers of the elastin molecule, which are covalently bound through cross-links. These cross-links are very flexible so that the polymer can stretch quite significantly without breaking, thus giving rise to the elasticity of tissue. Elastin has a strong TPEF (probably linked to the molecular cross-links). Elastin is found in many important organs e.g. skin, arteries and some types of cartilage. Imaging of elastin fibres is especially striking in elastic arteries where such imaging has been used for biomechanical applications and atherosclerosis research [13].

3.3. Cellulose

Cellulose is the main component of the cell wall of green plants. In vascular plants, cellulose is a major component of the structural fibres that provide mechanical stability and are responsible for transporting water and nutrients. Cellulose, lignin and hemicellulose are the major components of wood fibres, or tracheids. The cellulose molecules form the elementary fibrils, which have been reported to be the structural unit of natural cellulose, with a diameter of 3.5 nm [14] - [16]. The cellulosic components of a wood fibre wall structure are the cellulose molecule, the elementary fibril, the microfibril, the macrofibril and the lamellar membrane [15]. The mentioned cellulosic constituents compose various layers of the fibre wall structure, including the primary wall (P) and secondary walls (S1, S2 and S3) [17].

4. Sample preparation

For nonlinear microscopy one typically wants to cut samples which are much thicker than samples used for conventional histology. This is to take advantage of the 3D imaging capability of the techniques. We have successfully cut fixed and embedded sections of breast tissue with a rotary vibratome up to 100 microns. The challenge with thicker samples is to remove all the paraffin, but we found that this is usually not a problem if slightly longer incubation for removal of paraffin is used.

For samples which are not embedded, a vibratome is good choice for cutting sections of desired thickness. Here the challenge is on the other end of the scale, that is, creating thin samples. We have found that sections down to 40 microns of unfixed and unembedded cartilage are easily cut. Thinner samples are not necessary, as imaging down to 40 micron is straightforward and typically thicker sections can be cut. A vibratome requires a certain rigidity of the sample, but we have found that cartilage is an ideal sample to cut with this technique.

Cryosections are also possible. There is some concern that the freezing procedure might cause structural changes but this depends heavily on the structure that is to be imaged. For example, cryosections of collagen fibres in cartilage have been shown to be well preserved. This might also be the case for other structural fibres due to their inherent mechanical strength [30]. For other, more delicate structures (e.g. cell structures), preparation artefacts and lost morphology might be a concern.

Of course, in addition to the endogenous signal, tissues might also be stained with various fluorochromes to access the distribution of other molecules of interest. For thick sections, the additional stain penetration is an issue. This is highly dependent on the size of the staining molecule and the type of tissue. For fixed and embedded samples from breast tissue we have been able to stain nuclei with DAPI down to about 50 micron, yielding striking images of the cellular distribution relative to the collagen network imaged with SHG (Figure 3).

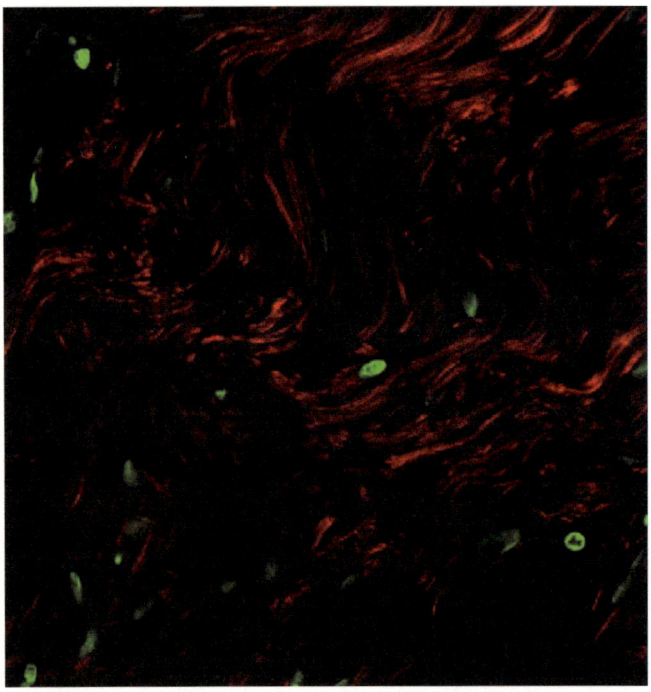

Figure 3. Breast cancer tissue imaged with SHG (red) to visualize collagen structure and TPEF (green) of DAPI stained cells. The collagen structure may serve as a prognostic marker for the degree of malignancy.

Since one of the main advantages of NLOM is the 3D imaging capability, enhancing this feature is often desirable. This leads to the concept of optical clearing. By substituting the water in the tissue with some other compound which has a refractive index comparable to cells and the collagen fibres, the scattering is greatly reduced. We have seen up to a two-fold increase in the penetration depth, based on this technique (typically 50 % increase) (see Figure 4). A useful compound is glycerol. It is nontoxic and has little interaction with the samples. Furthermore, the refractive index of glycerol can be tuned by using various mixing ratios of water and glycerol, to optimize the effect of the clearing.

5. Image analysis

The organization of structural fibres has a wide variety of effects on the biology of the components it supports. The primary property is structural, *i.e.* keeping the cells in appropriate relative positions. They also comprise a major part of the mechanical properties of tissue through load bearing, distribution of loads and mechanotransduction. It has also become clear recently, that structural fibres also have a profound biological role, acting both as a signalling molecule themselves, as well as determining the diffusion of other signalling molecules [18].

To be able to incorporate these parameters into mathematical models for biological systems, quantitative data needs to be derived from the images. Several image analysis methods have been developed for this purpose and a few of the most common ones are presented in this chapter.

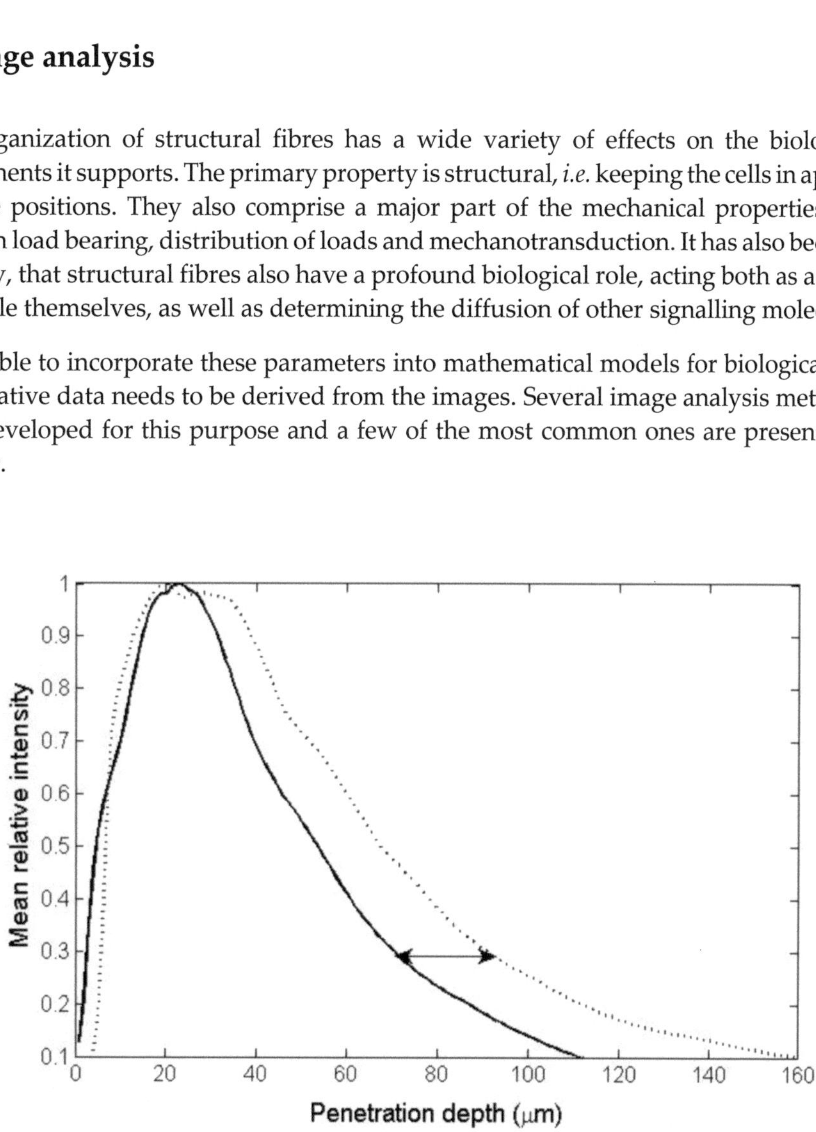

Figure 4. Change in penetration depth due to optical clearing. Mean SHG signal in an image frame as a function of depth for uncleared sample (solid line) and cleared with 70% glycerol (dashed line)

5.1. Fourier analysis

The linear arrangement of fibres in many types of structures and the resulting periodicity of the fibres in the direction perpendicular to the fibres immediately leads to the idea of using the Fourier transform of the image to determine the direction of the fibres (Figure 5) [19] - [21]. Even though the idea is simple, aligned fibres rarely yields a perfectly sinusoidal signals and significant processing is necessary to extract the desired information.

The one dimensional discrete Fourier transform $F(k)$ of a signal $f(x)$ of length N is given by

$$F(k) = \sum_{x=0}^{N-1} f(x) e^{-\frac{i2\pi kx}{N}} \tag{1}$$

where k is the spatial frequency of the signal. A periodic signal will give an impulse response in the frequency spectrum at the frequency of the periodic signal. Due to the linearity of the transform, a signal consisting of many frequency components will yield distinct peaks in the Fourier space. As the Fourier transform is complex, the power spectrum (magnitude of the Fourier spectrum) is typically employed for interpretation of the signals. In graphical representations of the power spectrum the DC component is usually shifted to the center of the image. An important point to note is that even though the Fourier transform is linear (Fourier transform of a sum of signals is equal to the sum of the Fourier transform of the individual signals) this is not true for the power spectrum. Hence, care should be taken when interpreting the power spectrum as a sum of individual components.

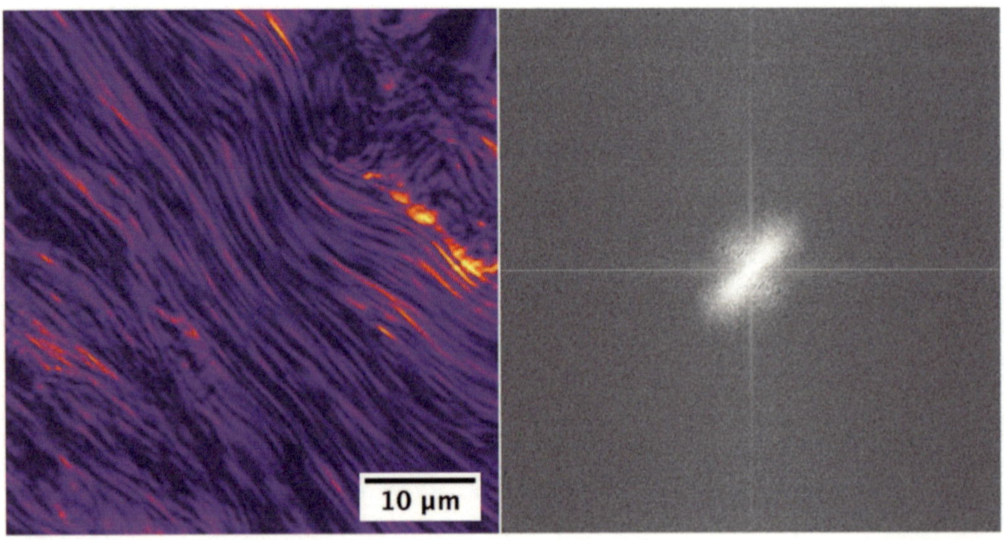

Figure 5. Left) SHG image of collagen fibres from chicken cartilage. Pseudocolours have been applied for better visualization. Right) The corresponding Fourier transform of the SHG image.

The most common way to interpret Fourier transform images is to assume that the frequency components are distributed on an ellipse (or ellipsoid in three dimensions). The long axis is perpendicular to the direction of the fibre, which corresponds to strong frequency components (periodicity) in this direction. Again, it must be noted that an ellipse is not the resulting frequency pattern of a specific ordered fibre distribution (as seen in Figure 5, right), but rather represents some of the inherent variability of the systems which are analyzed. There are several methods for fitting an ellipsoid to data. See for example [34].

To avoid aliasing in the image due to edge effects, it is important to add a windowing function to the image. The edge effect can be seen in Figure 5 (right), as straight lines through the center of the image. These are however easily removed by a windowing function. Several options are possible, with the Blackman-Harris function and the Hann window being popular candidates [19].

Also, as can be seen in Figure 5 (right) the pure Fourier image is quite noisy, and there is a lot of signal which is not relevant to the structure of interest. Before extracting the directional information it is often necessary to filter the data to get a good result. Typically, a band pass filter can be employed which lets through the frequencies associated with the fibrils. The approach removes random high frequency components and low frequency components which represent slow variations in intensity across the image. The frequency limits of the band pass filter must be determined from the size of the structures of interest. An example can be found in [34].

Due to the linearity of sums (or integrals for the continuous case) used in the Fourier transform, the extension of the one dimensional transform to two, three and higher dimensions is trivial. Hence, the 3D image volumes generated by CLSM and NLOM may be analyzed by the 3D Fourier transform. An important point to keep in mind when analyzing 3D volumes with the Fourier transform, is that the point spread function of the microscope is not equal in all three directions. This will cause an elongation of signals in the z-direction, thereby reducing the higher order frequency components in this direction. This must be taken into account when trying to extract quantitative information.

5.2. Gradient techniques

During the years several methods for characterization texture have been developed, including autocorrelation, Fourier transform, gradient analysis and quadtree decomposition. Gradient analysis is an interesting approach, as this technique can be applied for estimating the orientation of fibrous structure, *i.e.* collagen and cellulose nanofibrils (Figure 6). Several gradient definitions have been proposed [22]. However, Sobel operators are particularly attractive due to the noise enhancing effect of derivatives [23]. Examples are the orientation of short-fibre composites [24] and actin fibres in cytoskeletal structures [25].

A relatively simple approach for quantification of structural orientation has been implemented in the SurfCharJ plugin for ImageJ [26]. The plugin plots the frequency of azimuthal angles for estimating the preferred orientation of a given structure. In addition, an estimation of the structural orientation based on the mean resultant vector is provided [27].

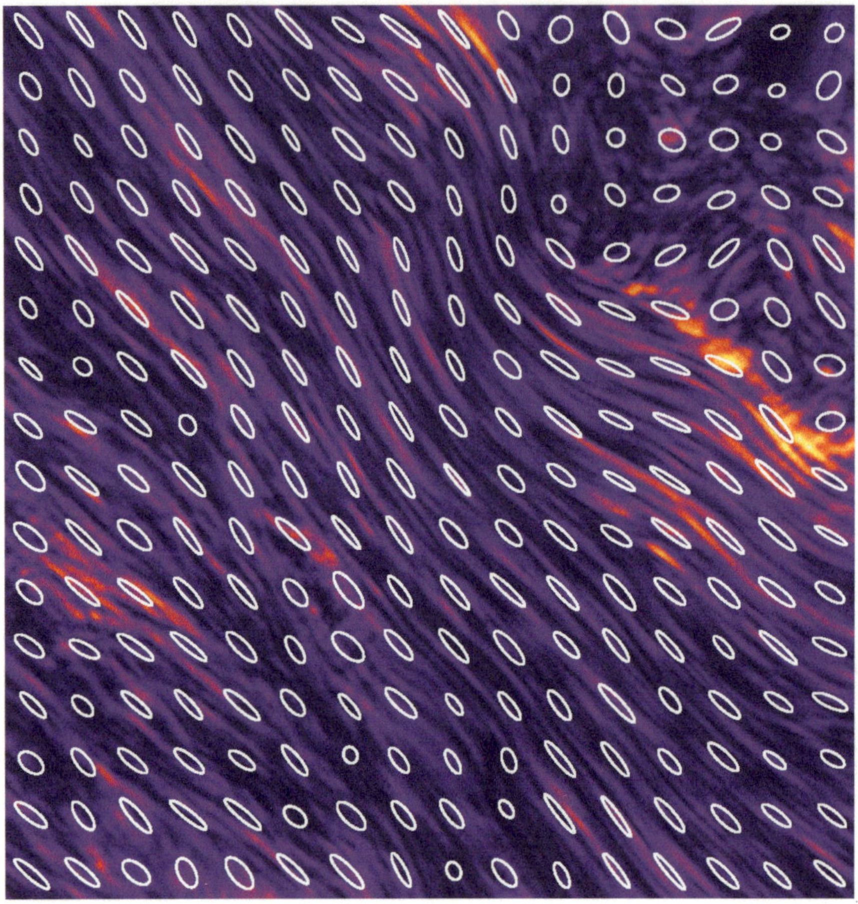

Figure 6. SHG image of collagen fibres analyzed with gradient techniques. Ellipses have been superimposed to exemplify the local direction of the fibres.

5.3. Distance transforms

A distance transform converts a binary image to a grey level image, where all pixels have a value corresponding to the distance to the nearest feature pixel [28]. The lateral size of fibrous structures can thus be estimated, provided that the previous segmentation is adequate (Figure 7). Some common metrics for estimating the distance between pixels are the City-block (*Eq. 2),* Chessboard (*Eq. 3)* and Euclidean distance (*Eq. 4)*:

$$d_b = |x_1 - x_2| + |y_1 - y_2| \tag{2}$$

$$d_c = [\,|x_1 - x_2|,\ |y_1 - y_2|\,] \tag{3}$$

$$d_e = \sqrt{(x_1 - x_2)^2 + (y_1 - y_2)^2} \tag{4}$$

where d_b, d_c, and d_e represent the distance between pixel (x_1, y_1) and pixel (x_2, y_2) as computed by the City-block, Chessboard and Euclidean distance transforms, respectively.

A distance transform can be efficiently applied for estimating the thickness of collagen and cellulose fibres, as exemplified in Figure 7.

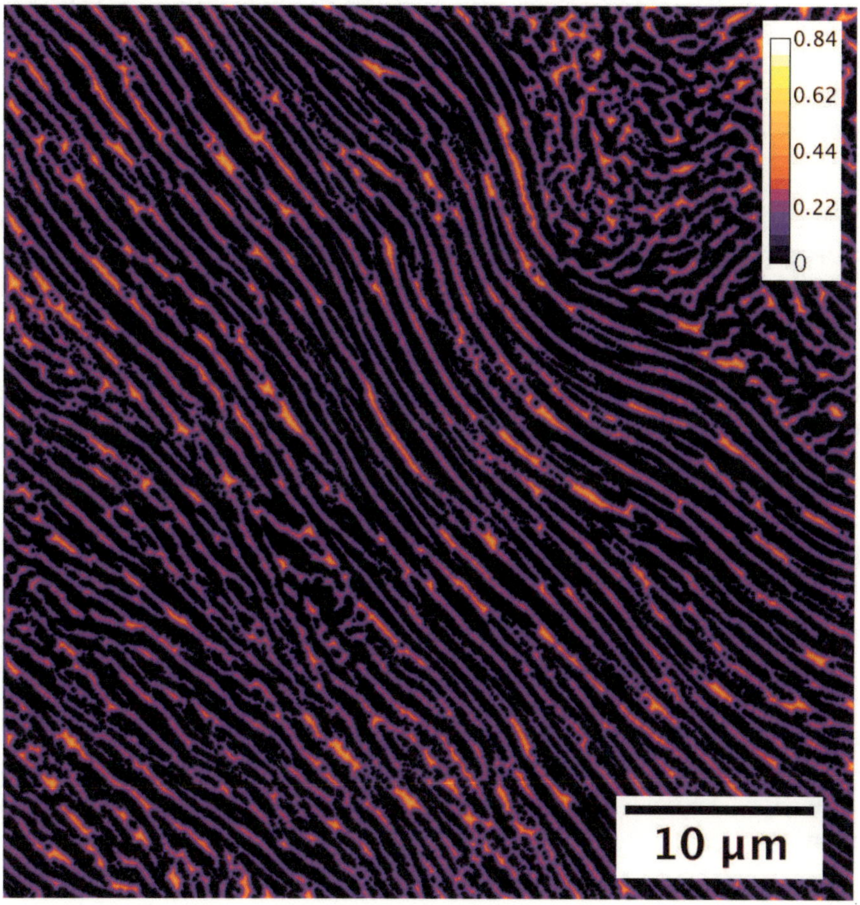

Figure 7. SHG image of collagen fibres. The distance transform map exemplifies the approximate lateral dimensions of the fibres. The calibration and scale bars are given in micrometers.

6. Image acquisition in NLOM

Optimizing a nonlinear optical system is demanding, especially for more complex techniques like SRS and CARS. We will in this section address a few points, which must be considered. To improve the generated signal, the first step is to maximize the photon density, while keeping the average power constant. As the femtosecond laser pulses undergo significant dispersion as they propagate through the optics of the microscope, pre-chirping the laser pulse is necessary to achieve a transform limited pulse. The most recent generation of Ti:Sapphire lasers have such a system integrated.

The next thing to consider is the objective, which should have a high numerical aperture (NA). High NA usually means large magnification which is typically a disadvantage, as a large field of view is also desirable. The latest high-end objectives can achieve about 20-25x magnification with a NA of around unity. The objective should also have good transmission and achromaticity in the NIR spectra region. Another important point to consider, especially when imaging in deep tissue, is that there might be significant spherical aberration. Therefore a correction collar might be useful.

Often the excitation light wavelength is governed by the fluorophore used. However, if the application allows a choice (e.g. with SHG), longer wavelengths are scattered less and will result in deeper penetration depths [29]. It should also be kept in mind that many molecules, especially structural ones which have a preferred orientation, have a highly polarization dependent response [6]. To avoid bias in the images due to this effect, circularly polarized excitation should be used or ideally a variable polarization of the laser beam and an analyzer on the detection side. This polarization dependence can also be used to derive additional information, as already explained.

7. Applications

Collagen, ubiquitous in the vertebrate body, has been imaged in a host of different tissues using SHG. Several proof-of-principle studies exist, but there are few that actually have used the signals to derive clinical, biological or quantitative data. However, as the instrumentation becomes more user-friendly and wide spread, we will likely see an even further increase in the use of NLOM for more clinical and biological applications. We will here present a few model systems where nonlinear and confocal microscopy can provide relevant structural information.

7.1. Cartilage

Cartilage consists of a highly hydrated (70% water) proteoglycan gel, which is reinforced with collagen fibres, primarily type II collagen (Figure 8). The tissue provides both distribution of load across the joint surface, and a low friction surface for joint articulation. Degradation and traumatic injuries to cartilage are major health challenges, especially with an increasingly old and obese population. Osteoarthritis (OA), characterized by a degradation of the collagen matrix is one such disease. The development of OA is linked both to biological and biomechanical causes, and there is interplay between these effects. To understand the biological effects, the biomechanical effects of OA must be understood. The biomechanics is tightly connected to the collagen matrix which not only imparts tensile strength to the tissue, but also affects diffusion and fluid flow, which affects the biomechanical response. SHG, being able to characterize the collagen structure in great detail, is an ideal tool for generating high fidelity structural input parameters for biomechanical models.

SHG from cartilage was first described by Yeh et al. which imaged canine cartilage and showed that several structural features could be discerned [31]. Mansfield used polarization sensitive SHG to derive directional information from the cartilage [32], [33]. We have taken a different approach and used Fourier transforms of images where individual fibrils are resolved to extract 3D structural information. As a model tissue, we use excised cartilage from chicken. These samples were sectioned in 100 μm sections using a vibratome, placed on objective slides and subsequently imaged. We have shown that this information can be directly employed in biomechanical models to improve the fidelity of such models. This has been applied to assess how the tissue responds to mechanical loading and how the mechanical properties are altered as the tissue undergoes pathological changes [34].

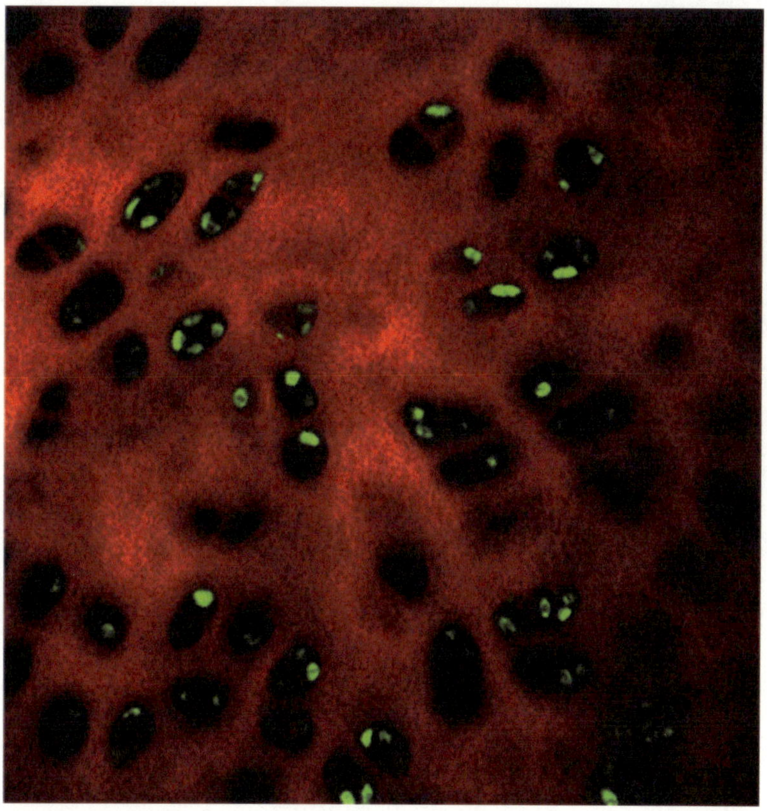

Figure 8. Collagen and chondrocytes in the transitional layer of cartilage imaged with SHG (red) and TPEF (green), respectively.

7.2. Atherosclerosis

Atherosclerosis is a disease characterized by an inflammatory process where monocytes are recruited to the vessel and differentiated into macrophages (Figure 9). As the diseases progress, a necrotic core of lipids and cellular debris develops, which is covered by a collagenous cap holding the pathological tissue, called a plaque. It is believed that a large part of heart attacks are caused by the rupture of these plaques which exposes the blood stream to the thrombogenic substances in the plaque and causes an occlusion of the artery. Plaques, which are prone to rupture, are denoted vulnerable plaques.

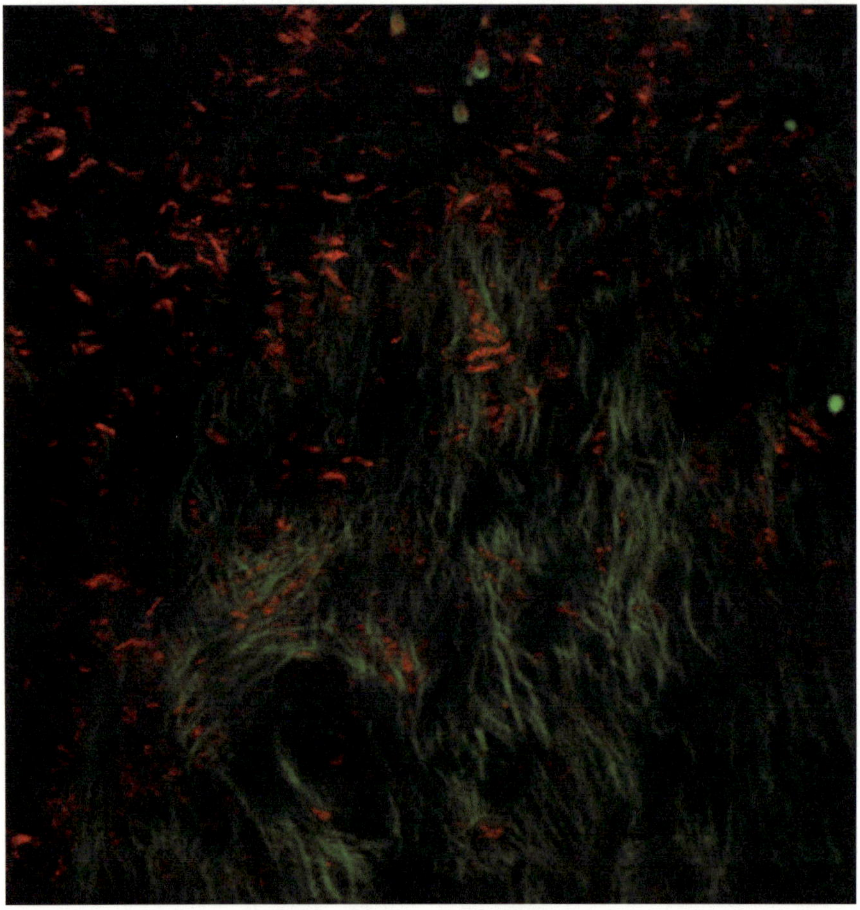

Figure 9. Collagen and elastin in arterial wessel imaged with SHG (red) and TPEF (green), respectively

There are several highly important molecules related to atherosclerosis, which can be imaged label-free using NLOM. Firstly, the collagenous cap of the atherosclerotic plaques are easily imaged with SHG. The cap is a very important structure as it is really the mechanical degradation (through the degradation of collagen by matrix metalloproteinases) of this cap, which ultimately leads to the disruption. The elastic fibres of the media are readily imaged using

TPEF. As the plaque remodels and develops, smooth muscle cells are recruited through the media and any disruption or increased permeability may be monitored using TPEF. Using samples from human autopsies, we have studied how different plaque structures can be characterized using NLOM [17]. In this study, unprepared samples were placed on a cover glass and imaged from the lumen side. Hence, similar data could potentially be extracted using an *in-vivo* fibre based probe. Finally CARS and SRS are ideal for imaging the lipids in the necrotic core. They can be used to differentiate between many different lipid types (triglycerides, cholesterol and its esters) as well as the degree of crystallinity through SHG [13], [35]. The development of cholesterol crystals have been linked to an increased vulnerability of the plaques.

7.3. Drug delivery through the extracellular matrix

In cancer therapy, encapsulation of drugs into nanoparticles is a promising strategy to enhance the accumulation of the drugs in tumor tissue and reduce the exposure to healthy tissue (Figure 10). This is due to the leaky blood vessels in tumor tissue, allowing the extravasation of nanoparticles from the blood vessels to the ECM, whereas the nanoparticles are constrained to the vessel in normal tissue. However, a major obstacle for successful delivery of drugs and nanoparticles to cells is the poor penetration through the ECM. The ECM consists of a network of the structural protein collagen embedded in a gel of hydrophilic glycosaminoglycans. Both the hydrophilic gel and the collagen network represent a hindrance for delivery of nanoparticles. To study the barrier represented by the collagen network, imaging of the collagen by the SHG signal gives valuable information to improve drug delivery [36], [37]. Especially intravital microscopy of tumors growing in dorsal window chambers on the back of athymic mice combined with SHG provides a valuable tool to image collagen fibres *in vivo* (Figure 10). The structure of the collagen network can be characterized by various parameters such as the anisotropy parameter [38], the second order nonlinear optical susceptibility [39], fibre angle [40], skewness and entropy [41].

7.4. Cellulose nanofibrils

Being the most abundant organic polymer on Earth, cellulose generates a wide range of research worldwide. Additionally, cellulose is exploited industrially, including various applications, *e.g.* a major component in paper, as thickeners in tooth paste, in paints and a series of pharmaceutical applications. During the last decades major research has been performed on the production and applications of cellulose nanofibrils [42] - [45]. This nanostructural component is commonly produced from wood pulp fibres, but other agricultural waste products have also been assessed as potential raw materials for their production [46].

Recently, cellulose nanofibrils from wood have been proposed as a major component of cryo-gels for wound healing applications. Cryo-gels based on cellulose are macro-porous and have wall structures composed of nano-sized fibrils (Figure 11) [47]. Due to this multi-dimensionality, proper characterization of porous cellulose nanofibril structures is demanding. Considering the limitations with respect to resolution, and provided an adequate preparation of the sample, the walls of cellulose macro-porous structures can be visualized. Two-photon fluorescence microscopy may be used for this purpose by staining the cellulose with for example calcofluor white. The 3D capabilities provided by nonlinear microscopy can provide enhanced understanding of the water retaining properties of cryo-gels, which is most important (Figure 12). CARS has also been used for imaging cellulose fibres and using polarization sensitive measurements, sub-resolution about molecular orientation may be derived [11]. This label-free imaging technique is thus expected to be useful for characterizing samples, which are meant for future use (quality assurance) as well as for monitoring the development of nanofibrils as they are processed using various chemical pre-treatments [48].

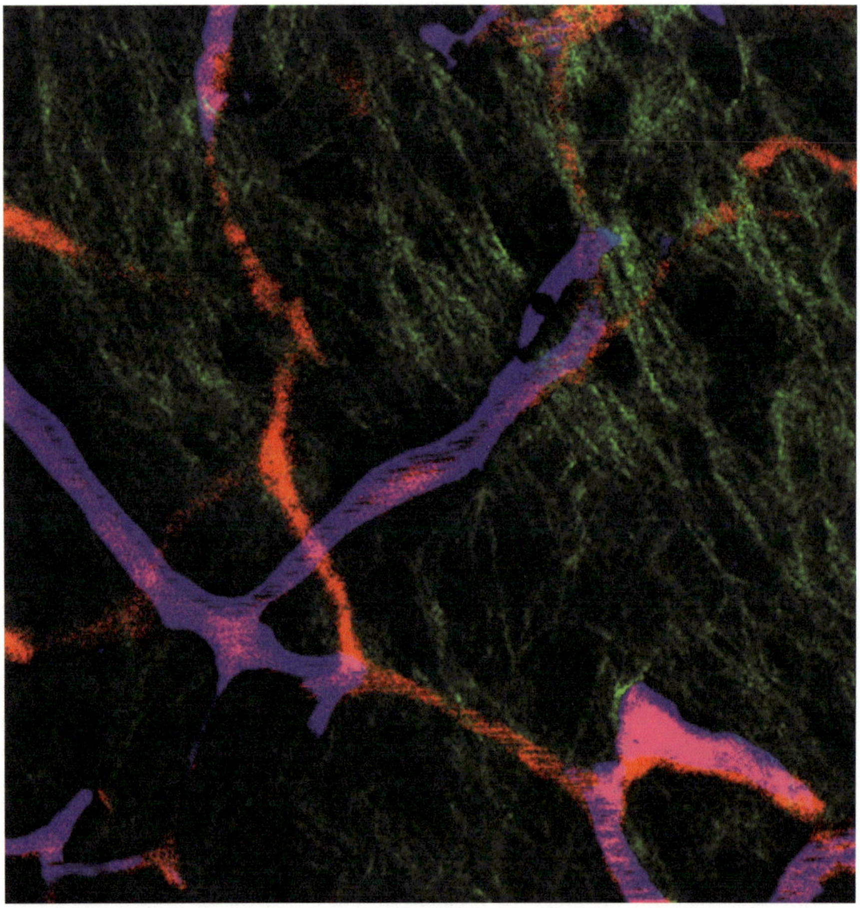

Figure 10. Tumor tissue growing in a dorsal window chamber. Collagen visualized by SHG (green), blood vessels (red) and nanoparticles (blue).

http://dx.doi.org/10.5772/55891

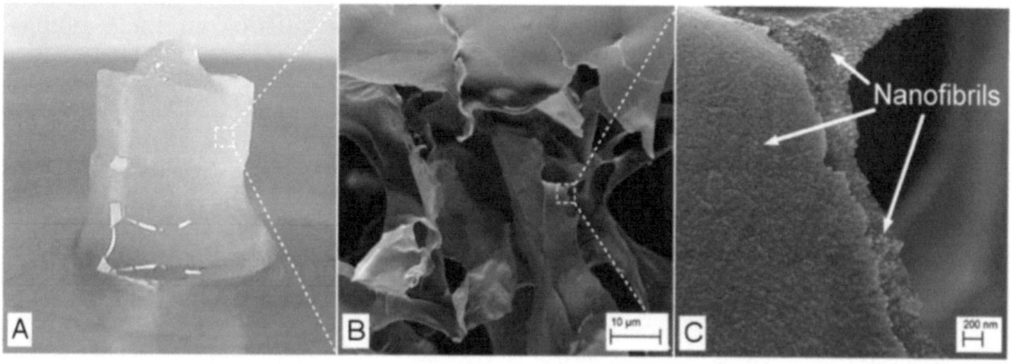

Figure 11. A) Elastic cryo-gel prepared from cross-linked cellulose nanofibrills. B) The macro-porous structure is exemplified. C) The nanofibrils are visualized. Reproduced from [47].

Figure 12. Three-dimensional rendering of a macro-porous cellulose structure in a wet state, observed from different angles. The sample was stained with Calcofluor White (CFW) before imaging.

8. Conclusion

Structural fibres are ubiquitous in biology were they play a wide array of roles. They provide support to the cells to develop complex organs. Not only do the fibres provide mechanical support but they are also modifiers of cellular behaviour, as they control the diffusion of signalling molecules and act as signalling molecules themselves, for example through mechanobiological transduction. The structural fibres included in this book chapter are cellulose and collagen fibres, from plant and animal tissue, respectively. We have in this chapter shown that nonlinear optical microscopy is an important extension of basic CLSM providing both label-free and three dimensional imaging. Adequate image acquisition and relevant image analysis techniques widen the possibilities for structural characterization. The ability to adequately characterize structural fibres thus expands our understanding of biological processes, which can be most important in medical applications.

Acknowledgements

This work has been funded in part by the NanoHeal project - Bio-compatible cellulose nanostructures for advanced wound healing applications, Grant 219733. Anna Bofin provided the breast cancer samples. The images of the breast cancer samples were acquired by Anders Brabrand and Ian Kariuki.

Author details

Magnus B. Lilledahl[1], Gary Chinga-Carrasco[2] and Catharina de Lange Davies[1]

1 Dept. of Physics, Norwegian University of Science and Technology, Trondheim, Norway

2 Paper and Novel Materials, PFI, Trondheim, Norway

References

[1] Brown, E. B.; Campbell, R. B.; Tsuzuki, Y.; Xu, L.; Carmeliet, P.; Fukumura, D.; Jain, R. K. *Nat Med* 2001, *7*, 864.

[2] Helmchen, F.; Denk, W. *Nature Mehods* 2005, *2*, 932.

[3] Patterson, G. H.; Piston, D. W. *Biophys. J.* 2000, *78*, 2159.

[4] Theer, P.; Hasan, M. T.; Denk, W. *Optics Letters* 2003, *28*, 1022.

[5] Xu, C.; Zipfel, W.; Shear, J. B.; Williams, R. M.; Webb, W. W. *P Natl Acad Sci USA* 1996, *93*, 10763.

[6] Stoller, P.; Kim, B.-M.; Rubenchik, A. M.; Reiser, K. M.; Da Silva, L. B. *J Biomed Opt* 2002, *7*, 205.

[7] Campagnola, P. J.; Loew, L. M. *Nat Biotech* 2003, *21*, 1356.

[8] Williams, R. M.; Zipfel, W. R.; Webb, W. W. *Biophysical journal* 2005, *88*, 1377.

[9] Zumbusch, A.; Holtom, G. R.; Xie, X. S. *Physical Review Letters* 1999, *82*, 4142.

[10] Freudiger, C. W.; Min, W.; Saar, B. G.; Lu, S.; Holtom, G. R.; He, C.; Tsai, J. C.; Kang, J. X.; Xie, X. S. *Science* 2008, *322*, 1857.

[11] Zimmerley, M.; Younger, R.; Valenton, T.; Oertel, D. C.; Ward, J. L.; Potma, E. O. *J Phys Chem B* 2010, *114*, 10200.

[12] Psilodimitrakopoulos, S.; Santos, S. I. C. O.; Amat-Roldan, I.; Thayil, A. K. N.; Artigas, D.; Loza-Alvarez, P. *J Biomed Opt* 2009, *14*, 014001.

[13] Lilledahl, M. B.; Davies, C. d. L.; Haugen, O. A.; Svaasand, L. O. *J Biomed Opt* 2007, *12*, 044005.

[14] Frey-Wyssling, A. *Science* 1954, *119*, 80.

[15] Meier, H. *Pure and applied chemistry* 1962, *5*, 37.

[16] Heyn, A. N. J. *Journal of Ultrastructure Research* 1969, *26*, 52.

[17] Brandström, J. *IAWA Journal* 2001, *22*, 333.

[18] Vogel, W. F. *Eur J Dermatol* 2001, *11*, 506.

[19] Chaudhuri, S.; Nguyen, H.; Rangayyan, R. M.; Walsh, S.; Frank, C. B. *Biomedical Engineering, IEEE Transactions on* 1987, *BME-34*, 509.

[20] Xia, Y.; Elder, K. *J Microsc-Oxford* 2001, *204*, 3.

[21] Brezinski, M. E.; Tearney, G. J.; Bouma, B. E.; Izatt, J. A.; Hee, M. R.; Swanson, E. A.; Southern, J. F.; Fujimoto, J. G. *Circulation* 1996, *93*, 1206.

[22] Tovey, N.; Krinsley, D. *Bulletin of Engineering Geology and the Environment* 1992, *46*, 93.

[23] Gonzalez, R.; Woods, R. E. *Digital Image Processing*; Addison-Wesley Publishing Company, Inc., 1993.

[24] Gadala-Maria, F.; Parsi, F. *Polymer Composites* 1993, *14*, 126.

[25] Yoshigi, M.; Clark, E. B.; Yost, H. J. *Cytometry Part A* 2003, *55A*, 109.

[26] Chinga, G.; Johnsen, P. O.; Dougherty, R.; Berli, E. L.; Walter, J. *Journal of Microscopy* 2007, *227*, 254.

[27] Curray, J. R. *Bulletin of the american association of petroleum geologists* 1953, *40*, 2440.

[28] Borgefors, G. *Computer Vision, Graphics, and Image Processing* 1986, *34*, 344.

[29] Kobat, D.; Horton, N. G.; Xu, C. *J Biomed Opt* 2011, *16*, 106014.

[30] Brockbank, K.; MacLellan, W.; Xie, J.; Hamm-Alvarez, S.; Chen, Z.; Schenke-Layland, K. *Cell and Tissue Banking* 2008, *9*, 299.

[31] Yeh, A. T.; Hammer-Wilson, M. J.; Van Sickle, D. C.; Benton, H. P.; Zoumi, A.; Tromberg, B. J.; Peavy, G. M. *Osteoarth Cartilage* 2005, *13*, 345.

[32] Mansfield, J.; Yu, J.; Attenburrow, D.; Moger, J.; Tirlapur, U.; Urban, J.; Cui, Z. F.; Winlove, P. *J Anat* 2009, *215*, 682.

[33] Mansfield, J. C.; Winlove, C. P.; Moger, J.; Matcher, S. J. *J Biomed Opt* 2008, *13*, 044020.

[34] Lilledahl, M. B.; Pierce, D. M.; Ricken, T.; Holzapfel, G. A.; de Lange Davies, C. *Medical Imaging, IEEE Transactions on* 2011, *30*, 1635.

[35] Suhalim, Jeffrey L.; Chung, C.-Y.; Lilledahl, Magnus B.; Lim, Ryan S.; Levi, M.; Tromberg, Bruce J.; Potma, Eric O. *Biophysical Journal* 2012, *102*, 1988.

[36] Brown, E.; McKee, T.; diTomaso, E.; Pluen, A.; Seed, B.; Boucher, Y.; Jain, R. K. *Nature Medicine* 2003, *9*, 796.

[37] McKee, T. D.; Grandi, P.; Mok, W.; Alexandrakis, G.; Insin, N.; Zimmer, J. P.; Bawendi, M. G.; Boucher, Y.; Breakefield, X. O.; Jain, R. K. *Cancer Research* 2006, *66*, 2509.

[38] Hompland, T.; Erikson, A.; Lindgren, M.; Lindmo, T.; de Lange Davies, C. *J Biomed Opt* 2008, *13*, 054050.

[39] Erikson, A.; Örtegren, J.; Hompland, T.; Davies, C. d. L.; Lindgren, M. *J Biomed Opt* 2007.

[40] Erikson, A.; Andersen, H. N.; Naess, S. N.; Sikorski, P.; Davies, C. d. L. *Biopolymers* 2008, *89*, 135.

[41] Erikson, A.; Tufto, I.; Bjønnum, A. B.; Bruland, Ø. S.; De Lange Davies, C. *Anticancer Res* 2008, *28*, 3557.

[42] Turbak, A. F.; Snyder, F. W.; Sandberg, K. R. *Microfibrillated cellulose, a new cellulose product: properties, uses, and commercial potential*, 1983.

[43] Siró, I.; Plackett, D. *Cellulose* 2010, *17*, 459.

[44] Klemm, D.; Kramer, F.; Moritz, S.; Lindström, T.; Ankerfors, M.; Gray, D.; Dorris, A. *Angewandte Chemie International Edition* 2011, *50*, 5438.

[45] Chinga-Carrasco, G. *Nanoscale Research Letters* 2011, *6*, 417.

[46] Alila, S.; Besbes, I.; Vilar, M. R.; Mutjé, P.; Boufi, S. *Industrial Crops and Products* 2013, *41*, 250.

[47] Syverud, K.; Kirsebom, H.; Hajizadeh, S.; Chinga-Carrasco, G. *Nanoscale Research Letters* 2011, *6*, 1.

[48] Saar, B. G.; Zeng, Y.; Freudiger, C. W.; Liu, Y.-S.; Himmel, M. E.; Xie, X. S.; Ding, S.-Y. *Angewandte Chemie International Edition* 2010, *49*, 5476.

Chapter 10

Confocal Microscopy as Useful Tool for Studying Fibrin-Cell Interactions

Rita Marchi and Héctor Rojas

Additional information is available at the end of the chapter

http://dx.doi.org/10.5772/55586

1. Introduction

Fibrinogen is a 340 kDa plasma protein that circulates in blood at 2 - 4 g/L. It is composed by two set of polypeptide chains: Aα, Bβ, and γ. Fibrinogen plays diverse roles in coagulation and inflammation. Thrombin is formed after sequential steps of proteases activation, cleaving fibrinogen at the N-terminal of Aα and Bβ chains. This partially degraded fibrinogen starts to polymerize forming at the end a tridimensional meshwork, the fibrin clot that is further stabilized by the transglutaminase FXIIIa (reviewed in [1]). The fibrin clot together with aggregated platelets form a plug that halt blood extravasation, and simultaneously became a temporal extracellular matrix for wound healing [2] and angiogenesis [3]. Fibrin(ogen) achieve these physiologic functions through its binding to several receptors, integrin and no integrin type, present in platelets (αII_b-β_3), endothelial cells ($\alpha_v\beta_3$, ICAM-1, and VE-cadherin), and leucocytes ($\alpha_M\beta_2$ and $\alpha_x\beta_2$).

Integrins are one of the major families of cell adhesion receptors composed by an α and β subunit, non-covalently linked [4]. Both subunits are type I transmembrane proteins, containing large extracellular domains and, in general, a short cytoplasmic domains. Integrin affinities for their extracellular ligands, such as fibronectin, fibrinogen and collagen are regulated by cellular signaling that results in "inside-out" integrin activation. The αII_b-β_3 and $\alpha_v\beta_3$ recognize ligands containing an RGD sequence [5]. Immobilized fibrinogen binds to the platelet receptor αII_b-β_3 through the sequence γ 400-411 (HHLGGAKQ**AGD**V) [6] and to $\alpha_v\beta_3$ through the C-terminal part of the Aα chain RGD (572-574) [7, 8]. More recently, it has been found that the fibrinogen segment γ 365-383 may function as the binding site for αII_b-β_3 during clot retraction [9], and additionally new fibrinogen $\alpha_v\beta_3$ binding sites have been discovered: γ 190-202 (GWTVFQKRLDGSV) and γ 346-358 (GVYYQGGTYSKAS), and in the α_E (a minor 420 kDa form of fibrinogen, which has a C-terminal highly homologous to the C terminal of the β and

INTECH
open science | open minds

γ chain). It seems that the Aα RGD 572-574 binds to $\alpha_v\beta_3$ at early stage of cell attachment [10]. All these binding sites are inhibited in the presence of the synthetic peptide RGD and antibodies directed against the integrins.

Fibrin(ogen) bridge inflammatory cells to endothelium promoting migration and inflammation by interacting with the leukocyte receptor $\alpha_M\beta_2$ and endothelial cell receptor ICAM-1 [11, 12]. Fibrin may also bridge leukocytes to the endothelium through the interaction with the endothelial cell receptor VE-cadherin [13]. The fibrin binding to the non integrin receptor VE-cadherin take place at the N-terminal part of fibrin (β15-42)$_2$ sequences [14, 15] that become exposed after fibrinopeptides B released. Further, it was demonstrated that fibrin promotes the formation of capillaries after binding to VE-cadherin [16].

Several studies have shown that near the endothelial cells surface the fibrin clot is organized into a dense structure, while farther from the cell´s surface, at approximately 30 to 50 μm the fibrin fibers are looser [17, 18]. This fibers arrangement proximal to the cells is lost in the presence of antibodies against α_v and/or β_3 integrin subunits [17] or the synthetic RGD peptide [18]. In contrast, other studies have concluded that the fibrin structure near the cell surface was related to the thrombin generated on the cells surface, and in their experiments the addition of RGD peptide did not affect the fibrin structure [19, 20]. They did not add antibodies against the integrin subunits that would be more conclusive.

Dysfibrinogenemias are inherited fibrinogen disorders that affect fibrin(ogen) functions according to the localization of the mutation. The affected patients are diagnosed due to a prolonged thrombin time, decreased functional fibrinogen and normal antigenic levels [21].

Abnormal fibrinogens with mutations located in the integrin/non integrin binding sites have been reported. Fibrinogen Kaiserslautern (γ Lys380Asn) and Osaka V (γ Arg375GLy) had delayed clot retraction and impaired αII_b-β_3 binding [22, 23]. Fibrinogen Nieuwegein, with a truncated Aα chain (stretch 454-610 deleted), was less supportive for endothelial cells invasion that was related to the abnormal clot structure rather than to the absence of the RGD integrin $\alpha_v\beta_3$ binding site (α 572-574).

In our laboratory we have examined the clot structure of several dysfibrinogenemias near the cell surface of human dermal microvascular endothelial cells (HMEC-1), forming clots with plasma and purified fibrinogen. We have included fibrinogen Caracas I (Aα 466-610 deleted) and fibrinogen Caracas V (Aα Ser532Cys), with mutations in or near α 572-574, and fibrinogen Caracas VIII (Bβ Tyr41Asn), all of them heterozygous for the mutation. Furthermore, for comparative purposes, clots were formed with fragment X, a degraded form of fibrinogen with around 10% of αC domain remaining, and partially degraded Bβ chain (1-42 residues) [24].

2. Materials and methods

2.1. Fibrinogen coupling to Alexa Fluor® 488

Human plasma fibrinogen was purified in the laboratory by standard protocol [25]. The fibrinogen was coupled to Alexa Fluor 488® (Invitrogen, Molecular Probes, Rochester, NY,

USA) according to manufacturer guidelines. Briefly, purified protein was diluted to 2 mg/ml with phosphate buffered saline (PBS), and 1M sodium bicarbonate solution was added (1:10). This mixture was transferred to the dye vial, gently mixed and stirred during 1 h at room temperature. The coupled fibrinogen to Alexa 488 was separated from the uncoupled dye using a column provided by the manufacturer. The absorbance of the conjugate solution was read at 280 and 494 nm and protein concentration was calculated using the following equation:

Protein concentration (mg/ml) = $[A_{280} - (A_{494} \times 0.11) \times \text{dilution factor}]/1.51$

2.2. Cell culture and clot assembly

The human dermal microvascular endothelial cells (HMEC-1) were kindly donated by Dr. Edwin Ades and Mr. Francisco J. Candal of the Centers for Disease Control and Prevention (CDC, Atlanta, GA, USA), and Dr. Thomas Lawley of Emory University (Atlanta, GA, USA). HMEC-1 were cultured in MCDB 131 medium supplemented with 10 % fetal bovine serum, epidermal growth factor (10 ng/ml), penicillin (100 U/ml), streptomycin (100 μg/ml), fungizone 0.25 μg/ml and 2 mM L-glutamine. Cells (120,000) were seeded in Lab Tek glass chamber slides and maintained at 37 °C in a humid atmosphere with 95 % air and 5 % CO_2 up to reach approximately 80 % confluence. The cells were stained with 4 μM di-8-anepps (Invitrogen, Molecular Probes, Rochester, NY, USA) for 15 min and washed three times with PBS. Apart, 200 μl of plasma (2 mg/ml fibrinogen concentration) or purified fibrinogen (1 mg/ml) were mixed with fibrinogen - Alexa 488 (95 μg/ml), and clotted with a thrombin - $CaCl_2$ solution, 2.7 and 1 nM of human thrombin (American Diagnostica, Greenwich, CT, USA), and 20 mM and 10 mM $CaCl_2$, respectively. The mixture was immediately transferred on the top of the cells. Simultaneously, clots were formed in other wells in the absence of cells. Clot formation was allowed to progress for 2 h in a tissue culture incubator at 37 °C. Then, the top of the clot was loaded with 200 μl of MCDB 131 medium with not supplement.

2.3. Laser scanning confocal microscopy

The fibrin clots was observed in a laser scanning confocal microscopy (LSCM) system with 2 lasers, an argon ion laser (488 nm excitation and 515/530 LP filter for emission) and a HeNe laser (543 nm) mounted on one Nikon Eclipse TE 2000-U microscope. The objective used was Plan Apo VC 60X water immersion with a work distance of 0.27. The acquisition pinhole was set to 60 μm. The confocal was controlled through the software EZ-C1 from Nikon. The images were acquired with a field of view of 212x212 μm. A z- and orthogonal 3D projection of 10 μm thick (0.5 μm/slice) were done near the cell surface, far from the cell surface (20 to 30 μm), and without cells, using ImageJ 1.46 and Olympus FV10ASW (version 0.2010404) softwares, respectively. For images analysis, the thickness chosen was 10 μm near the cell surface, 5 μm at 25 μm far from the cell surface, and 10 μm without cells. The histogram of the frequency *vs* intensity (gray scale) of z projections were calculated with the Olympus FV10ASW (version 0.2010404), which gave the number of pixels associated to a gray value (0-255; the 0 and the saturation gray values were omitted); these quantities were multiplied and normalized (0-1) using the Origin 8.1 SR2 software, with the following equation:

$$F = (F_0 - Fmin) / (Fmax - Fmin)$$

F: normalized fluorescence

F_0: frequency of each gray-value

Fmin: frequency of the lowest gray value

Fmax: frequency the highest gray value

The mean of F was calculated (F_{mean}), and results represented as column graph, $F_{mean} \pm$ SEM. Statistical analysis was performed using the t-student, and $p<0.05$ was considered statistically significant.

3. Results and discussion

Fibrin clots anchor to different receptors of different cell types through specific fibrin recognition sites present in the three fibrinogen chains. The clot structure changes according to the distance to the cell surface. Near the cell the fibrin density is increased, and pores filling space are smaller compared to greater distance, at approximately 30 μm. This particular fibrin architecture is related to the presence of cells since when clots were formed in its absence the clot structure was uniform, resembling that observed far from the cell surface. The figure 1 shows the column graph of the F_{mean} for Alexa 488 of clots performed with plasma, and figure 2 for clots done with purified fibrinogen. For patients and controls, the fluorescence intensity was greater near the cell surface compared to that far from it, and without the presence of cells, which were statistically significant at $p<0.05$. Furthermore, fibrinogens Caracas I, Caracas V, and Caracas VIII had less fluorescence near the cell surface compared to their respective controls ($p<0.05$).

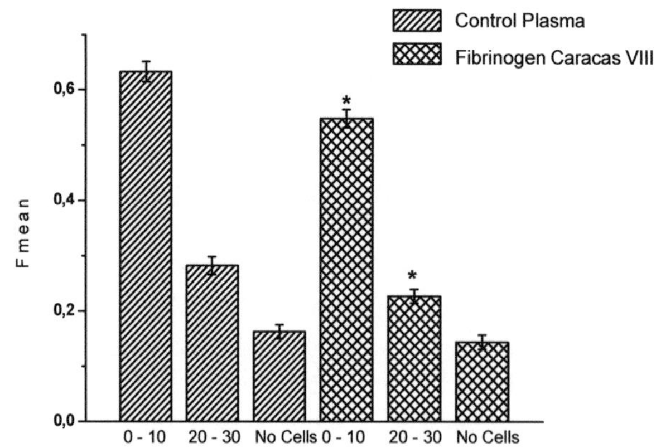

Figure 1. Column graph of Alexa 488-fibrin fluorescence of clot done with plasma. The first and second column of Control plasma and Fibrinogen Caracas VIII represent the distance from the cell surface in μm. * $p < 0.05$.

http://dx.doi.org/10.5772/55586

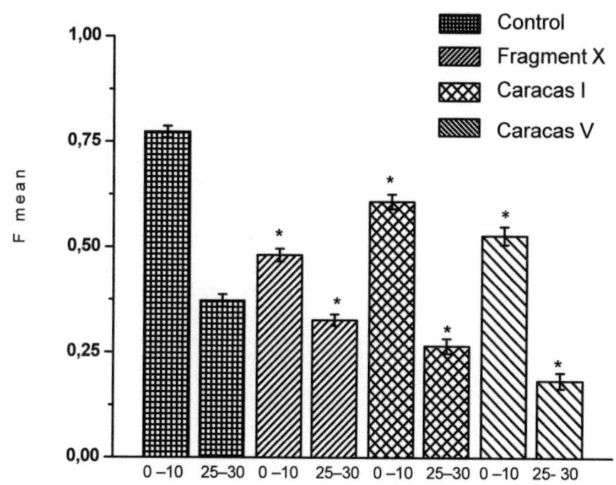

Figure 2. Column graph of Alexa 488-fibrin fluorescence of clot done with purified fibrinogen. The first and second column of each sample represents the distance from the cell surface in µm. * $p < 0.05$, compared to control.

The figure 3 shows a Z projection of the region close to the cell surface, 10 µm thick, when clots were formed with plasma (figure 3a), with purified fibrinogen (figure 3b), and with fibrinogen fragment X (figure 3c). The fields where fibers interacted with cell receptors are brighter compared to that without interactions. Clots formed with plasma and purified fibrinogen had 1.3 and 1.6 times more interaction with the cell surface compared to fragment X, differences statistically significant. It was common to observe stressed fibers as consequence of fibrin binding to cell receptors (signaled with arrow in figure 3b). The interaction of purified fibrinogen with cells was approximately 1.2 times higher than plasma ($p<0.05$). Although plasma had more fibrinogen (0.4 mg total) compared to purified fibrinogen (0.2 mg), the presence of other proteins that compete with fibrin for receptor binding, mainly fibronectin and vitronectin, could decrease fibrin interaction.

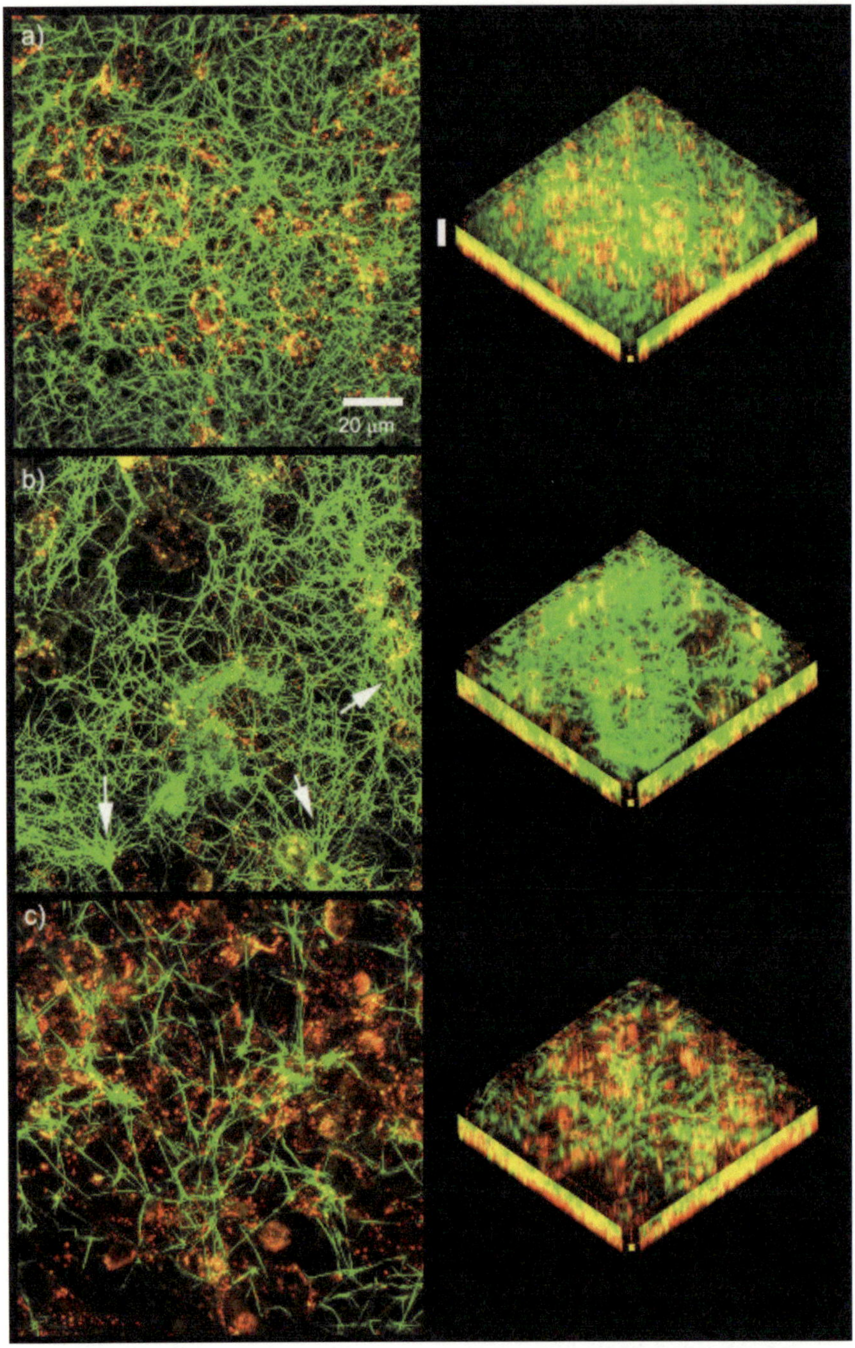

Figure 3. Z- projection of control fibrin clots performed on the top of HMEC-1 cells. Fibrin was labeled with Alexa 488 (green), and cells with di- 8-anepps (red). The thickness of each clot corresponded to 10 µm (from the bottom of the dish). a) Control plasma, b) Normal purified fibrinogen, c) Fibrinogen Fragment X. Next to each Z projection is shown the orthogonal 3D projection. The vertical bar represents 10 µm.

When Z projections were done farther from the cell surface (from 20 to 30 μm), figure 4, the fibrin structure became more homogeneous, the fibrin density decreased and the pores filling space were larger both in plasma (figure 4a) and purified fibrinogen clots (figure 4b), similar to what was observed when clots were formed without cells (figure 4d and 4e); however, the appearance of the fragment X - fibrin almost did not change (figure 4c), similar to the condition without cells (compare figure 3c with 4c and 4f).

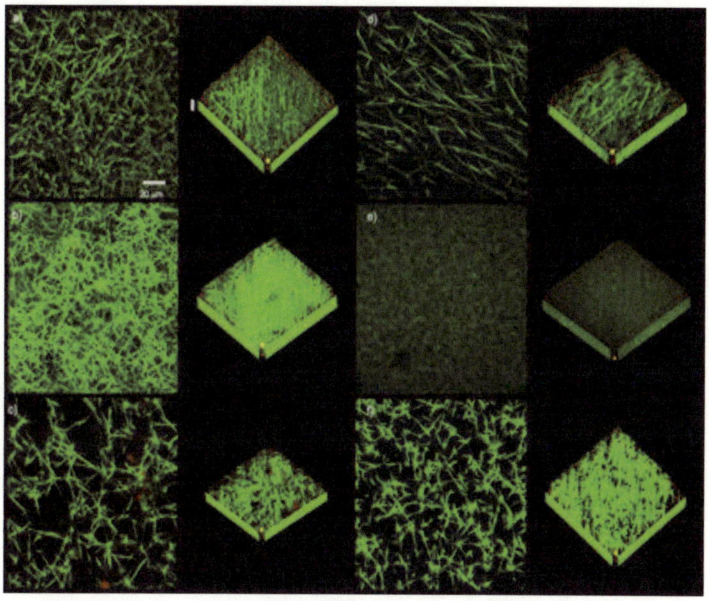

Figure 4. Z projection of control fibrin network located at 20 to 30 μm from the cell surface, and of clots performed in the absence of cells. Fibrin was labeled with Alexa 488 (green), and cells with di-8-anepps (red). The cells were not visualized at this distance. a) Control plasma, b) Normal purified fibrinogen, c) Fibrinogen Fragment X. Clots without cells: d) Control plasma, e) Normal purified fibrinogen, f) Fibrinogen Fragment X. Next to each Z projection is shown the orthogonal 3D projection. The vertical bar represents 10 μm.

Our results confirmed previous work performed by Jerome et al [17]. Other authors have found that the clot organization was related to the thrombin generated on the cell surface, and when they added exogenous thrombin no change was observed in the fibrin morphology [19].

In order to study the effect of certain fibrinogen mutations on fibrin organization on the cell surface, we have chosen dysfibrinogenemias with mutations in or close to the $\alpha_V\beta_3$ integrin binding site (α572-574): fibrinogen Caracas I (Aα Ser466Stop) and Caracas V (Aα Ser532Cys); and in the VE-Cadherin binding site $(B\beta15\text{-}66)_2$ [26]: fibrinogen Caracas VIII (Bβ Tyr41Asn).

Fibrinogen Caracas I and V formed tight clots composed by very thin fibers (figure 5 and 6). The figure 2 shows that fibrin Caracas I and V have less interaction with the receptors of HMEC-1 compared to control (1.3 and 1.5, respectively). The fibrin architecture near the cell surface is shown in figure 5a (Caracas I) and 5b (Caracas V). This pattern was lost far from the cell surface, similar to that without cells (figure 6a and 6b, and 6d and 6e, respectively). Both fibrinogen Caracas I and V have decreased interaction to the $\alpha_v\beta_3$ receptor, due to that fibrinogen Caracas I has decreased population of fibrinogen molecules with α572-574, while the point mutation of Caracas V at α Ser532 probably is closer to the integrin binding site.

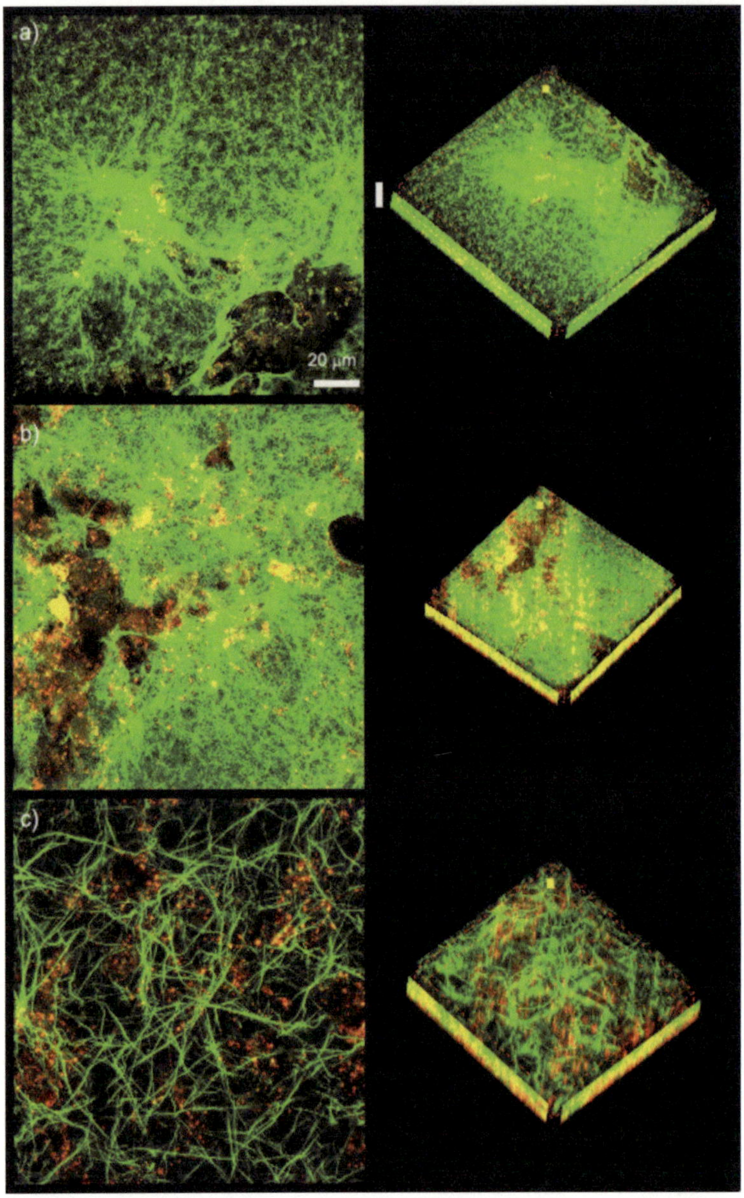

Figure 5. Z-projection of fibrin clots of dysfibrinogenemic patients performed on the top of HMEC-1 cells. Fibrin was labeled with Alexa 488 (green), and cells with di-8-anepps (red). The thickness of each clot corresponded to 10 µm (from the bottom of the dish). a) Purified fibrinogen Caracas I, b) Purified fibrinogen Caracas V, and c) Plasma from fibrinogen Caracas VIII. Next to each Z projection is shown the orthogonal 3D projection. The vertical bar represents 10 µm.

http://dx.doi.org/10.5772/55586

The clinical implications of these results are far to be understood, while fibrinogen Caracas I outcome was mild bleeder, fibrinogen Caracas V was thrombotic. It is more likely that less interaction of the thrombus with the endothelium would embolize the vessel blood, since less adhesion could imply that the force needed to sweep it away are less. Fibrinogen Caracas V had pulmonary embolism but not Caracas I. If there is a threshold value, since fibrinogen Caracas V had less interaction, is subject of future investigation.

The fibrin network of fibrinogen Caracas VIII was done with plasma. The meshwork was composed by thick fibers and large pores (figure 5 and 6). The interaction of patient's fibers with the cell receptors was approximately 1.2 times less than control plasma (figure 1); in spite the fact that the C-terminal region of the α-chain was normal. The figure 5c shows fibrin Caracas VIII organization on the cell surface, compared to fibrinogen Caracas I and V, the fibrin density around the cells was very decreased. The architecture of Caracas VIII fibrin far from the cell surface was similar to that performed without cells (figure 6c and 6f, respectively).

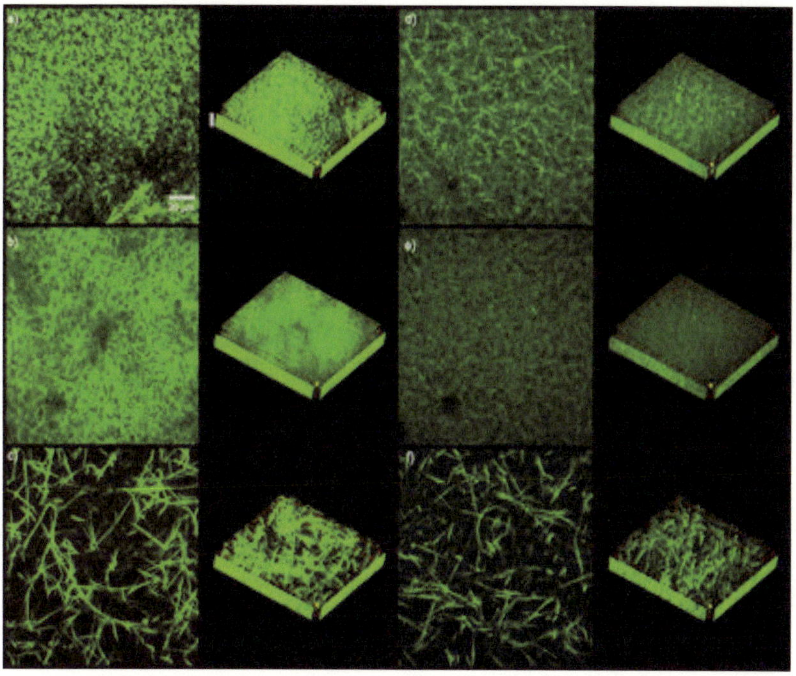

Figure 6. Z projection of dysfibrinogenemic fibrin network located at 20 to 30 μm from the cell surface, and of clots performed in the absence of cells. Fibrin was labeled with Alexa 488 (green), and cells with di- 8-anepps (red). The cells were not visualized at this distance. a) Purified fibrinogen Caracas I, b) Purified fibrinogen Caracas V, and c) Plasma from fibrinogen Caracas VIII. Clots without cells: d) Purified fibrinogen Caracas I, e) Purified fibrinogen Caracas V, and f) Plasma from fibrinogen Caracas VIII. Next to each Z projection is shown the orthogonal 3D projection. The vertical bar represents 10 μm.

One possible explanation is that fiber thickness influence the quantity of ligands per unit of fibrin surface. This would partially explain why Fibrinogen Caracas VIII with normal αC had less interaction with the cells. The contribution of fibrin binding to VE-cadherin to fibrin organization is less clear, since the VE-cadherin contributes to cell-cell and not to cell-extracellular matrix interaction [27].

Fibrinogen Caracas VIII was asymptomatic, and the fibrinolysis process was much faster compared to the above mentioned fibrinogens. Probably this abnormal fibrinogen is protected against embolization due to the fast lysis rate.

The fibrin formed with fibrinogen fragment X had approximately 90% less αC domains and degraded N-terminal Bβ chain, and the least interaction with the cell surface, highlighting the importance of the α572-574 for fibrin organization on the cell surface.

In summary, these preliminary data indicated that the fibrin structure is important for its interaction with cell receptors and that the α572-574 stretch seemed to be the main responsible of fibrin organization on cell surface.

Acknowledgements

We are grateful to Dr. Edwin Ades and Mr. Francisco J. Candal of the Centers for Disease Control and Prevention, and Dr. Thomas Lawley of Emory University (Atlanta, GA, USA) for the donation of the human microvascular endothelial cells (HMEC-1). We want to thank M.Sc. Jenny Dimelza Gómez from the University of Pamplona of Colombia for the fragment X preparation.

Author details

Rita Marchi[1] and Héctor Rojas[2]

*Address all correspondence to: rmarchi@ivic.gob.ve

1 Venezuelan Research Institute, Experimental Medicine Department, Caracas, Venezuela

2 Immunology Institute, Central University of Venezuela, Caracas, Venezuela

References

[1] Stief, T. W. ed. Fibrinogen and Fibrin: structure and functional aspects., ed. R. Marchi. (2012). NOVA.

[2] Laurens, N, Koolwijk, P, & De Maat, M. P. Fibrin structure and wound healing. Journal of thrombosis and haemostasis, (2006). , 932-939.

[3] Van Hinsbergh, V. W, Collen, A, & Koolwijk, P. Role of fibrin matrix in angiogenesis. Annals of the New York Academy of Sciences, (2001). , 426-437.

[4] Hynes, R. O. Integrins: bidirectional, allosteric signaling machines. Cell, (2002). , 673-687.

[5] Humphries, J. D, Byron, A, & Humphries, M. J. Integrin ligands at a glance. Journal of cell science, (2006). Pt 19): , 3901-3903.

[6] Kloczewiak, M, et al. Platelet receptor recognition site on human fibrinogen. Synthesis and structure-function relationship of peptides corresponding to the carboxy-terminal segment of the gamma chain. Biochemistry, (1984). , 1767-1774.

[7] Cheresh, D. A, et al. Recognition of distinct adhesive sites on fibrinogen by related integrins on platelets and endothelial cells. Cell, (1989). , 945-953.

[8] Smith, J. W, et al. Interaction of integrins alpha v beta 3 and glycoprotein IIb-IIIa with fibrinogen. Differential peptide recognition accounts for distinct binding sites. The Journal of biological chemistry, (1990). , 12267-12271.

[9] Podolnikova, N. P, et al. Identification of a novel binding site for platelet integrins alpha IIb beta 3 (GPIIbIIIa) and alpha 5 beta 1 in the gamma C-domain of fibrinogen. The Journal of biological chemistry, (2003). , 32251-32258.

[10] Smith, R. A, et al. Evidence for new endothelial cell binding sites on fibrinogen. Thrombosis and haemostasis, (2000). , 819-825.

[11] Altieri, D. C. Regulation of leukocyte-endothelium interaction by fibrinogen. Thrombosis and haemostasis, (1999). , 781-786.

[12] Languino, L. R, et al. Fibrinogen mediates leukocyte adhesion to vascular endothelium through an ICAM-1-dependent pathway. Cell, (1993). , 1423-1434.

[13] Petzelbauer, P, et al. The fibrin-derived peptide Bbeta15-42 protects the myocardium against ischemia-reperfusion injury. Nature Medicine, (2005). , 298-304.

[14] Bach, T. L, et al. Endothelial cell VE-cadherin functions as a receptor for the beta15-42 sequence of fibrin. The Journal of biological chemistry, (1998). , 30719-30728.

[15] Gorlatov, S, & Medved, L. Interaction of fibrin(ogen) with the endothelial cell receptor VE-cadherin: mapping of the receptor-binding site in the NH2-terminal portions of the fibrin beta chains. Biochemistry, (2002). , 4107-4116.

[16] Martinez, J, et al. Interaction of fibrin with VE-cadherin. Annals of the New York Academy of Sciences, (2001). , 386-405.

[17] Jerome, W. G, Handt, S, & Hantgan, R. R. Endothelial cells organize fibrin clots into structures that are more resistant to lysis. Microscopy and microanalysis : the official journal of Microscopy Society of America, Microbeam Analysis Society, Microscopical Society of Canada, (2005). , 268-277.

[18] Marchi, R, et al. Structure of fibrin network of two abnormal fibrinogens with mutations in the alphaC domain on the human dermal microvascular endothelial cells 1. Blood coagulation & fibrinolysis : an international journal in haemostasis and thrombosis, (2011). , 706-711.

[19] Campbell, R. A, et al. Cellular procoagulant activity dictates clot structure and stability as a function of distance from the cell surface. Arteriosclerosis, thrombosis, and vascular biology, (2008). , 2247-2254.

[20] Campbell, R. A, et al. Contributions of extravascular and intravascular cells to fibrin network formation, structure, and stability. Blood, (2009). , 4886-4896.

[21] Hill, M, & Dolan, G. Diagnosis, clinical features and molecular assessment of the dysfibrinogenaemias. Haemophilia : the official journal of the World Federation of Hemophilia, (2008). , 889-897.

[22] Ridgway, H. J, et al. Fibrinogen Kaiserslautern (gamma 380 Lys to Asn): a new glycosylated fibrinogen variant with delayed polymerization. British journal of haematology, (1997). , 562-569.

[23] Yoshida, N, et al. Characterization of an abnormal fibrinogen Osaka V with the replacement of gamma-arginine 375 by glycine. The lack of high affinity calcium binding to D-domains and the lack of protective effect of calcium on fibrinolysis. The Journal of biological chemistry, (1992). , 2753-2759.

[24] Marder, V. J, & Budzynski, A. Z. The structure of the fibrinogen degradation products. Progress in hemostasis and thrombosis, (1974). , 141-174.

[25] Jacobsen, E. K.P., A modified beta-alanine precipitation procedure to prepare fibrinogen free of antithrombin III and plasminogen. Thrombosis research, (1973). , 145-148.

[26] Yakovlev, S, & Medved, L. Interaction of fibrin(ogen) with the endothelial cell receptor VE-cadherin: localization of the fibrin-binding site within the third extracellular VE-cadherin domain. Biochemistry, (2009). , 5171-5179.

[27] Alghisi, G. C, Ponsonnet, L, & Ruegg, C. The integrin antagonist cilengitide activates alphaVbeta3, disrupts VE-cadherin localization at cell junctions and enhances permeability in endothelial cells. PLoS One, (2009). , e4449.

Section 3

Applications in Food Science

Chapter 11

Applications of Confocal Laser Scanning Microscopy (CLSM) in Foods

Jaime A. Rincón Cardona, Cristián Huck Iriart and
María Lidia Herrera

Additional information is available at the end of the chapter

http://dx.doi.org/10.5772/55653

1. Introduction

Much of the work in the area of physical properties of fats is aimed at determining the relationship among triglyceride structure, crystal properties, crystallization conditions, and macroscopic properties of fats. In finished product containing fat, some of these many macroscopic properties include spredability of margarine, butter and spreads; snap of chocolate; blooming of chocolate; and graininess, smoothness, mouthfeel, water binding, and emulsion stability of spreads [1]. Plastic fats consist of a crystal network in a continuous oil matrix. Many articles in the past have been focused on establishing relationships between lipid composition or polymorphism and macroscopic properties of fats without much consideration of the microstructure of the fat crystal network. Germane to a thorough understanding of plastic fat rheology is a characterization of its microstructure. Not including microstructure as a variable will lead to failure in the prediction of macroscopic properties. In many other non fat or low fat products macroscopic properties depend on their structural organization. Emulsion stability, which is one of the most important physical properties of multiple-phase systems, is strongly determined by oil droplet size and interactions among components that determine spatial distribution of lipid and aqueous phases. Thus, control of food properties for various applications requires a better understanding of the relationships between the food microstructure and macroscopic properties.

Light microscopy is a well-developed and increasingly used technique for studying the microstructure of food systems in relation to their physical properties and processing behavior. To obtain good-quality, high-resolution images of the internal structures of foods it is necessary to cut thin sections of the sample. Procedures that applied substantial shear and compressive forces may destroy or damage structural elements, and sectioning is time consuming and involves chemical processing steps that may introduce artifacts and make image interpretation difficult. Investigations of microstructural changes in foods are increasingly common especially with the growing availability of new microscopic techniques such as the confocal laser scanning microscopy (CLSM) that can probe in situ changes in the microstructure without disturbing the sample. In this instrument, image formation does not depend on transmitting light through the specimen, and, therefore, bulk speciments can be used for the first time in light microscopy. Thus, the CLSM does not require sample fixation and/or dehydration. In addition, when combined with three-dimensional (3-D) reconstruction techniques, optical sectioning may be sufficient to reveal novel information typically unobtainable via traditional two-dimensional (2-D) micrographs since in CLSM information from regions distant from the plane of focus does not blur the image of the focal plane [2]. The primary value of the CLSM to research is its ability to produce optical sections through a three-dimensional (3-D) specimen, for example a thick object such as cheese, yogurt or chocolate. The instrument uses a focused scanning laser to illuminate a subsurface layer of the specimen in such a way that information from this focal plane passes back through the specimen and is projected onto a pinhole (confocal aperture) in front of a detector. Only a focal plane image is produced, which is an optical slice of the structure at some preselected depth within the sample. By moving the specimen up and down relative to the focused laser light, a large number of consecutive optical sections with improved lateral resolution (compared with conventional light microscopy) can be obtained with a minimum of sample preparation. A further advantage of CLSM is the possibility to follow in situ the dynamics of processes such as phase separation, coalescence, aggregation, coagulation, solubilization, etc. Specially designed stages, which allow heating, cooling or mixing of the sample, give the possibility to simulate food processing under the microscope [3].

CLSM has been used in food science since the eighties. Several reviews discussing the application of this technique in microstructural studies of food products have been reported in literature [1, 4, 5]. These reviews have shown the advantages of using CLSM over conventional techniques for studying the relation between the composition, processing, and final properties of food products [6]. The optical sectioning capability of CLSM has proved very useful in the examination of high-fat foods, which are difficult to prepare using the conventional microscopy without the loss or migration of fat globules [7]. In systems like cheese, one of the advantages of this technique is that it can both visualize and chemically differentiate cheese components through the use of protein and lipid specific stains. As internal probes cannot be excited using the lasers typically installed on commercial confocal microscopes, extrinsic fluorechromes must be used such as acridine orange, Nile blue, fluorescein isothiocyanate (FITC), rhodamine B and fast green FCF to stain protein and Nile red to stain fat [8]. An even more detail analysis of food structure may be achieved by the simultaneous labeling of two or more components of foods with probes which are specific for each component. In applications of CLSM in food science there is a minimum sample preparation. A detailed protocol which described how to stain a fat system with Nile red was reported by Herrera and Hartel [6]. Besides, there are no special techniques or modifications to the equipment for food science. The techniques of microscopy are identical as for the life science. The possibility to combine CLSM with rheological measurements, light scattering and other physical analytical techniques in the same experiments with specially designed stages allows obtaining detailed structural information of complex food systems. This chapter reviews applications of CLSM in food systems. Results on bulk fat, emulsions, gels, and a variety of products are reported.

http://dx.doi.org/10.5772/55653

2. Description of fats in bulk

Milk fat has a unique taste and is therefore an important ingredient in many food products. The composition of milk fat is complex, e.g., it contains at least 400 different fatty acids of which 12 are present in proportions greater than 1%. Therefore, the diversity of triglycerides species in milk fat is enormous, resulting in broad crystallization and melting ranges. Furthermore, the breed and feeding of the cow have influence on the milk fatty acids composition, which has impact on the crystallization behavior for anhydrous milk fat and cream. The macroscopic properties of a fat are influenced by a hierarchy of factors. The solid-like behavior in particular is influenced by the amount of solid fat present, the type of crystals, and the interactions among crystals leading to the formation of a fat crystal network. Lipid composition and crystallization conditions will influence crystal habit upon crystallization. Thus, different polymorphic forms and crystal morphologies are possible. Crystallization of milk fat affects several properties that are important for product quality, such as texture, mouthfeel, and rheology. Processing factors such as cooling rate, final temperature and agitation rate are very relevant to crystallization behavior and will also affect product quality. Several authors have studied the microstructure and rheology of fractions of milk fat. Herrera and Hartel [9] have used CLSM to describe the microstructure of blends of 30, 40, and 50% high-melting fraction [Mettler dropping point (MDP) = 47.5°C] in the low-melting fraction (MDP = 16.5°C) of milk fat. The effect of cooling and agitation rates, crystallization temperature, chemical composition of the blends, and storage time on crystalline microstructure (number, size, distribution, etc.) was investigated by CLSM. Samples were crystallized at the selected crystallization temperatures and then were stored at 10°C/min. Therefore, two levels of structure were found: the primary crystals which were quantified (number and size) by using other type of microscopy (polarized light microscopy, PLM) and the final structure whose distribution was qualitatively described by CLSM. This distribution was very relevant to understand rheological properties. To improve resolution, a mix of Nile blue and Nile red dyes was dissolved in the melted samples in proportions that did not modify the nucleation kinetics. When the high melting fraction of milk fat was studied the effect of processing conditions on microstructure was evident. Figure 1 reports the microstructure obtained when milk fat was crystallized at 0.1°C/min and 5.5°C/min. The dark elements are crystals. The liquid oil is bright. Slowly crystallized samples (0.2°C/min) formed different structures from rapidly crystallized samples (5.5°C/min). When slow cooling was used, crystals were sometimes diffuse and hard to distinguish from the liquid. Samples were darker as a result of this solid-mass distribution. However, rapidly crystallized samples had well-defined crystals and seemed to be separated by a distinct liquid phase. These crystals were not in touch with each other as was the case for slowly crystallized samples. Figure 1 showed that slow cooling promoted crystal growth. Fewer crystals with bigger size were obtained, the size of which was quantified by PLM. These microstructures are expected to be related to different mechanical properties, being structures with small crystals harder samples than the ones with greater crystals.

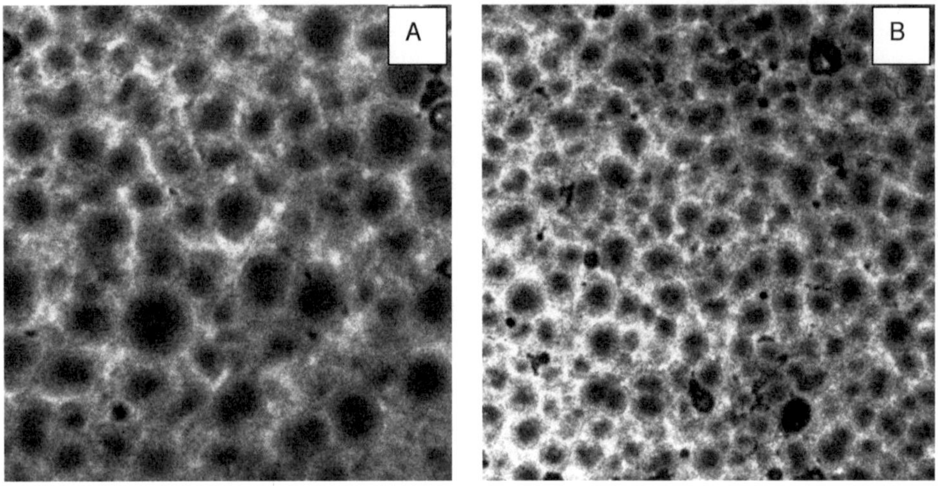

Figure 1. Effect of cooling rate on crystal size for milk fat after 90 min at 35°C and 24 h at 10°C/min. A) slow cooling (0.1°C/min) and B) fast cooling (5.5°C/min).

When agitation rate was tested, higher agitation rates led to smaller crystal size due to enhanced nucleation. Temperature also had a strong impact in crystal size. Larger crystals were formed when crystallization occurred at higher temperatures. High-melting fraction of Milk fat crystals grew during storage: the longer storage time the greater crystal size. The blends of different fractions (high and low melting) of milk fat showed the same microstructural behavior as high melting fraction of milk fat. Greater crystals were found when samples were crystallized at slow cooling rate, slower agitation rate (50 rpm) and higher temperatures (30°C), [9].

Wiking et al. [10] performed further studies to investigate the relationship of microstructure and macroscopic properties analyzing non-fractioned anhydrous milk fat. Using differential scanning calorimetry, synchrotron time resolve X-ray diffraction and pulsed nuclear magnetic resonance, crystallization mechanisms of milk fat were elucidated. Under the same crystallization conditions, the microstructure of the milk fat was analyzed with CLSM and oscillatory rheology. The milk fat was cooled without agitation to 20°C at two different cooling rates, i.e., 0.1 and 10°C/ min. Thereafter, the isothermal crystallization at 20°C was monitored. In agreement with the results shown in Figure 1, faster cooling resulted in a two-step crystallization, and a microstructure that comprised smaller and more uniform crystals than was the case with slower cooling. Consequently, the final texture of the faster cooled milk fat was firmer, i.e., higher complex modulus, than that of the slower cooled milk. X-ray diffraction showed that the two-step crystallization involved a polymorphic transition from α to β' phase.

Milk fat fractions have found application in a variety of food products. The high melting point stearins are useful in puffy pastry, whereas the mid fractions are useful in Danish cookies. Stearins are also used in the reduction in blooming properties of chocolate. The modification by blending of milk fat stearins is an interesting and important approach for utilization of milk fat fractions in a number of edible fat products. Spreadable butter products based on milk fat blended with vegetable oils are becoming increasingly popular and a growing market motivates the industry to develop novel products in this category. Martini et al. [11, 12] studied the effect of blending sunflower oil with a high-melting fraction of milk fat using PLM and CLSM. Figure 2 shows two PLM micrographs of a blend of 90 wt.% high-melting fraction of milk fat and 10 wt.% sunflower oil (A) and 60 wt.% high-melting fraction of milk fat and 40 wt.% sunflower oil (B) after 90 min of crystallization at 40°C. Samples were crystallized with a cooling rate of 0.1°C/min. As may be noticed in Figure 2, the addition of sunflower oil markedly increased crystal size and delayed crystallization kinetics. Less crystals (coming from a smaller number of nuclei) were formed when 40 wt.% sunflower oil was added to high-melting fraction of milk fat. A quantitative analysis of these systems may be found in Martini et al. [11, 12].

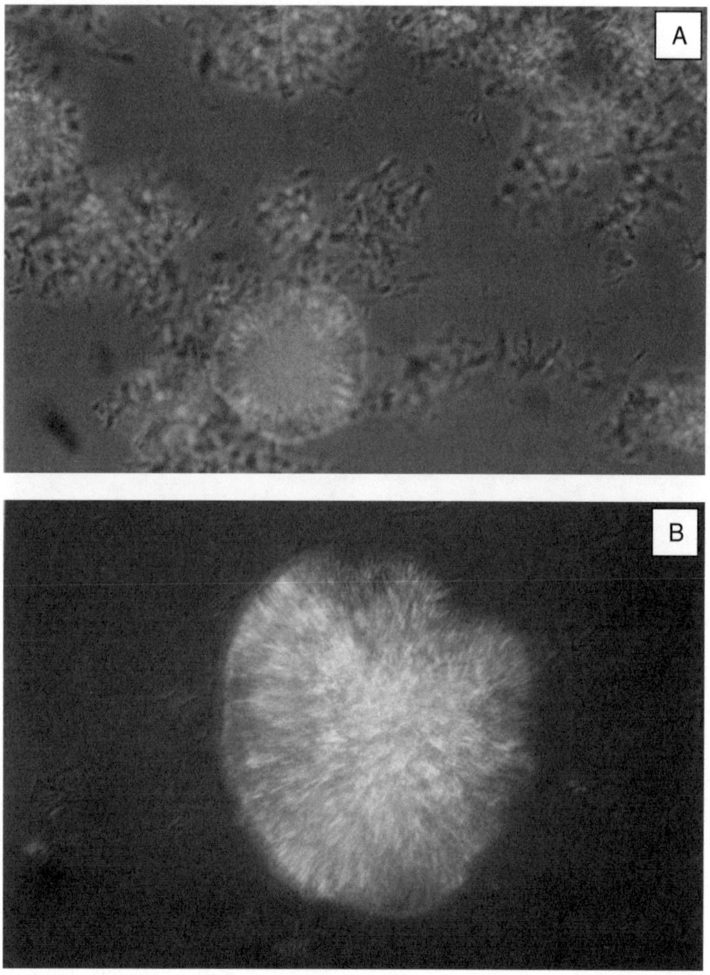

Figure 2. PLM images of two different blends of high-melting fraction of milk fat and sunflower oil crystallized at 40°C for 90 min: A) a blend of 90 wt.% high-melting fraction of milk fat and 10 wt.% sunflower oil and B) a blend of 60 wt. % high-melting fraction of milk fat and 40 wt.% sunflower oil.

Figure 3 shows CLSM images of two different blends of high-melting fraction of milk fat and sunflower oil: A) a blend of 90 wt.% high-melting fraction of milk fat and 10 wt.% sunflower oil and B) a blend of 60 wt.% high-melting fraction of milk fat and 40 wt.% sunflower oil. These images correspond to the PLM images report in Figure 2. The blends were crystallized in two-steps: first they were kept at 40°C for 90 min (as in samples in Figure 2) and secondly they were stored at 10°C for 24 h. Slow cooling rate (0.1°C/min) was used in both cases. In the first step big dark crystals were formed. Then, smaller crystals appeared on the background when samples were cooled to 10°C. The formed structures were too opaque to study by PLM, therefore they were analyzed by CLSM. The advantage of this approach is that the real microstructure can be described without diluting the system. It may be noticed in Figure 3 that addition of sunflower oil increased primary crystals size. In addition, more separated crystals and a clear liquid phase were formed in the background when blends were stored at 10°C.

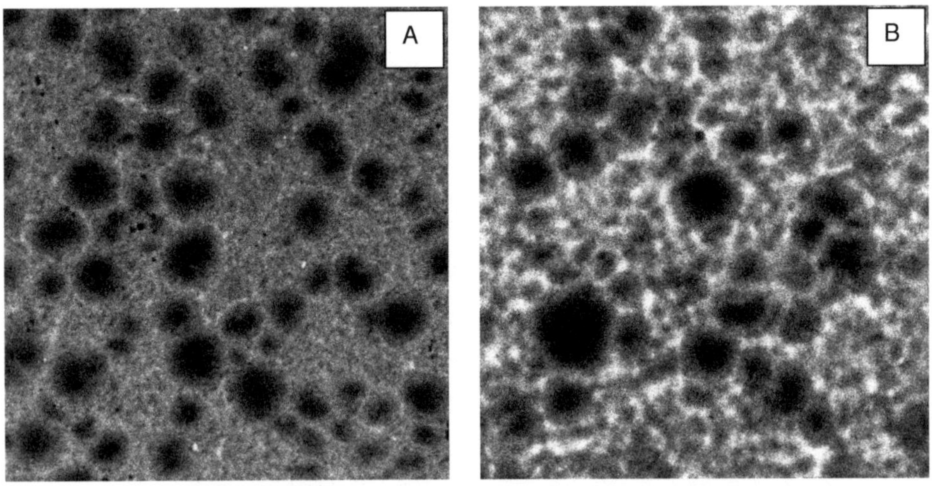

Figure 3. CLSM images of two different blends of high-melting fraction of milk fat and sunflower oil crystallized at 40°C for 90 min and then stored at 10°C for 24 h: A) a blend of 90 wt.% high-melting fraction of milk fat and 10 wt.% sunflower oil and B) a blend of 60 wt.% high-melting fraction of milk fat and 40 wt.% sunflower oil.

Martini et al. [11, 12] also investigated the effect of cooling rate on microstructure of the blends. Cooling rate is known to affect the polymorphic state of the fat crystals as well as its microstructure and texture. Slow cooling usually promotes stable polymorphic forms such as the β' or β forms while less stable polymorphs are obtained using fast cooling, typically the α form. The β form generally leads to greater crystals or harder texture than the other polymorphs. Thus, the different polymorphic forms have different microstructure and different textures. Fast cooling causes rapid development of the thermodynamic driving force for crystallization compared with slow cooling and therefore more and smaller crystals are formed. Figure 4 shows the effect of cooling rate on a 20 wt.% sunflower oil in high-melting fraction of milk fat blend microstructure. The blend was crystallized in two-steps. The first step took place in dynamic conditions with an agitation rate of 50 rpm. After 90 min at 35°C samples were cooled

http://dx.doi.org/10.5772/55653

to 10°C and stored in quiescent conditions for 24 h before CLSM images were taken. Big crystals were obtained in the first step while small crystals were obtained during quiescent crystallization at 10°C. It may be noticed that when fast cooling was applied, nucleation predominated over crystal growth, resulting in more crystals with smaller crystal size. During fast cooling, triacylglycerols were forced to adopt a crystal structure at conditions far from equilibrium, forming mixed or compound crystals. Fast cooling has been shown to yield a higher solid fat content than slow cooling proving that lower melting point triacylglycerols are included in higher melting point triacylglycerols network. Slow cooling allows triacylglycerols to organize in the right conformation to form crystals and more pure crystals with triacylglycerols of higher melting point are obtained resulting in a lower solid fat content and a higher melting point. In all cases the selected blends crystallized in the same polymorphic form, the β′ form. Although cooling rate had a strong influence in microstructure of high-melting fraction of milk fat/sunflower oil blends, it did not have much influence in polymorphism of these systems.

Kaufmann et al. [13] studied anhydrous milk fat and its blends with rapeseed oil containing up to 40% (w/w) of the oil. Samples were crystallized using a slow (0.05 °C/min) or fast (5 °C/min) cooling rate. Melting behavior was examined using differential scanning calorimetry, texture was analyzed by parallel plate compression tests, microstructure was studied by confocal laser scanning microscopy, and solid fat content was measured by pulse-Nuclear Magnetic Resonance. In slow cooled samples addition of rapeseed oil decreased the hardness, which was ascribed to an increase in crystal cluster size. Similar systems were studied by Buldo and Wiking [14]. In agreement with Kaufmann et al. [13], the microstructure found for the blends of butter and rapeseed oil, as analyzed with CLSM, explained the effect on the rheological behavior. The microstructure analysis showed that a high content of rapeseed oil and high processing temperatures produce a less dense crystal network and a change in protein/water distribution.

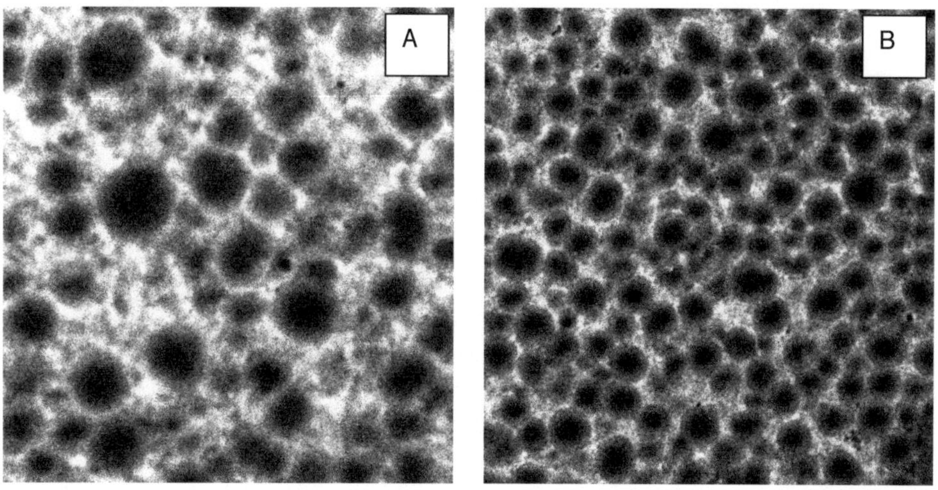

Figure 4. Effect of cooling rate on crystal size for a blend of sunflower oil/high-melting fraction of milk fat after 90 min at 35°C and 24 h at 10°C/min. A) slow cooling (0.1°C/min) and B) fast cooling (5.5°C/min).

3. Biopolimer mixtures

Over the last two decades there has been a resurgence in the study of biopolymer mixtures and in the development of "novel" processing technologies such as High-Pressure Processing. While, the former is motivated by the fact that polysaccharides and proteins almost always co-exist in food products, the later is motivated by the need to preserve food products and extend their shelf-life, without affecting their organoleptic and nutritional properties. To explore both aspects, Hemar et al. [15] performed an experimental study on the effect of High Hydrostatic Pressure on the flow behavior of skim milk–gelatin mixtures. The mixtures which contained 0–1 wt.% gelatin were subjected to different pressures (0–600 MPa) for 15 min at an initial temperature of 5 °C. The behavior of these mixtures was compared to that of aqueous gelatin solutions subjected to the same pressure treatments. Microstructural observations using confocal scanning laser microscopy, small-deformation oscillatory rheology and particle size measurements were performed in an attempt to relate the microstructural properties of these skim milk–gelatin mixtures to their flow behavior. According to the authors this fundamental work, dealing with the effect of high pressure on the physicochemical properties skim milk–gelatin mixtures could be relevant to the industry in several ways. Firstly, skim milk–gelatin mixtures are widely used in the dairy industry, particularly in yoghurt manufacture, where gelatin is used as a stabilizer. In addition the application of High Hydrostatic Pressure to such a system is also relevant, as this technology could be used as a substitute to the conventional heat treatment processes. Secondly, an important finding of this study is that under certain conditions of high pressure and gelatin concentration, an increase in viscosity is observed at intermediate shear-rate (between 10 and 100 s−1). This is highly relevant to industry if the system requires subsequent pumping. Thirdly, from a sensory view point, this range of shear rates (10 and 100 s−1) is comparable to that experienced by a food bolus during swallowing. Thus, this effect of high pressure on the viscosity can influence sensory attribute of the skim milk–gelatin food system.

4. Emulsion systems

Oil-in-Water emulsions consist of small lipid droplets dispersed within an aqueous environment. The lipid droplets are normally coated by a thin layer of emulsifier molecules to prevent them from aggregating. In the food industry, a variety of different kinds of surface-active molecules are used as emulsifiers, including small molecule surfactants, phospholipids, proteins, and polysaccharides. Among proteins, sodium caseinate is widely used as an ingredient in the food industry because its functional properties include emulsification, water-binding, fat-binding, thickening, and gelation. It contains a soluble mixture of surface active caseins ($\alpha s1$-, $\alpha s2$-, β-, and κ-). The caseins adsorb rapidly at the oil–water interface during emulsification and provide long-term stability to oil-in-water emulsions due to a combination of electrostatic and steric stabilization. In milk protein-stabilized emulsions, the presence of ethanol can confer increased stability by reducing the interfacial tension between oil and aqueous phases and so producing a lower average droplet size during emulsification. Motivated by separate observations of the sensitivity of emulsion flocculation to Ca^{2+} content and ethanol concentration, Radford et al. [16] investigated the *combined* roles of ionic calcium content and ethanol concentration as variables controlling depletion-induced flocculation in caseinate emulsion systems. Construction of the global stability diagram was based on a combination of techniques, including rheology, particle-size distribution analysis, visual creaming observations, and CLSM. The emulsion microstructure at various caseinate–calcium–ethanol compositions has been observed by CLSM using the fluorescent dye Rhodamine B. This dye stains only the protein. On the addition of a moderate concentration of ethanol (15 wt%), the dense flocculated protein network appeared to be completely broken down. At that point the microstructure was made up of discrete flocculated droplets separated by relatively large distances. This difference in microstructure with 15 wt.% alcohol addition suggested that alcohol enhance stability at this concentration. A single narrow stable (noncreaming) region was identified, indicating limited cooperation between calcium ions and ethanol.

Physical instability results in an alteration in the spatial distribution or structural organization of the molecules. Creaming, flocculation, coalescence, partial coalescence, phase inversion, and Ostwald ripening are examples of physical instability. The development of an effective strategy to prevent undesirable changes in the properties of a particular food emulsion depends on the dominant physicochemical mechanism(s) responsible for the changes. In practice, two or more of these mechanisms may operate in concert. It is therefore important for food scientists to identify the relative importance of each mechanism, the relationship between them, and the factors that influence them, so that effective means of controlling the stability and physicochemical properties of emulsions can be established. Emulsions have been studied by numerous techniques, such as particle sizing, microscopy, rheology, among others, to characterize their physical properties. Most of these techniques involve some form of dilution. This dilution disrupts some structures that contribute to destabilization. The ability to study the stability of food emulsions in their undiluted forms may reveal subtle nuances about their stability. That is the case of CLSM and another new techniques such as Turbiscan. The Turbiscan method, allows scan the turbidity profile of an emulsion along the height of a glass tube filled with the emulsion, following the fate of the turbidity profile over time. The analysis of the turbidity profiles leads to quantitative data on the stability of the studied emulsions and allows making objective comparisons between different emulsions. Turbiscan measurements together with dynamic light scattering measurements are two techniques that allow quantifying the microstructures described by CLSM. Alvarez Cerimedo et al. [17] studied emulsions stabilized by sodium caseinate. In that article the effect of trehalose on emulsion stability was followed by Turbiscan, the microstructure of emulsions was described by CLSM using Nile red as fluorechrome, and the particle size distribution of emulsion droplets was studied by dynamic light scattering. The fat phase appears in red in CLSM images. The volume-weighted mean diameter ($D_{4,3}$), volume percentage of particles exceeding 1 μm in diameter ($\%V_{d>1}$), and width of the distribution (W) of emulsions formulated with 10 wt.% fish oil as lipid phase and different concentrations of trehalose or NaCas are summarized in Table 1.

Sample		$D_{4,3}$	$\%V_{d>1}$	W
NaCas	T	(μm)	(%)	(μm)
4.5	0	0.49	8.22	0.85
5.0	0	0.37	2.71	0.52
5.5	0	0.46	7.18	0.80
6.0	0	0.47	7.21	0.80
6.5	0	0.46	6.20	0.75
7.0	0	0.45	6.01	0.74
4.5	40	0.24	0.00	0.23
5.0	40	0.29	0.00	0.33
5.5	40	0.22	0.00	0.15
6.0	40	0.21	0.00	0.15
6.5	40	0.21	0.00	0.15
7.0	40	0.21	0.00	0.14
4.5	20	0.37	2.83	0.52
4.5	30	0.30	0.79	0.35
7.0	20	0.32	1.12	0.39
7.0	30	0.27	0.51	0.29

Table 1. Volume-weighted mean diameter ($D_{4,3}$, μm), volume percentage of particles exceeding 1 μm in diameter ($\%V_{d>1}$), and width of the distribution (W) of emulsions formulated with fish oil (10 wt.%) as lipid phase and different concentrations of trehalose or sodium caseinate (NaCas) immediately after preparation.

In all cases, fish oil emulsions showed a monomodal distribution regardless of sodium caseinate concentration. However, $D_{4,3}$,$\%V_{d>1}$, and W significantly decreased as sodium caseinate concentration increased indicating that protein concentration limited the fat globule size in this concentration range. Emulsions with trehalose also showed a monomodal distribution. $D_{4,3}$ was always smaller than for the emulsions without sugar in the aqueous phase showing that trehalose had strong interactions with the protein influencing droplet size. Besides, distribution for emulsions with trehalose (specially 40 wt.%) were very narrow.

The Turbiscan equipment has a reading head which is composed of a pulsed near-IR light source (λ= 850 μm) and two synchronous detectors. The transmission detector receives the light, which goes through the sample (0°), while the back-scattering detector receives the light back-scattered by the sample (135°). The curves obtained by subtracting the BS profile at t= 0 from the profile at t, that is $\Delta BS=BS_t - BS_0$, display a typical shape that allows a better quantification of creaming, flocculation and other destabilization processes. Creaming was detected using the Turbiscan as it induced a variation of the concentration between the top and the bottom of the cell. The droplets moved upward because they had a lower density than the surrounding liquid. When creaming take place in an emulsion, the ΔBS curves show a peak at heights between 0–20 mm. Flocculation was followed by measuring the BS_{av} as a function of storage time in the middle zone of the tube. The optimum zone was the one no affected by creaming (bottom and top of the tube), that is, the 20–50 mm zone. Figure 5 reports as an example, changes in back scattering (BS) profiles (expressed in reference mode, ΔBS) as a function of the tube length with storage time (samples were stored for 1 week, arrow denotes time) in quiescent conditions for emulsions with fish oil as fat phase, no trehalose added to the aqueous phase, and different concentrations of sodium caseinate: A) 4.5 wt.%, B) 5.0 wt.%, C) 5.5 wt.%, D) 6.0 wt.%, E) 6.5 wt.%, and F) 7.0 wt.%. All emulsions but the one stabilized with 6.0 wt.% of sodium caseinate destabilized by flocculation as indicated by the decrease in BS in the central part of the tube (20-50 mm). The 6.0 wt.% emulsion was more stable. However, a slightly creaming was detected by Turbiscan as evidenced by the decreased of the profile at the bottom of the tube (zone 0-20 mm) and the increase at the top of the tube. The stability zone is very narrow for emulsions without trehalose.

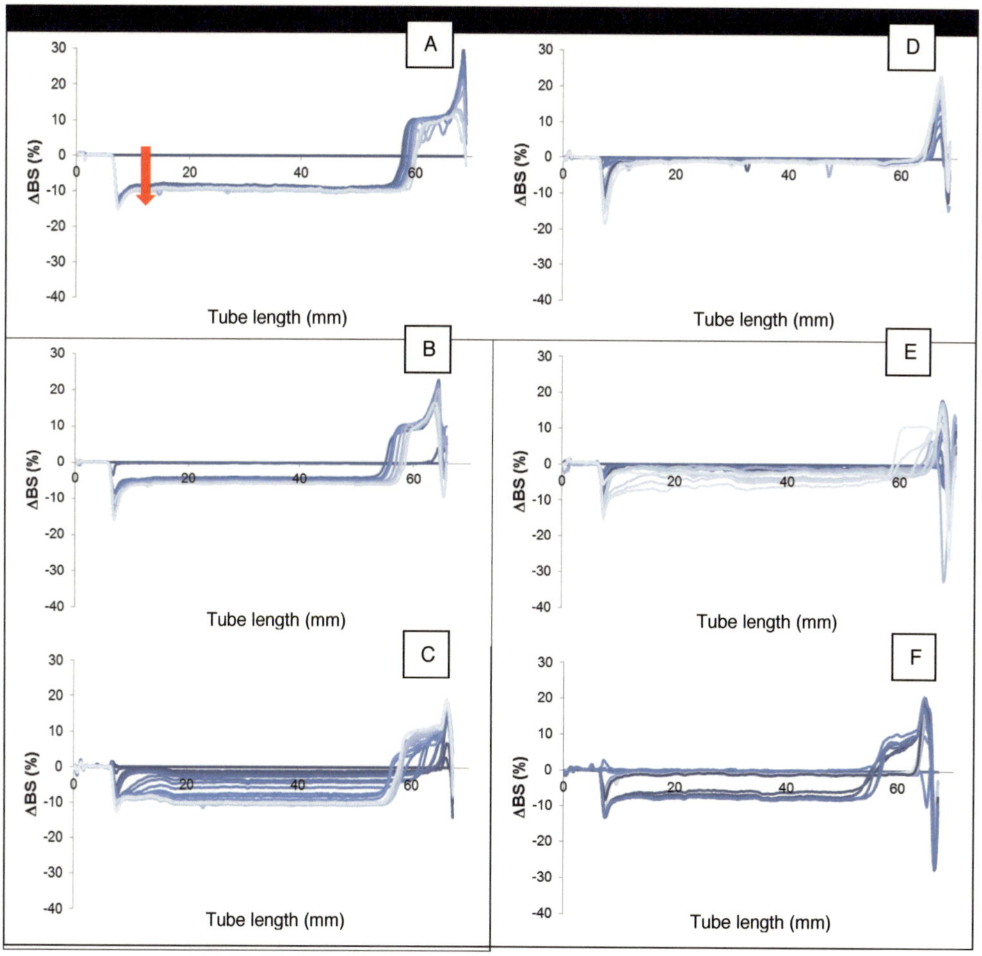

Figure 5. Changes in back scattering profiles in reference mode, as a function of the tube length with storage time (the emulsion was stored for 1 week, arrow denotes time) in quiescent conditions. It was formulated with 10 wt. % fish oil as fat phase, no trehalose added to the aqueous phase, and a concentration of sodium caseinate of A) 4.5, B) 5.0, C) 5.5, D) 6.0, E) 6.5, and F) 7.0 wt.%. Tube length 65 mm.

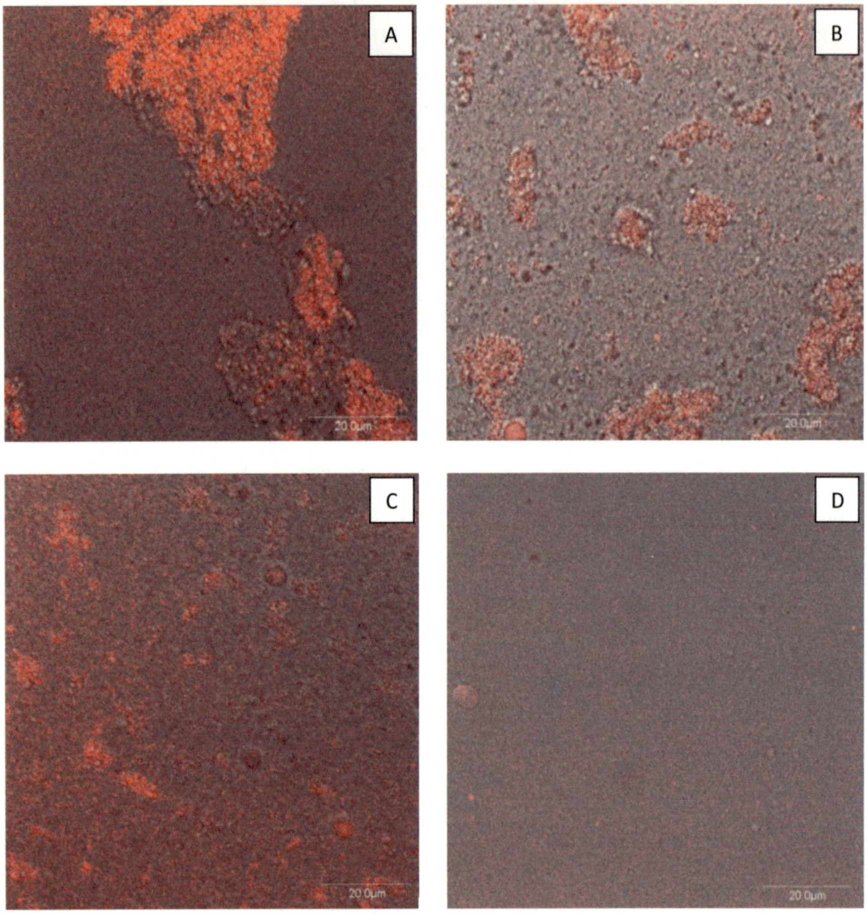

Figure 6. CLSM images of emulsions formulated with 10 wt.% fish oil, 4.5 wt.% sodium caseinate and different concentrations of trehalose: A) 0, B) 20, C) 30, and D) 40 wt.%.

When 40 wt.% trehalose was added to the aqueous phase of emulsions in Figure 5 the rate of destabilization was markedly lower. The Turbiscan analysis showed profiles similar than the one reported in Figure 5 D. The emulsions stabilized with 4.5 wt.% sodium caseinate or more did not flocculate during a week at 22.5 °C. ΔBS profiles in the central zone of the tube (20-50 mm) did not change during that time. These results were in agreement with the decrease in $D_{4,3}$ showed in Table 1. It was reported that polysaccharides which are not particularly surface active such as xanthan gum and which are usually added to the aqueous phase of emulsions as thickening agents to retard instability mechanisms did not affect the size of the emulsion droplets. On the contrary, the fact that particle size diminished for sugar addition does not allow disregarding interactions. In some systems droplet size can be smaller if polysaccharides are present with the protein during homogenization, so the rate of creaming can be reduced as long as there is no bridging flocculation. Interactions between polysaccharides and proteins are based on hydrogen bond, and dipole-dipole associations, in which the presence of OH-groups plays a predominant role. Beside, in the Biochemistry field, strong interactions between peptides and short oligosaccharides have been recently reported as playing an important role in protein recognition. All the above indicates that sugar-protein interactions are very common. Regarding trehalose special properties described in literature it is not unreasonable to think that trehalose may work as coadjutant in micelle formation and, also, may change the quality of water as a solvent, improving the solvent-protein interaction, leading to smaller particles sizes and more stable emulsions.

To further explore this hypothesis, emulsions structure were studied by CLSM. Figure 6 reports CLSM images of emulsions formulated with 10 wt.% fish oil, 4.5 wt.% sodium caseinate and different concentrations of trehalose: A) 0, B) 20, C) 30, and D) 40 wt.%. As may be noticed from the CLSM images of Figure 6 A, when the aqueous phase did not contain trehalose the resulting emulsion destabilized by flocculation. The Turbiscan profile of this emulsion was similar than the one showed in Figure 5 A. In agreement with the Turbiscan profile, CLSM detected flocs formation in this sample. When sugar was added microstructure changed being the changes more evident as trehalose concentration increased. Flocs diminished their size with sugar addition (Figure 6 B and C). When trehalose concentration was 40 wt.%, single drops with a smaller size were noticeable in CLSM image (Figure 6 D). This more uniform microstructure was indicative of a greater stability. These changes in microstructure, qualitatively detected by CLSM, were quantified by measuring droplet size distribution by dynamic light scattering techniques (Table 1) and evaluating the Turbiscan profiles. These quantitative results were in agreement with the description provided by CLSM images, i.e., image reported in Figure 6 A corresponds to the Turbiscan in Figure 5 A. It is clear in Figure 6 A that emulsion structure is formed by flocs. This was the expected result from the profile in Figure 5 A that corresponds to an emulsion which main mechanism of destabilization is flocculation. The Turbiscan profile corresponding to the CLSM image in Figure 6 D was similar to the one reported in Figure 5 D. In agreement with Turbiscan measurements the microstructure of this emulsion (Figure 6 D) shows a homogeneous structure.

Figure 7 reports CLSM images of emulsions formulated with 10 wt.% fish oil, 7 wt.% sodium caseinate and different concentrations of trehalose: A) 0, B) 20, C) 30, and D) 40 wt.%. All these microstructures corresponded to homogeneous emulsions with small droplets evenly distributed. As expected from CLSM images, the quantification of all systems by Turbiscan showed that they were very stable. Besides, in agreement with their microstructures, these emulsions showed a monomodal and very narrow distribution with a small $D_{4,3}$ when analyzed by dynamic light scattering (Table 1).

In order to slow down the destabilization of emulsions, thickening agents such as polysaccharides and hydrocolloids are frequently used. In Álvarez Cerimedo et al. study [17] it was shown that the effect of trehalose was further than the ability to form viscous solutions since it diminished average particle size values for the same processing conditions. The interactions between protein and sugar also played an important role in stabilization although was not enough to suppress the depletion effect that led to instability of the emulsions formulated with sodium caseinate concentrations from 2 to 4 wt.%. Although sucrose has a different structure than trehalose, with the monosaccharides units bound C1–C4, according to our results, it seems

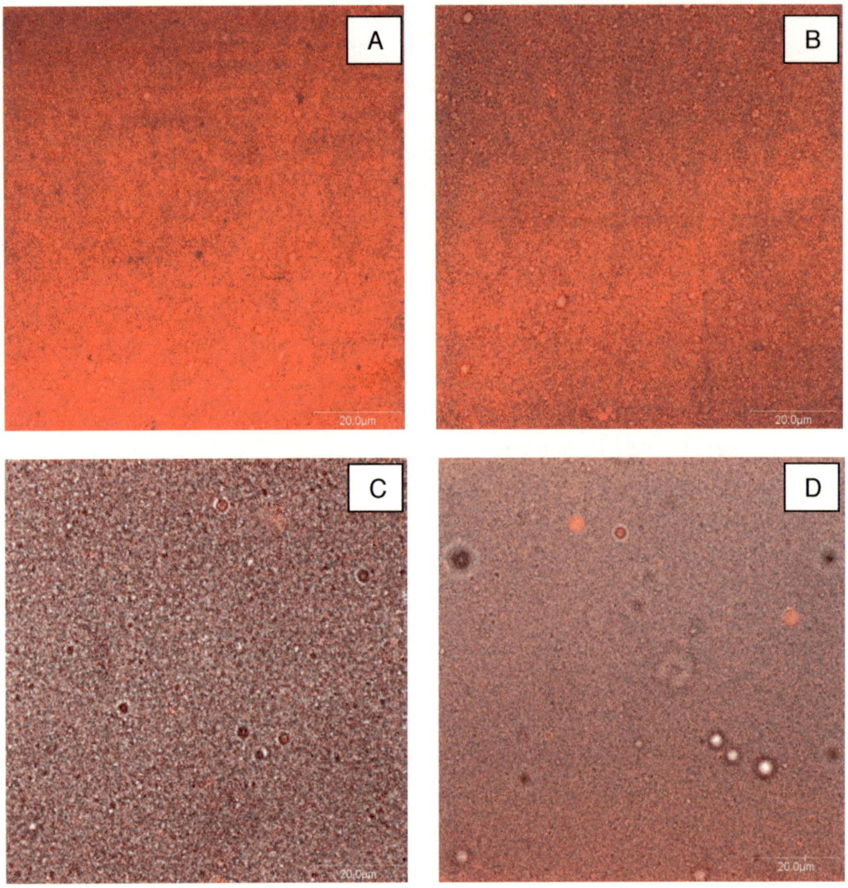

Figure 7. CLSM images of emulsions formulated with 10 wt.% fish oil, 7 wt.% sodium caseinate and different concentrations of trehalose: A) 0, B) 20, C) 30, and D) 40 wt.%.

to have the same microstructural properties. It was reported that there was a pronounced dissociation of sodium caseinate sub-micelles in the presence of sucrose at a pH above the protein´s isoelectric point due, most likely, to direct hydrogen bonding between sodium caseinate and sucrose. The dissociation of sodium caseinate sub-micelles was in excellent agreement with the more homogeneous microstructure and the formation of smaller compact protein structures as detected by CLSM.

4.1. Combination of CLSM and other techniques

Other applications of CLSM in emulsions in combination with other techniques are as follows:

Emulsions were also used to encapsulate lipids. Emulsion-based delivery systems have been developed to increase the oral bioavailability of lipophilic compounds within the gastrointestinal tract. They have also been used to control the release of lipophilic agents at specific locations within the gastrointestinal tract, such as the mouth, stomach, small intestine, or colon.

Emulsion-based delivery systems typically utilize relatively small lipid droplets (d = 10 nm–100 μm) to contain the lipophilic components. The functionality of this type of delivery system can be tailored by controlling their composition and structural organization. This has led to the development of a number of different categories of emulsion-based delivery systems, including conventional emulsions, nanoemulsions, multilayer emulsions, solid lipid nano-particles, multiple emulsions, microcluster emulsions, and filled hydrogel particles. Each of these systems has its own specific advantages and disadvantages for controlling lipid digestion and for releasing lipophilic components. Li et al. [18] evaluated the performance of four emulsion-based delivery systems with different structures: (A) conventional emulsions; (B) small microcluster emulsions; (C) large microcluster emulsions; (D) filled hydrogel beads. The fate of the delivery systems within the gastrointestinal tract was ascertained by introducing them into rat stomachs. Confocal microscopy showed that system D remained intact in the stomach, but systems A, B and C exhibited considerable disruption leading to droplet coalescence. This study showed that an in vitro digestion model is a useful predictive tool for in vivo feeding studies, and that encapsulation is an effective strategy to control the fate of lipids within the gastrointestinal tract. As mentioned, oil-in-water emulsions are widely used in the food industry to encapsulate lipophilic functional components, such as vitamins, colors, flavors, nutraceuticals, and antimicrobials. They are also used to provide desirable optical, rheological and sensory characteristics to many types of food products, including beverages, sauces, dips, dressings and deserts. Ziani et al. [19] investigated on the interaction of lipid droplets of emulsions with the surfaces of packaging materials, an area in which much less research was done. An improved understanding of the interactions of lipid droplets with packaging materials is important for a number of reasons. If lipid droplets are strongly attracted to the surface of a packaging material, then they may form a coating on the internal walls of the package. This coating may give an undesirable appearance to the product, particularly if the lipid droplets contain colorants. In addition, the lipid droplet concentration within the product will be depleted, which could also change the overall product appearance. The deposition of lipid droplets onto the container walls may also mean that any bioactive components encapsulated within them (such as vitamins or nutraceuticals) are not ingested when the product is consumed. On the other hand, the ability of lipid droplets to become attached to packaging surfaces may be desirable in other applications. For example, one or more layers of lipid droplets containing active ingredients could be deposited onto the surfaces of a packaging material to modify its functional characteristics, such as charge, optical properties, and antimicrobial activity. Ziani et al. [19] examined specifically the interaction between lipid droplets coming from a corn oil/water emulsion and polyethylene surfaces. They prepared lipid droplets with various surface charges in the presence of surfactants with different electrical characteristics: non-ionic (Tween 80), cationic (lauric arginate), and/or anionic (sodium dodecyl sulfate). The ionic properties of polyethylene surfaces were modified by UV treatment. Stable emulsions containing small droplets (d < 200 nm) with nearly neutral (Tween 80), cationic (Tween 80: lauric arginate), and anionic (Tween 80: sodium dodecyl sulfate) charges were prepared by adding different levels of the ionic surfactants to Tween 80 stabilized emulsions. Scanning electronic microscopy, confocal fluorescence microscopy, and ATR-FTIR showed that the number of droplets attached to the polyethylene surfaces depended on the droplet charge and the polyethylene surface characteristics. The greatest degree of droplet adsorption was observed for the cationic droplets to the UV–ozone treated polyethylene surfaces, which was attributed to electrostatic attraction. These results were a contribution to the understanding of the behavior of encapsulated lipophilic components in food containers. From Ziani et al. results [19], it seems to be possible to create novel functional properties of packaging materials by depositing lipid droplets containing active agents onto their surfaces, such as antimicrobials or antioxidants.

4.2. Applications of CLSM in emulsion-based products

Almonds are considered to be a great source of proteins, dietary fiber, health-promoting unsaturated fatty acids, vitamin E, other vitamins, minerals; they are also low in saturated fats, and contain no cholesterol. As a result, almonds are used in several food products, such as almond-based beverages, pastes, butter, snacks and baking goods. Lipids in almonds are present as oil bodies in the nut. These oil bodies are surrounded by a membrane of proteins and phospholipids and are a delivery vehicle of energy in the form of triglycerides, similarly to the more studied bovine milk fat globule membrane. Gallier et al. [20] performed chemical, physical and microscopic analyses of these systems. Their results revealed major differences in the composition and structure of almond oil bodies and bovine milk fat globules. The lipids of both natural emulsions differed in degree of unsaturation, chain length, and class. The almond oil body membrane does not contain any cholesterol or sphingomyelin unlike the bovine milk fat globule membrane. The membranes, a monolayer around almond oil bodies and a trilayer around bovine fat globules, may affect the stability of the lipid droplets in a food matrix and the way the lipids are digested.

The conversion of an emulsion into a stable foam requires the formation of a structure capable of retaining bubble integrity. It is well known that controlled destabilization of an emulsion by partial coalescence of fat droplets can provide the necessary means of foam stabilization. Dairy colloids such as whipped cream and ice cream are important examples of such aerated emulsions, being primarily stabilized by a matrix of partially aggregated fat globules at the air–water interface. Allen et al. [21] compared the aeration properties of acidified casein-stabilized emulsions containing liquid oil droplets to the whipping of dairy cream. The foam systems were characterized in terms of overrun, microstructure, drainage stability, and rheology. CLSM studies were performed staining protein and oil with Nile Blue and Nile Red, respectively. These dyes were incorporated into the emulsion immediately prior to whipping. Nile Red was used to stain the oil in whipped cream. Allen et al. [21] have found that a stable foam of high overrun can be made by aerating a sodium caseinate-stabilized emulsion which is undergoing gradual acidification. Such a foam, similar in fat content to whipped cream, yet containing only liquid oil droplets, is stabilized not by partial coalescence, but by aggregation of the protein-coat surrounding the droplets. Heuer et al. [22] performed further investigations on foam microstructure when subject to pressure change. They studied the static and dynamic effects of pressure on the stability of air bubbles to coalescence in process and formulation regimes that are relevant to much of the food industry. Analysis of the digital images obtained was then used to quantify the effects of pressure on bubble size distribution. According to

http://dx.doi.org/10.5772/55653

these authors, the foam under study was relatively stable under quiescent conditions but could destabilize when subject to a pressure change similar to that of a typical industrial process.

Emulsions can be stabilized not only by surfactants, but also by solid particles, whence they are termed Pickering emulsions. Depending on their surface hydrophobicity, particles can stabilize Oil-in-Water, Water-in-Oil, or multiple emulsions. It is seen that the particles must be reasonably hydrophobic in order to adsorb but this also means that they will also have a tendency to aggregate in the aqueous phase. This aggregation may slow down the rate of droplet coverage by solid particles and curtail stable emulsion formation. However, for Oil-in-Water emulsions there is the option of reducing the aggregation of the particles by dispersing them in the hydrophobic (oil) phase before emulsification. Yusoff and Murray [23] created hydrophobic starch particulates from starch granules (i.e., not free hydrophobic starch molecules) and tested their ability to stabilize model Oil-in-Water emulsions, optimizing the modification required for the formation of new and improved surface active agents. The physical properties (particle size and surface activity) and emulsifying properties (emulsion microstructure and stability) of the starch particulates were investigated. Conventional light transmission microscopy, CLSM, scanning electron microscopy, multi-angle light scattering and laser Doppler light scattering all suggested that a wide range of starch particle sizes was produced. Some particles were considerably smaller than the original starch granule sizes, but a large proportion appeared to be above several microns in size. The prepared emulsions creamed readily, but they were extremely stable to coalescence with no significant change in the emulsion droplet-size distributions for over 3 months.

Controlling the rate of droplet growth is critical for emulsion based products and processes. Particles added to Oil-in-Water emulsion formulations may attach to drops and impart kinetic stability for significant periods (weeks or longer). There is growing interest in how attached particle layers alter the rates of interfacial reactions and release of materials from drops by dissolution or evaporation. Little is known, however, about how particle-stabilized (Pickering) emulsions break. Predicting the lifetime of an emulsion and the processes by which it destabilizes remains challenging, even for surfactant-stabilized emulsions. Avendaño Juárez and Whitby [24] studied droplet evolution in unstable, dilute Oil-in-Water Pickering emulsions with the aim of knowing the processes of destabilization of an emulsion at low particle concentrations that do not kinetically stabilize emulsions. The systems under studied were toluene-in-water emulsions formed at low concentrations of silanised fumed silica nanoparticles. The emulsions were characterized using a combination of light scattering, CLSM and rheology. Their results showed that the particles coagulate in the aqueous phase of the emulsions, undergoing reaction limited, or slow, aggregation. The main cause of destabilization arises from aggregation of the particles into compact clusters that do not protect newly formed drops against flocculation. Transfer of oil between flocculated drops occurs by permeation of toluene molecules through the pores of the close packed particulate shells coating the drops. In contrast, particles form loose networks that keep the drops well separated in emulsions formed at higher particle concentrations. These findings show that the minimum particle concentration required to kinetically stabilize a Pickering emulsion is the concentration required for the particle aggregates to be connected into a single expanded network structure.

Among proteins of vegetable origin, soy protein is an abundant byproduct of soybean oil industry and has good functional properties for food processing because of high nutritional value and the contribution to food texture and emulsifying properties. The isoelectric point of soy protein is about pH 4.8, therefore, soy protein emulsions are unstable at pH around 5. As most foods and beverages are acidic, the poor emulsion stability at isoelectric point limits the applications of soy protein emulsions in food and beverage industries. The stability of protein emulsion can be improved by protein-polysaccharide conjugate produced via covalent bond or protein/polysaccharide complex formed by electrostatic attraction. Yin et al. [25] used a simple, green and effective strategy to produce long-term stable oil in water emulsion from soy protein and soy polysaccharide. Soy protein and soy polysaccharide formed dispersible complexes at pH around 3.25. A high pressure homogenization produced the protein/polysaccharide complex emulsion having a droplet size about 250 nm. A heat treatment of the emulsion resulted in the protein denaturation, forming irreversible oil–water interfacial films composed of soy protein/soy polysaccharide complexes. The droplets of the emulsion were characterized by dynamic light scattering, ζ-potential, transmission electron microscopy, polysaccharide digestion via pectinase, and CLSM observation via dual fluorescence probes. As a result of the polysaccharide being fixed on the droplet surface, the emulsions exhibited long-term stability in the media containing pH values of 2–8 and 0.2 mol/L NaCl. According to Yin et al. [25], the stable soy protein/soy polysaccharide complex emulsion is a suitable food-grade delivery system in which lipophilic bioactive compounds can be encapsulated.

The bioavailability of lipid components depends on their chemical structure, physicochemical properties, and the nature of the food matrix that surrounds them. In some situations it may be advantageous to increase the bioavailability of an ingested lipid, e.g., highly non-polar and crystalline bioactive components, such as carotenoids or phytosterols. In other situations, it may be more beneficial to decrease the bioavailability of an ingested lipid, e.g., saturated fats or cholesterol that can have a negative impact on human health. An improved understanding of the factors that impact the bioavailability of dietary lipids would enable the food industry to design foods to increase, decrease or control lipid digestion and absorption within the human gastrointestinal tract. Hur et al. [26] examined the impact of emulsifier type on the micro-structural changes that occur to emulsified lipids as they pass through a model gastro-intestinal system. Lipid droplets initially coated by different kinds of emulsifiers (lecithin, Tween 20, whey protein isolate and sodium caseinate) were prepared using a high speed blender. The emulsified lipids were then passed through an in vitro digestion model that simulated the composition (pH, minerals, surface active components, and enzymes) of mouth, stomach and small intestine juices. The change in structure and properties of the lipid droplets were monitored by CLSM, conventional optical microscopy, light scattering, and micro-electrophoresis. Nile red (a fat soluble fluorescent dye) was excited with 488 nm argon laser line. The general shape of the particle size distributions of all the emulsions remained fairly similar from initial-to-mouth-to-stomach, exhibiting a major population of large droplets and a minor population of smaller droplets. The largest change in mean droplet size occurred when the emulsions moved from the simulated stomach to small intestine, which might be attributed to digestion of the emulsified lipids by lipase and the incorporation of the lipid digestion products within mixed micelles and vesicles. In general, there was a decrease in mean droplet diameter (d_{32}) as the droplets moved from mouth to stomach to small intestine.

5. Gel-like emulsions

Whey protein is a milk protein widely used in the food industry for gelation and texture formation. The aggregation and network formation of this globular protein have been previously well investigated in the bulk phase. Manoi and Rizvi [27] studied the emulsifying activity and emulsion stability indices of texturized whey protein concentrate and its ability to prevent coalescence of Oil-in-Water emulsions and compared its functionality with the commercial whey protein concentrate 80. The cold, gel-like emulsions were prepared at different oil fractions ($\phi = 0.20–0.80$) by mixing oil with the 20% (w/w) texturized whey protein concentrate dispersion at 25 ºC and evaluated using a range of rheological techniques. Microscopic structure of cold, gel-like emulsions was also observed by CLSM. The results revealed that the texturized whey protein concentrate showed excellent emulsifying properties compared to the commercial whey protein concentrate 80 in slowing down emulsion breaking mechanisms such as creaming and coalescence. Very stable with finely dispersed fat droplets, and homogeneous Oil-in-Water gel-like emulsions could be produced. The emulsion products showed a higher thermal stability upon heating to 85 ºC and could be used as an alternative to concentrated Oil-in-Water emulsions and in food formulations containing heat-sensitive ingredients. Liu and Tang [28] also studied cold gel-like whey protein concentrate emulsions at various oil fractions (ϕ: 0.2–0.6). Emulsions were formed through thermal pretreatment (at 70°C for 30 min) and subsequent microfluidization. The rheogical properties and microstructures, as well as emulsification mechanism of these emulsions were characterized. CLSM analyses confirmed close relationships between rheological properties and gel network structures at various ϕ values. Chen and Dickinson [29] investigated surface properties of heat-set whey protein gels (14 wt%) using surface friction measurement and CLSM. The aims of that work was to establish reliable techniques for surface characterization of aggregated whey protein particle gels and to understand how the gel surfaces are affected by various influencing factors. In their studies, a He-Ne laser was used as the light source and fluorescence from the sample was excited at 543 nm. Rhodamine B was used to stain protein and it was added after gelation. CLSM observations revealed that the protein gel without salt addition has a smooth and flat surface, while the gel with added salt has a much rougher surface.

Mixed biopolymer gels are often used to model semi-solid food products. Understanding of their functional properties requires knowledge about structural elements composing these systems. Thus, it is of prime importance to establish an in-depth understanding of the detailed nature and spatial distribution of structural elements included in the gels. Such information may also allow understanding the formation of protein/polysaccharide mixed gels and developing food products with tailored well-defined textural properties. Plucknett et al. [30] used CLSM to follow the dynamic structural evolution of several phase-separated mixed biopolymer gel composites. Two protein/polysaccharide mixed gel systems were examined: gelatin/maltodextrin and gelatin/agarose. These materials exhibit emulsion-like structures, with included spherical particles of one phase (i.e. polymer A) within a continuous matrix of the second (i.e. polymer B). Compositional control of these materials allows the phase order to be inverted (i.e. polymer B included and polymer A continuous), giving four basic variants for the present composites. For the micrographs presented in the current paper, the gelatin-rich phase appears light, whereas both the maltodextrin-rich and agarose-rich phases appear dark. Tension and compression mechanical tests were conducted dynamically on the CLSM. This technique allowed describing microstructure under deformation. As a conclusion of the study it is postulated that polymer interdiffusion occurred across the interface for the gelatin/agarose system, to a significantly greater extent than for interfaces between gelatin and maltodextrin, resulting in higher interfacial fracture energy. van den Berg et al. [31] also investigated mixed biopolymer gels. They described the structural features of mixed cold-set gels consisting of whey protein isolate and different polysaccharides (gellan gum, high methyl pectin or locust bean gum) at different length scales by using CLSM and scanning electron microscopy. Whey protein cold-set gels were prepared at different concentrations to emulate stiffness of various semi-solid foods. The gels were stained with an aqueous solution of Rhodamine B prior to gelation and allowed to gel inside a special CLSM cuvette. The dye binds non-covalently to the protein network. van den Berg et al. [31] described the structure of their systems with CLSM, at the micrometer length scale, showing that the gels had an homogeneous nature with a maximum resolution of around 1 µm. The CLSM images indicated that density of the protein network was proportionally related to the protein concentration. Going from low concentration to high concentration the pores within the protein network became smaller and disappeared at a protein concentration of 9% (w/w). With decreasing pore size the density of the protein network increases. However, the morphology of whey protein isolates aggregates (i.e. size, shape and their connectivity) could not be described as it was below the resolution of CLSM. During the gel preparation (acidification), the presence of polysaccharides in the whey protein gels led to an initial phase separation into a gelled protein phase and a polysaccharide/serum phase at a micrometer length scale. At the final pH of the gels (pH 4.8, i.e. below the pI of whey proteins), the negatively charged polysaccharides interacted with the protein phase and their spatial distribution was effected by charge density. Polysaccharides with a higher charge density were more homogeneously distributed within the protein phase. Neutral polysaccharide, locust bean gum, did not interact with the protein aggregates but was present in the serum phase. Whey protein gels were also analyzed by scanning electron microscopy. According to van den Berg et al. [31], a combination of the results of two microscopic techniques, CLSM and scanning electron microscopy, appeared to offer unique possibilities to characterize the structural elements of whey protein/polysaccharide cold-set gels over a wide range of length scales. Gaygadzhiev et al. [32, 33] also studied mixed biopolymer gels with focus on the impact of the concentration of casein micelles and whey protein-stabilized fat globules on the rennet-induced gelation of milk. The effect of different volume fractions of casein micelles and fat globules was investigated by observing changes in turbidity, apparent radius, elastic modulus and mean square displacement, in addition to CLSM imaging of the gels. Increasing the volume fraction of fat globules showed a significant increase in gel elasticity, caused by flocculation of the oil droplets. The presence of flocculated oil globules within the gel structure was confirmed by CLSM observations. Moreover, a lower degree of κ-casein hydrolysis was needed to initiate casein micelles aggregation in milk containing whey protein-stabilized oil droplets compared to skim milk. Gaaloul et al. [34] characterized the behavior of mixtures containing κ-carrageenan (κ-car) and whey protein isolate using image analysis method. Heat treatment of whey protein isolate/κ-car mixtures induces protein gelation and a phase-separation process. This is due to a modification in protein conformation. CLSM observations revealed the appearance of protein aggregate domains when phase separation occurred, with microgel droplets of whey protein isolate included in a continuous κ-car phase.

6. Applications of CLSM to different products

6.1. Low fat products

Industry is following the present trend of the market in developing low-fat products without changing the sensory and techno-functional properties of semi-solid milk products. Initially, hydrocolloids and stabilizers were used to imitate the fat perception and to enhance the stability of yoghurt regarding syneresis during storage. A reduced fat content caused a loss in viscosity and structure resulting in an altered appearance, texture, and mouthfeel. An alternative to hydrocolloids and stabilizers is the use of milk ingredients like whey proteins arising as a coproduct in the dairy industry. Krzeminski et al. [35] studied the effect of whey protein addition on structural properties of stirred yoghurt systems at different protein and fat content using laser diffraction spectroscopy, rheology and CLSM. CLSM images illustrated that the presence of large whey protein aggregates and lower number of fat globules lead to the formation of an interrupted and coarse gel microstructure characterized by large interstitial spaces. The higher the casein fraction and/or the fat level, the less interspaced voids in the network were observed. However, according to Krzeminski et al. [35], it is evident that the addition of whey proteins reinforces firmness properties of low-fat yoghurts comparable to characteristics of full-fat yoghurt. Celigueta Torres et al. [36] also studied differences in the microstructure of low fat yoghurt manufactured with microparticulated whey proteins used as fat replacer. One commercial and three experimental microparticulated ingredients with different chemical characteristics were used in the yoghurt formulations and compared to both full and low fat yoghurts without fat replacer. The results showed that the amount of native and soluble whey proteins present in the microparticles had a positive influence on the structure of the formed gel. The created structure, dominated by dense aggregates and low amount of serum, had an increased degree of self similarity or fractality with yoghurts in which fat was present. Gel-like emulsions are very interesting systems since they may be applied in food formulations as a kind of carrier for heat-labile and active ingredients. Studying similar systems, Le et al. [37] described the physical properties and microstructure of yoghurt enriched with milk fat globule membrane material. In milk, lipids are present in the form of globules with a diameter varying from 0.1 to 10 mm. These fat globules have a surrounding thin membrane called the milk fat globule membrane. Fat in the core of the globules is mainly composed of neutral triacylglycerides whereas the lipid fraction in the milk fat globule membrane mainly comprises polar lipids. In view of its nutritional and technological properties, milk fat globule membrane material has a high potential to act as a novel ingredient for the development of new functional food products. From an economic point of view, by-products in dairy processing still have a lower price compared with main-stream products. The utilization of these sources to isolate the functional milk fat globule membrane material, and then apply it in the production of new products, may bring great benefit. Le et al. [37] studied applications of milk fat globule membrane in food product development such as enriched yoghurt. Milk fat globule membrane material was isolated from industrial buttermilk powder and both were used as a supplement for the production of yoghurt. Milk fat globule membrane isolated from BMP contained a high concentration of skim milk proteins. Based on confocal laser scanning microscopy, the microstructure of milk fat globule membrane-enriched yoghurts was denser from that of plain skim milk and buttermilk powder-enriched yoghurts. These results indicate the high potential of milk fat globule membrane material to be used as a novel ingredient for the development of new functional products, utilizing both the technological functionalities as well as the nutritional properties of the material.

6.2. Starch-based products

Starch represents the major storage product of most plants. In contrast to the transient starch found in photosynthetic tissues, storage starch accumulates in the plastids of starch storing tissues such as tubers and seeds over long periods of time to form large, well-organized granular structures. Starch granules are made up of two structurally distinct polymers of glucose. Amylose constitutes about 20-40% of typical storage starch and amylopectin constitutes the remaining 60-80% of the granule and is a much larger and highly branched polymer. The branch points are not randomly distributed but are clustered into an ordered arrangement allowing adjacent linear segments to form double helices. It is now widely accepted that the double helices formed by interacting chains of amylopectin form the basis for the crystalline structure of starch and are ordered into concentric crystalline lamellae interrupted by amorphous lamellae containing the branch points. Glaring et al. [38] used CLSM imaging and 8-amino-1,3,6-pyrenetrisulfonic acid (APTS) as a probe for starch molecule distribution and Pro-Q Diamond as a specific probe for phosphate. In conjunction with CLSM they described their systems with scanning electron microscope. The aim of their work was to study the relationship between internal and external structural features of starches extracted from different botanical sources and genetic backgrounds. The investigation has focused on characterizing the genotype-specific internal deposition of starch molecules and how this is manifested at the surface of the granule. By using a combined surface and internal imaging approach, interpretations of a number of previous structural observations is presented. In particular, internal images of high amylose maize and potato suggest that multiple initiations of new granules are responsible for the compound or elongated structures observed in these starches. CLSM optical sections of rice granules revealed an apparent altered distribution of amylose in relation to the proposed growth ring structure, hinting at a novel mechanism of starch molecule deposition. Well-described granule features, such as equatorial grooves, channels, cracks, and growth rings were documented and related to both the internal and external observations. Ishihara et al. [39] used CLSM to visualize the degree of starch gelatinization and the distribution of gum Arabic and soybean soluble polysaccharide in the starch/polysaccharide composite system using fluorescein isothiocianate and rhodamine B as fluorescent dye. The aim of their work was to improve the quality, overall palatability and commercial value of rice-based food products by using novel texture modifiers. CLSM allowed them to see both the starch granules and the glutinous layer around the surface of the starch granules as a result of gelatinization in the presence and the absence of gum Arabic and soybean soluble polysaccharide. According to their results, soybean soluble polysaccharide was more effective in lowering the degree of starch gelatinization than gum Arabic in terms of granular size and the volume of the glutinous layer. Their images showed that the distribution pattern was different between these polysaccharides. Gum Arabic was present in a dispersed and scattered manner while soybean soluble polysaccharide covered in a continuous way the surface of the starch granules. Trivedi et al. [40, 41] processed a range of commercial cheese samples containing starch on a Rapid Visco Analyser (RVA) and on a pilot plant scale. This work clearly demonstrated that it was possible to manufacture processed cheese with part of the protein replaced with potato starch, while maintaining similar rheological attributes (firmness) to those of the control and an acceptable melt index. The confocal micrographs of the processed cheeses prepared in the pilot plant suggested that the fat particle size decreased as the starch levels increased. The reduction in fat particle size would then contribute to increase firmness and decrease melt. Sensory evaluation showed that, although the reduced-protein cheese samples had a good, clean, fresh flavour that was comparable with that of the control, at high starch concentrations the starch-containing processed cheese had a pasty texture and tended to stick to the wrapper.

6.3. Processed potatoes

Potatoes (Solanum tuberosum L.) are an important source of carbohydrates and consumed widely in developing, as well as the developed world. Morphologically, a potato tuber is usually oval to round in shape with white flesh and a pale brown skin, although variations in size, shape, and flesh color are frequently encountered. The color, size, and cooked potato texture are the main quality attributes assessed by consumer for the acceptability of potatoes at domestic scale. However, a quality screening for the industrial processing of potatoes include several parameters, such as dry matter, starch content and characteristics, post harvest, and post processing shelf stability. Texture is one of the most essential technological quality attributes in processed potatoes. It is affected by raw material properties and processing conditions including salting of potatoes and cooking conditions. As salting of potatoes is an important issue for industrially processed potatoes, Straadt et al. [42] investigated the influence of salting on changes in texture, microstructure and water mobility and distribution in two raw material qualities of potatoes presented by two dry matter fractions within one variety. This study was the first report of the combined use of low-field nuclear magnetic resonance and CLSM in the study of the effects of salting on low and high dry matter potato tissue. The simultaneous use of CLSM and low-field nuclear magnetic resonance resulted in important information in relation to the interpretation of the origin of the low-field nuclear magnetic resonance water populations. Salting caused the raw potato cells to loose weight, which in the microscopic images was observed as loss of turgor pressure still further away from the edges of the samples with increased salting time. The paper illustrates the aptitude of low-field nuclear magnetic resonance and CLSM to determine and elucidate structural changes and associated changes in water mobility in potato tissue. Bordoloi et al. [43] described cooking, microstructural and textural characteristics from four New Zealand potato cultivars (Agria, Nadine, Moonlight, and Red Rascal). Potatoes from the waxy cultivar, Nadine, showed lowest dry matter and starch content and also had highest cooking time compared to the other cultivars. CSLM micrographs revealed Moonlight and Red Rascal raw potato parenchyma cellular structure to be well integrated, showing compact hexagonal cells. Raw tubers from these cultivars also exhibited higher hardness and cohesiveness, as observed using texture profile analysis. Moonlight potato parenchyma retained cell wall outline after cooking and its cells were observed to be completely filled with gelatinized starch matrix, whereas the cellular structure of Nadine potato parenchyma was completely disintegrated after cooking.

6.4. Chocolate

In addition to milk fat, another possible source of trapped fat in chocolate is from the cocoa ingredients, which may be in the form of cocoa mass, often also referred to as cocoa liquor, or cocoa powder. The total fat content of cocoa mass is typically about 55 g/100 g, while it is between 11 and 22 g/100 g for standard cocoa powders and less than 1 g/100 g for a highly defatted cocoa powder. The fat content of the cocoa ingredient counts towards the fat content of chocolate, which is usually around 30 g/100 g. A chocolate formulated with a cocoa solid of high trapped fat content contains less free fat which affects the viscosity behavior of the chocolate in its molten state since only the free fat 'contributes' to the chocolate's flowability. A certain level of flowability is required to facilitate processing such as pumping, conching, molding and enrobing as well as to impart an acceptable mouthfeel. Therefore, in an attempt to formulate chocolate with a lower fat content, simply decreasing the level of fat is not a valid approach. Aiming at the manufacture of reduced fat chocolates Do et al. [44] developed a novel method of trapped fat reduction: manipulation of the cocoa ingredient. Cocoa mass was replaced with cocoa powder (11 g/100 g or <1 g/100 g fat) and added 'free' cocoa butter. CLSM was used to visualize the internal microstructure of cocoa particles paying particular attention to fat partitioning. The approach was to let fluorescently stained oil diffuse into the particles to identify (i) the porous microstructure and (ii) the presence of fat inclusions on the CLSM micrographs acquired. Imaging of the cocoa solids was performed on the raw material as obtained. Results showed that the cocoa mass chocolate had a higher viscosity than cocoa powder chocolates of the same total fat content due to the presence of trapped fat globules as identified by CLSM. Based on the evidence presented, it can be concluded that standard defatted cocoa powder, as widely used by the industry, is indeed the best compromise in terms of free fat, particle size and morphology attempting to formulate fat reduced chocolate of acceptable molten state viscosity. Svanberg et al. [45, 46] also studied changes in chocolate structure with formulation. They described the effect of major chocolate ingredients (sugar, cocoa particles and lecithin), in combination with the two pre-crystallization techniques, seeding and non-seeding, on the kinetics of cocoa butter crystallization and the resulting microstructure. CLSM was used to monitor microstructural evolution under dynamic thermal conditions. Differential scanning calorimetry measurements and image analysis were also applied in order to quantify the impacts of processing and formulation on microstructure. Segmentation between solid particles (i.e., sugar and cocoa particles, crystalline and liquid

http://dx.doi.org/10.5772/55653

cocoa butter) was made through manual thresholding, which shows solid particles as black and liquid fat as significantly bright, allowing it to be distinguished from the crystals. Both the pre-crystallization techniques and the ingredients proved to have a significant impact on the kinetics and morphology of cocoa butter crystallization, and substantially different microstructures were observed. Complementary differential scanning calorimetry measurements were performed to estimate the amount of stable versus non-stable crystals present in the samples. These calculations also displayed large variations between seeded and non-seeded samples, as well as between the additions of ingredients, especially lecithin.

6.5. Cheese

Many of the properties of cheese such as cheese texture and flavour are determined by the spatial arrangement of components including: the casein particles that form a protein matrix, the fat globules, dispersed water and minerals. The arrangement of these components on the micron scale is known as the microstructure of cheese. In Cheddar cheese, this microstructure develops when milk coagulates to form a gel, usually through the enzymatic action of chymosin. Changes in process conditions or the choice of ingredients can alter the microstructure of the gel, curd and cheese considerably and thus the functional properties of the final cheese product. The ability to image and characterize these changes will provide an important tool for the quality control of cheese and other dairy products. Ong et al. [47] used cryo scanning electron microscopy and CLSM to visualize changes in the microstructure of milk, rennet-induced gel and curd during the manufacture of Cheddar cheese. When the structure of the gel, curd and cheese was observed using CLSM, spherical fat globules were mostly present in the serum pores of the gel prepared from unhomogenized milk but were found embedded in the aggregated chains of the casein network within the gel when prepared from homogenized milk. The porosity measurements obtained using cryo SEM were similar to those obtained using CLSM. These two complementary techniques can potentially be used to assist studies for the control of cheese texture and functionality. Costa et al. [48] investigated the effect of an exopolysaccharide-producing starter culture on milk coagulation and curd syneresis during the manufacture of half-fat Cheddar cheese by comparing the effect of an exopolysaccharide-producing starter with its non exopolysaccharide-producing isogenic variant. Coagulation and syneresis were monitored using light backscatter sensor technologies and the charge of the exopolysaccharide was also established. The distribution of the exopolysaccharide within the cheese microstructure was investigated by staining with Concavalin A and using CLSM. The study indicated that the exopolysaccharide produced by this starter culture did not interfere with coagulation but had a significant effect on reducing syneresis shortly after cutting. The exopolysaccharide was observed to be uniformly distributed throughout the cheese matrix and specifically located near the aqueous pores, possibly binding moisture and causing the observed decrease in syneresis.

Oiling-off is a major phenomenon that occurs during heating of semi-hard and hard-cheeses. The formation of a thin layer of free oil on the surface of the pieces of cheeses is required to prevent moisture evaporation during cooking and skin formation that impairs flow and stretch. Free oil lubricates the displacement of the para-casein matrix and has been described as the driving force of cheese melting. Oiling-off is also involved in the browning properties of cheeses which are heated for culinary applications. Indeed oiling-off may modulate dehydration and water activity, which is well known to strongly influence Maillard reactions. However, excessive oiling-off leads to appearance defects such as bi-phasing and to excessive chewiness of partially defatted melted cheese. Richoux et al. [49] investigated the influence of temperature and pressing strength on the oiling-off of Swiss cheese. Twin-cheeses were submitted to various cooling rates after the curd have been dipped from the vat. These cheeses were either pressed or non-pressed. Samples have been characterized using CLSM and were double stained: the fat phase was coded in red and the protein phase was coded in green. Aqueous phase appears as black areas in the CLSM images. Using this strategy, authors showed that a higher heat load yields to larger inclusions of fat in the cheese matrix and to higher oiling-off. The heat load applied to the curd (integral of the temperature vs. time curve) and oiling-off of cheeses aged of 5 days were linearly related. In contrast, the pressing strength did not influence oiling-off. As a conclusion, the heat load applied during pressing and acidification could be a tool to modulate the oiling-off of Swiss cheese.

6.6. Fried food

The market for fried foods remains very large and it continues to increase although there are health concerns associated with consuming high calorie foods. The peculiar organoleptic properties of fried foods such as good mouth-feel, distinct flavor, unique taste and palatability make them irresistible. Application of batter and breading coating is one of the means devised to reduce fat uptake during the frying operation aside from the fact that they add more value such as improved texture, appearance, taste and volume, to fried foods. The structure that is developed during processing to a large extent, defines some of the quality attributes of the coating system. Food structure can be characterized in terms of density, porosity, pore size distribution and specific volume. Porosity and pore size distribution are very important microstructural properties of fried foods needed in process optimization and product development. Adedeji et al. [50] characterized the pore properties and quantify fat distribution in deep-fat fried chicken nuggets batter coating using CLSM. Images were obtained at fluorescence mode with excitation/emission wavelengths of 488/570 nm to show fat distribution, and reflection/trans mode to obtain grayscale images which show structure of the sample that include pore distribution. Fat distribution obtained from image analysis was significantly ($P < 0.05$) affected by the frying temperature and time, and it decreased within the depth of the sample thickness. Pore size varied approximately between 1.20 and 550 μm. Frying process led to the formation of more micropores (pores < 40 μm) and bigger (pore ≥ 216 μm) pores.

6.7. High-protein snack bars

Many processed foods are multicomponent heterogeneous systems that are far from thermodynamic equilibrium. There is often a considerable time lag between manufacture and consumption, during which a product is transported and stored. During this storage time, multiple chemical, physical, and biological reactions occur serially and simultaneously. Some of these reactions lead to the development of desirable attributes, such as good flavors in aged wines and cheeses, but others create flavors, colors, or textures that impact negatively on quality, as perceived by consumers. The shelf life can be defined as the length of time for which a product can be stored before the appearance of the first characteristic that consumers find unappealing, e.g., texture that is too tough. High-protein snack bars (hereafter called 'protein bars') are a convenient and nutritious food format that was originally developed for athletes but is now formulated to appeal to a wide range of health-conscious consumers. These bars contain 15–35% protein, which consists almost exclusively of dairy or soy proteins because of their health benefits and cost effectiveness. Chocolate, sugars, and flavorings create an appealing taste and flavor. Nuts, wafers, nuggets, etc. may be added for novel texture. Vitamins, minerals, and/or fiber are often added for enhanced nutritional value. Loveday et al. [51] examined the contributions of various chemical and physical reactions to the hardening of a model protein bar stored for up to 50 days at 20 °C. During manufacture of the protein bars, a subsample was withdrawn after the final mixing and a few drops of dye were added. The dye was a mixture of Nile Blue and Fast Green FCF dissolved at 0.2% w/v in a commercial antifading mountant medium, Citifluor (Citifluor Ltd., Leicester, UK). The dye was mixed into the protein bar material with a knife until even coloring was achieved, as judged by eye, and then a drop of the mixture was placed on a glass cavity slide and a coverslip was applied. The slides were stored at 20 °C. Over the first 17 h after manufacturing, protein particles became more clustered, and soluble protein appeared to precipitate, as shown by CLSM. Moisture migration induced important changes in molecular mobility and in microstructure.

7. Conclusions

Based on the above presented examples it can be concluded that CLSM broadens the application of conventional light microscopy. Combined with other techniques it is a valuable tool to describe a variety of food systems because it gives the possibility to examine the internal structure of thick samples in three dimensions. Samples may be analyzed without perturbing the system and food transformations during processing may be followed *in situ* using specially designed stages. By using CLSM it is possible the simultaneous labeling of two or more components of food with probes which are specific for each component and therefore food structure may be described in more detail. A variety of model food systems was successfully analyzed, i.e., bulk fats, emulsions and gels. Besides, real products such as yoghurt, potatoes, chocolate, cheese, fried food and protein bar were also studied by CLSM in combination with other techniques given valuable data of food structure and its relationship with macroscopic properties.

Acknowledgements

This work was supported by CONICET through Project PIP 11220080101504, by the National Agency for the Promotion of Science and Technology (ANPCyT) through Project PICT 0060, and by the University of Buenos Aires through Project 20020100100467.

Author details

Jaime A. Rincón Cardona[1], Cristián Huck Iriart[2,3] and María Lidia Herrera[3*]

*Address all correspondence to: lidia@di.fcen.uba.ar

1 School of Science and Technology, University of San Martín (UNSAM), San Martín City, Buenos Aires Province, Argentina

2 Institute of Inorganic Chemistry, Environmental Science and Energy (INQUIMAE), National Research Council of Argentina (CONICET), Buenos Aires City-State, Argentina

3 Faculty of Exacts and Natural Sciences (FCEN), University of Buenos Aires (UBA), Buenos Aires City-State, Argentina

References

[1] Marangoni A G, Hartel R W. Visualization and Structural Analysis of Fat Crystal Networks. Food Technology 1998;52(9) 46-51.

[2] Ding K, Gunasekaran S. Three-Dimensional Image Reconstruction Procedure for Food Microstructure Evaluation. Artificial Intelligence Review 1998;12(1-3) 245–262.

[3] Dürrenberger M B, Handschin S, Conde-Petit B, Escher F. Visualization of Food Structure by Confocal Laser Scanning Microscopy (CLSM). LWT - Food Science and Technology 2001;34(1) 11-17.

[4] Heertje I, van der Vlist P, Blonk J C G, Hendrickx H A C M, Brakenhoff G J. Confocal Scanning Laser Microscopy in Food Research: Some Observations. Food Microstructure 1987;6(2) 115-120.

[5] Blonk J C G and van Aalst H. Confocal Scanning Light Microscopy in Food Research. Food Research International 1993;26(4) 297-311.

[6] Herrera M L, Hartel R W. Unit D 3.2.1-6 Lipid Crystalline Characterization, Basic Protocole. In: Current Protocols in Food Analytical Chemistry (CPFA). New York: John Wiley & sons, Inc.; 2001.

[7] Ong L, Dagastine R R, Kentish S E, Gras S L. Microstructure of Milk Gel and Cheese Curd Observed Using Cryo Scanning Electron Microscopy and Confocal Microscopy. LWT - Food Science and Technology 2011;44(5) 1291-1302.

[8] Auty M A E, Twomey M, Guinee T P, Mulvihill D M. Development and Application of Confocal Scanning Laser Microscopy Methods for Studying the Distribution of Fat and Protein in Selected Dairy Products. Journal of Dairy Research 2001;68(3) 417-427.

[9] Herrera M L, Hartel R W. Effect of Processing Conditions on Physical Properties of a Milk Fat Model System: Microstructure. Journal of the American Oil Chemists' Society 2000;77(11) 1197–1204.

[10] Wiking L, De Graef V, Rasmussen M, Dewettinck K. Relations Between Crystallisation Mechanisms and Microstructure of Milk Fat. International Dairy Journal 2009;19(8) 424–430.

[11] Martini S, Herrera M L, Hartel R W. Effect of Cooling Rate on Crystallization Behavior of Milk Fat Fraction/Sunflower Oil Blends. Journal of the American Oil Chemists' Society 2002;79(11) 1055-1062.

[12] Martini S, Puppo M C, Hartel R W, Herrera M L. Effect of Sucrose Esters and Sunflower Oil Addition on Microstructure of a High-Melting Milk Fat Fraction. Journal of Food Science 2002;67(9) 3412- 3418.

[13] Kaufmann N, Andersen U, Wiking L. The Effect of Cooling Rate and Rapeseed Oil Addition on the Melting Behaviour, Texture and Microstructure of Anhydrous Milk Fat. International Dairy Journal 2012;25(2) 73-79.

[14] Buldo P, Wikin L. The Role of Mixing Temperature on Microstructure and Rheological Properties of Butter Blends. Journal of the American Oil Chemists' Society 2012;89(6) 787–795.

[15] Hemar Y, Liu L H, Meunier N, Woonton B W. The Effect of High Hydrostatic Pressure on the Flow Behaviour of Skim Milk–Gelatin Mixtures. Innovative Food Science Emerging Technology 2010;11(3) 432–440.

[16] Radford S J, Dickinson E, Golding M. Stability and Rheology of Emulsions Containing Sodium Caseinate: Combined Effects of Ionic Calcium and Alcohol. Journal of Colloids and Interface Science 2004;274(2) 673–686.

[17] Álvarez Cerimedo M S, Huck Iriart C, Candal R J, Herrera M L. Stability of Emulsions Formulated with High Concentrations of Sodium Caseinate and Trehalose. Food Research International 2010;43(5) 1482–1493.

[18] Li Y, Kim J, Park Y, McClements D J. Modulation of Lipid Digestibility Using Structured Emulsion-Based Delivery Systems: Comparison of in Vivo and in Vitro Measurements. Food & Function 2012;3(5) 528-536.

[19] Ziani K, Barish J A, McClements D J, Goddard J M. Manipulating Interactions Between Functional Colloidal Particles and Polyethylene Surfaces Using Interfacial Engineering. Journal of Colloid and Interface Science 2011;360(1) 31–38.

[20] Gallier S, Gordon K C, Singh H. Chemical and Structural Characterisation of Almond Oil Bodies and Bovine Milk Fat Globules. Food Chemistry 2012;132(4) 1996–2006.

[21] Allen K E, Dickindon E, Murray B. Acidified Sodium Caseinate Emulsion Foams Containing Liquid Fat: A Comparison With Whipped Cream. LWT - Food Science and Technology 2006;39(3) 225–234.

[22] Heuer A, Coxa A R, Singleton S, Barigou M, van Ginkel M. Visualisation of Foam Microstructure When Subject to Pressure Change. Colloids and Surfaces A: Physicochemical and Engineering Aspects 2007;311(1-3) 112–123.

[23] Yusoff A, Murray B S. Modified Starch Granules as Particle-Stabilizers of Oil-in-Water Emulsions. Food Hydrocolloids 2011;25(1) 42-55.

[24] Avendaño Juárez J, Whitby C P. Oil-in-water Pickering Emulsion Destabilisation at Low Particle Concentrations. Journal of Colloids and Interface Science 2012;368(1) 319–325.

[25] Yin B, Deng W, Xu K, Huang L, Yao P. Stable Nano-Sized Emulsions Produced From Soy Protein and Soy Polysaccharide Complexes. Journal of Colloid and Interface Science 2012;380(1) 51–59.

[26] Hur S J, Decker E A, McClements D J. Influence of Initial Emulsifier Type on Microstructural Changes Occurring in Emulsified Lipids During in Vitro Digestion. Food Chemistry 2009;114(1) 253–262.

[27] Manoi K, Rizvi S S H. Emulsification Mechanisms and Characterizations of Cold, Gel-Like Emulsions Produced From Texturized Whey Protein Concentrate. Food Hydrocolloids 2009;23(7) 1837–1847.

[28] Liu F, Tang C H. Cold, Gel-Like Whey Protein Emulsions by Microfluidisation Emulsification: Rheological Properties and Microstructures. Food Chemistry 2011;127(4) 1641–1647.

[29] Chen J, Dickinson E. Surface Texture of Particle Gels. A Feature in Two Dimensions or Three Dimensions?. Chemical Engineering Research and Design 2005;83(7) 866–870.

[30] Plucknett K P, Pomfret S J, Normand V, Ferdinando D, Veerman C, Frith W J, Norton I T. Dynamic Experimentation on the Confocal Laser Scanning Microscope: Application to Soft-Solid, Composite Food Materials. *Journal of Microscopy* 2001;201(2) 279-290.

[31] van den Berg L, Rosenberg Y, van Boekel M A J S, Rosenberg M, van de Velde F. Microstructural Features of Composite Whey Protein/Polysaccharide Gels Characterized at Different Length Scales. Food Hydrocolloids 2009;23(5) 1288–1298.

[32] Gaygadzhiev Z, Corredig M, Alexander M. The Impact of the Concentration of Casein Micelles and Whey Protein-Stabilized Fat Globules on the Rennet-Induced Gelation of Milk. Colloids and Surfaces B: Biointerfaces 2009a;68(2) 154–162.

[33] Gaygadzhiev Z, Hill A, Corredig M. Influence of the Emulsion Droplet Type on the Rheological Characteristics and Microstructure of Rennet Gels From Reconstituted Milk. Journal of Dairy Research 2009b;76(3) 349–355.

[34] Gaaloul S, Turgeon S L, Corredig M. Phase Behavior of Whey Protein Aggregates/κ-Carrageenan Mixtures: Experiment and Theory. Food Biophysics 2010;5(2) 103–113.

[35] Krzeminski A, Großhable K, Hinrichs J. Structural Properties of Stirred Yoghurt as Influenced by Whey Proteins. LWT - Food Science and Technology 2011;44(10) 2134-2140.

[36] Celigueta Torres I, Amigo Rubio J M, Ipsen R. Using Fractal Image Analysis to Characterize Microstructure of Low-Fat Stirred Yoghurt Manufactured with Microparticulated Whey Protein. Journal of Food Engineering 2012;109(4) 721–729.

[37] Le T T, van Camp J, Pascual P A L, Meesen G, Thienpont N, Messens K, Dewettinck K. Physical Properties and Microstructure of Yoghurt Enriched with Milk Fat Globule Membrane Material. International Dairy Journal 2011;21(10) 798-805.

[38] Glaring M A, Koch C B, Blennow A. Genotype-Specific Spatial Distribution of Starch Molecules in the Starch Granule: A Combined CLSM and SEM Approach. Biomacromolecules 2006;7(8) 2310-2320.

[39] Ishihara S, Nakauma M, Funami T, Nakaura Y, Inouchi N, Nishinari K. Functions of Gum Arabic and Soybean Soluble Polysaccharide in Cooked Rice as a Texture Modifier. Bioscience Biotechnology and Biochemistry 2010;74(1) 101-107.

[40] Trivedi D, Bennett R J, Hemar Y, Reid D C W, Lee S K, Illingworth D. Effect of Different Starches on Rheological and Microstructural Properties of (I) Model Processed Cheese. International Journal of Food Science & Technology 2008a;43(12) 2191–2196.

[41] Trivedi D, Bennett R J, Hemar Y, Reid D C W, Lee S K, Illingworth D. Effect of Different Starches on Rheological and Microstructural Properties of (II) Commercial Processed Cheese. International Journal of Food Science & Technology 2008b;43(12) 2197–2203.

[42] Straadt I K, Thybo A K, Bertram H C. NaCl-Induced Changes in Structure and Water Mobility in Potato Tissue as Determined by CLSM and LF-NMR. LWT – Food Science and Technology 2008;41(8) 1493–1500.

[43] Bordoloi A, Kaur L, Singh J. Parenchyma Cell Microstructure and Textural Characteristics of Raw and Cooked Potatoes. Food Chemistry 2012;133(4) 1092–1100.

[44] Do T A L, Vieira J, Hargreaves J M, Mitchell J R, Wolf B. Structural Characteristics of Cocoa Particles and Their Effect on the Viscosity of Reduced Fat Chocolate. LWT - Food Science and Technology 2011;44(4) 1207-1211.

[45] Svanberg L, Ahrné L, Lorén N, Windhab E. Effect of Sugar, Cocoa Particles and Lecithin on Cocoa Butter Crystallisation in Seeded and Non-Seeded Chocolate Model Systems. Journal of Food Engineering 2011a;104(1) 70–80.

[46] Svanberg L, Ahrné L, Lorén N, Windhab E. Effect of Pre-Crystallization Process and Solid Particle Addition on Microstructure in Chocolate Model Systems. Food Research International 2011b;44(5) 1339–1350.

[47] Ong L, Dagastine R R, Kentish S E, Gras S L. The Effect of pH at Renneting on the Microstructure, Composition and Texture of Cheddar Cheese. Food Research International 2012;48(1) 119–130.

[48] Costa N E, O'Callaghan D J, Mateo M J, Chaurin V, Castillo M, Hannon J A, McSweeney P L H, Beresford T P. Influence of an Exopolysaccharide Produced by a Starter on Milk Coagulation and Curd Syneresis. International Dairy Journal 2012;22(1) 48-57.

[49] Richoux R, Aubert L, Roset G, Briard-Bion V, Kerjean J R, Lopez C. Combined Temperature–Time Parameters During the Pressing of Curd as a Tool to Modulate the Oiling-Off of Swiss Cheese. Food Research International 2008;41(10) 1058–1064.

[50] Adedeji A A, Liu L, Ngadi M O. Microstructural Evaluation of Deep-Fat Fried Chicken Nugget Batter Coating Using Confocal Laser Scanning Microscopy. Journal of Food Engineering 2011;102(1) 49–57.

[51] Loveday S M, Hindmarsh J P, Creamer L K, Singh H. Physicochemical Changes in a Model Protein Bar During Storage. Food Research International 2009;42(7) 798–806.